FISH BREEDING & GENETICS

Editor

Venkat Kumar

scitus
academics

Fish Breeding & Genetics

Edited by **Venkat Kumar**

This Book is Printed on Acid-Free Paper. ♾

ISBN: 9781681179483

Library of Congress Control Number: 2017949096

SCITUS Academics LLC,

616, Corporate Way, Suite 2, 4766,

Valley Cottage, NY 10989

www.scitusacademics.com

Notice

Printed in United States of America.

Preface

Aquatic germplasm resources have enormous potential to contribute towards the economic well-being of the nation. The fisheries and aquaculture play a vital role in the social development by providing nutritional security and economic upliftment of fishers and fish farmers. General aspects of fish breeding, just like with fish keeping, include proper feeding and providing the right environment. The breeding environment needs to be maintained with proper water conditions and you need to watch for disease or other ailments. Aquaculture is the fastest growing food production industry, and the vast majority of aquaculture products are derived from Asia. The quantity of aquaculture products directly consumed is now greater than that resulting from conventional fisheries. The nutritional value of aquatic products compares favorably with meat from farm animals because they are rich in micronutrients and contain high levels of healthy omega-3 fatty acids. Compared with farm animals, fish are more efficient converters of energy and protein. If the aquaculture sector continues to expand at its current rate, production will reach 132 million tons of fish and shellfish and 43 million tons of seaweed in 2020. Future potential for marine aquaculture production can be estimated based on the length of coastline, and for freshwater aquaculture from available land area in different countries. In recent years there has been increasing research activity in genetics of farmed fish, especially relating to the salmonid species.

This book covers all aspects of breeding genetics which are of direct or potential relevance to improving the efficiency of commercial production of farmed fish. This guide accompanies a comprehensive review of the ways in which molecular genetic technologies can contribute to the management of wild fisheries and a critical evaluation of the role of population, evolutionary and molecular genetics in the management of wild fisheries. This brings a wealth of information, providing a book which should be bought by all fish biologists, fisheries scientists, geneticists and aquarists.

Table of Contents

CHAPTER 1

Domestication of Marine Fish Species: Update and Perspectives

Fabrice Teletchea

Research Unit Animal and Functionalities of Animal Products (URAFPA), University of Lorraine—INRA, 2 Avenue de la Forêt de Haye, BP 172, 54505 Vandoeuvre-lès-Nancy, France

ABSTRACT

Domestication is a long and endless process during which animals become, generations after generations, more adapted to both captive conditions and humans. Compared to land animals, domestication of fish species has started recently. This implies that most farmed marine fish species have only changed slightly from their wild counterparts, and production is based partly or completely on wild inputs. In the past decades, global marine fish production has increased tremendously, particularly since the 1990s, to reach more than 2.2 million tons in 2013. Among the 100 marine fish species listed in the FAO's database in 2013, 35 are no longer produced, and only six have a production higher than 100,000 tons. The top ten farmed marine species accounted for nearly 90% of global production. The future growth and sustainability of mariculture will depend partly on our ability to domesticate (*i.e.*, control the life cycle in captivity) of both currently farmed and new species.

Keywords: domestication level; wild; domesticated; marine fish species; capture-based aquaculture; bottlenecks

1. INTRODUCTION

Domestication is a long and endless process during which animals become, generations after generations, more adapted to both captive conditions and humans [1,2,3]. Therefore, domestication should not be confused with taming, which is conditioned behavioral modification of wild-born animals [3,4]. During domestication, captive animals are progressively modified from their wild ancestors and at a certain moment are considered domesticated. Nevertheless, it is difficult to determine when captive animals have become domesticated, and such a decision is subjective and arbitrary [1]. According to most authors, a domesticated animal is bred in captivity and thereby modified from its wild ancestors in ways making it more useful to humans who control its reproduction and its food supply [4,5,6].

Domestication on land started about 12,000 years ago [4]. Over millennia, animal populations were modified by humans and changes in behavior, physiology and morphology occurred [6,7]. At the beginning of the twentieth century, modern breeding programs were initiated leading to dramatic changes in productivity, e.g., increase laying rate for laying hens or improved feed conservation ratio, meat yield and growth rate in broiler chickens [2,8,9]. As a result, thousands of genetically distinct livestock breeds have been created, and there is an apparent dichotomy between domesticated species and their wild congeners, which have sometimes gone extinct [6,10,11].

Compared to the domestication of land animals, the domestication of aquatic animals is a recent phenomenon [5]. Except for few species, such as the common carp (*Cyprinus carpio*) and the goldfish (*Carassius auratus*), the bulk of farming has started in the past century [5,12,13]. Most fish species farmed today are not much different from their wild conspecifics [9,10,13,14,15]. It is estimated that 90% of the global aquaculture industry is based on wild, undomesticated or non-selectively bred stocks [2,16]. Conversely, less than 10% of aquaculture production comes from selectively bred farm stocks [2]. The Atlantic salmon (*Salmo salar*) is an outlier, as almost 100% of the total production is based on selectively bred stocks [8,16]. Consequently, depending on the species considered, the control over aquaculture production can vary from managing only a portion of the life cycle to managing the complete life cycle in captivity [11,14,17,18,19,20]. In order to better describe the various fish production strategies, Teletchea and Fontaine [20] proposed a new classification based on the level of human control over the life cycle of farmed species and independence from wild inputs. This classification comprises five levels of domestication with one being the least domesticated to five being the most domesticated. Among the 250 species recorded in the Food and Agriculture Organization (FAO) database in 2009, 39 belong to level 1 (first trials of acclimatization to the culture environment), 75 to the level 2 (part of the life cycle closed in captivity, also known as capture-based aquaculture), 61 to the level 3 (entire life cycle closed in captivity with wild inputs), 45 to the level 4 (entire life cycle closed in captivity without wild inputs) and 30 to the level 5 (selective breeding programs are used focusing on specific goals) [20].

The main goal of the present study is to analyze the evolution of the aquaculture production of marine fish species since 1950, while updating the number of species per domestication level since 2009.

2. MATERIALS AND METHODS

The central source of data about the world's fisheries and aquaculture operations is the United Nation's Food and Agriculture Organization (FAO). With fisheries catch and aquaculture production data going back to 1950, the FAO's database is an invaluable source of temporal information about the quantity, value, and geographic location of global seafood production [19]. Nevertheless, concerns have been raised in the past two decades about the quality of the data, mainly due a lack of clarity and transparency in terms of what is and is not being reported as "aquaculture", if reported at all [19,21]. However, a full discussion of these concerns lay outside the scope of this paper (instead, see [19,22]). In the present study, I choose to focus on marine fish species, thus excluding diadromous species, such as Atlantic salmon [23]. The domestication level for all marine fish species listed in the FAO in 2013 (n = 100 species) was determined based on [20] for species already in the database in 2009 (n = 87), and on the literature for "new" species (n = 17). Group of species were excluded.

3. RESULTS

3.1. Evolution of Global Marine Fish Aquaculture Production

Global marine fish production increased slightly from 1950 up to the beginning of the 1970s. Then, the production increased steadily up to the 1990s; thereafter, it rose tremendously (Figure 1). However, more than half of the production is not identified at the species level in the FAO database (Figure 1). One group called "Marine fish nei (not elsewhere included)" totaled 621,275 tons, which is more than one-quarter of the global production in 2013.

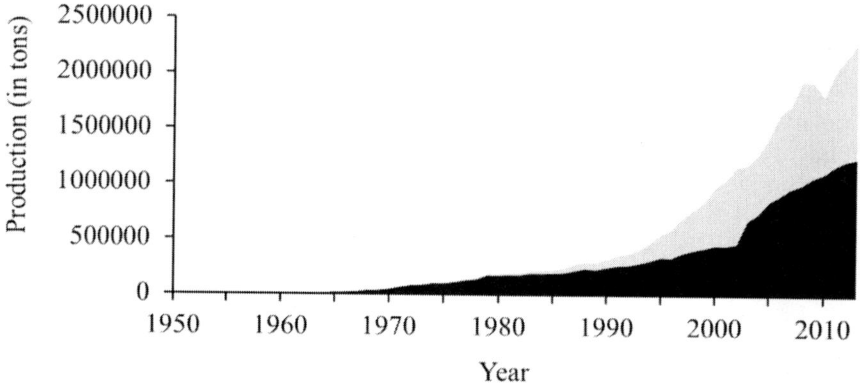

Figure 1. Evolution of global aquaculture production of marine fish species. Group of species, not identified at the species level (**upper** part, in grey), identified at the species level (**lower** part, in black).

In 2013, global marine fish production reached more than 2.2 million tons (Figure 1), which represents about half the production of diadromous fish and about 6% of freshwater fish production (Figure 2). This implies that despite its strong increase, marine fish aquaculture remains small compared to non-marine fish production.

Marine fish are mainly produced in Asia (83.1%), followed by Europe (9.2%) and Africa (7.1%) (Figure 3a). Asia is also the main producer of diadromous fish (39.0%); yet the production is more evenly distributed in the world, with Europe (37.8%) and the Americas (21.9%) (Figure 3b). For freshwater fish, almost the entire production (93.8%) is realized in Asia (Figure 3c).

3.2. Evolution of the Number of Farmed and Domesticated Fish Species

The number of farmed species has strongly increased in the past decades to reach 65 in 2013 (Figure 4). Since the mid-1990s, the number has doubled, despite slight decreases in 1997, 1998, 2006, 2007, 2009, 2011 and 2012.

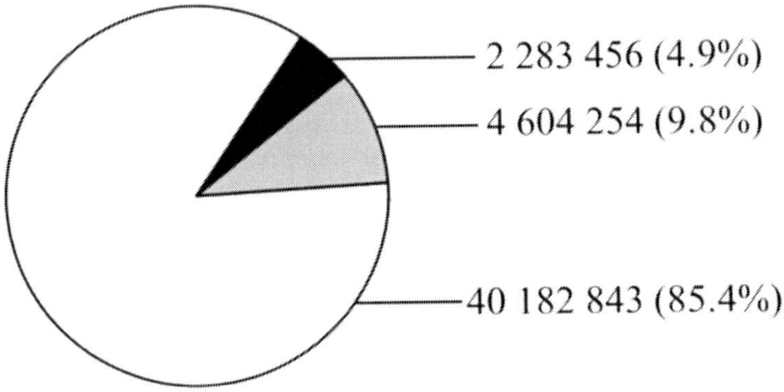

Figure 2. Comparison of the aquaculture production of marine fish species (**black**) with those of diadromous (**grey**) and freshwater (**white**) species in 2013. The first number is the total production followed by the percentage in parentheses.

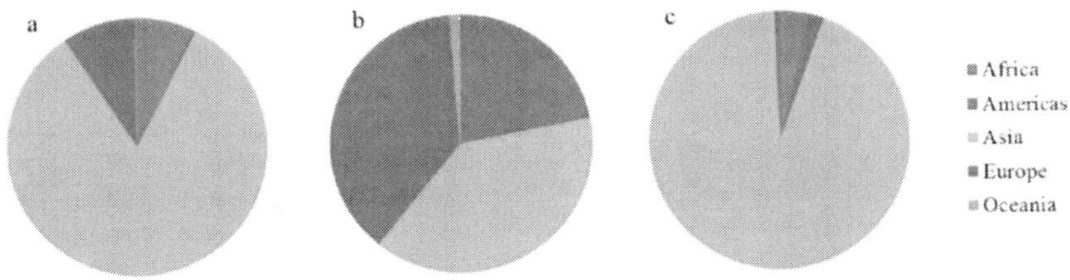

Figure 3. Main aquaculture regions of fish species in 2013: (**a**) marine fish species; (**b**) diadromous fish species; and (**c**) freshwater fish species.

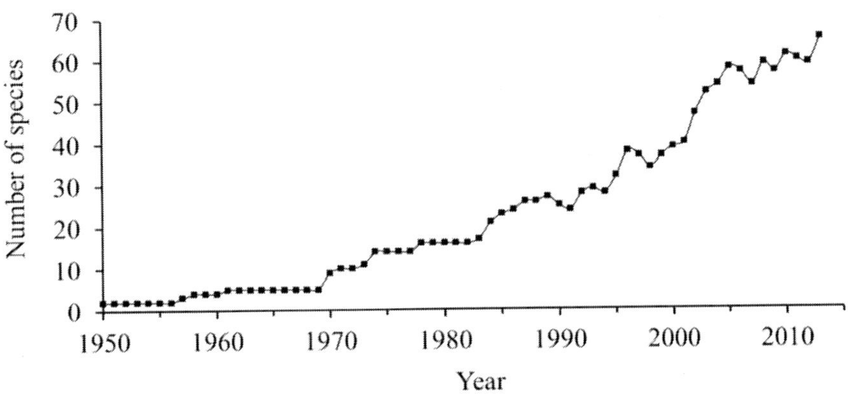

Figure 4. Evolution of the number of marine fish species farmed per year.

The comparison between 2009 and 2013 showed that 87 species were listed in 2009 and 100 in 2013. Four species listed in 2009 are no longer present in 2013, the striped weakfish (*Cynoscion striatus*), the eastern pomfred (*Schuettea scalaripinnis*), the streamerish (*Agrostichthys parkeri*) and the rice-paddy eel (*Pisodonophis boro*). Seventeen "new" species were present (in bold in Appendix). Among the 100 species listed in the FAO database, 35 were no longer produced in 2013, 24 had a production less than 100 tons, 13 had a production between 100 and 1000 tons, 13 species between 1001 and 10,000, nine species between 10,001 and 100,000, and only six have a production higher than 100,000 tons. The domestication level of marine species ranged from one to five (Figure 5). The domestication level for each species is provided in the Appendix. There are only slight differences between 2009 and 2013: only the numbers of species at levels 1, 2 and 3 have increased.

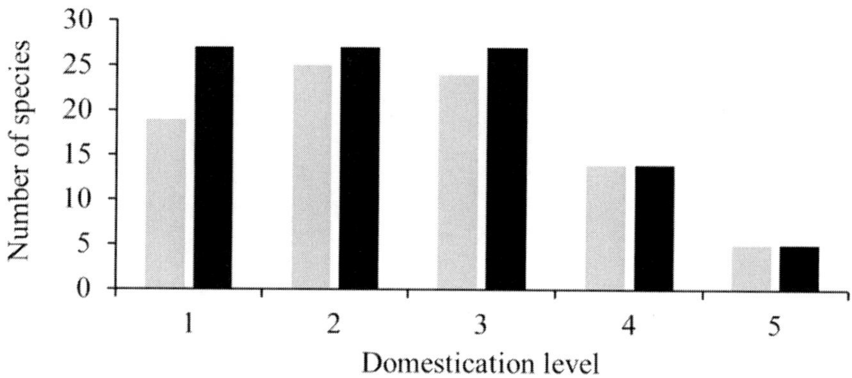

Figure 5. Domestication level of marine fish species in 2009 (**grey**) and 2013 (**black**).

The top ten marine farmed species in 2013 totaled 86.5% of the global production, which was 1,241,149 tons (when excluding groups not identified at the species level). These species are almost exclusively produced in Asia, except for the two leading species, gilthead seabream (*Sparus aurata*) and European sea bass (*Dicentrarchus labrax*), for which about half of the production is in Europe (Figure 6).

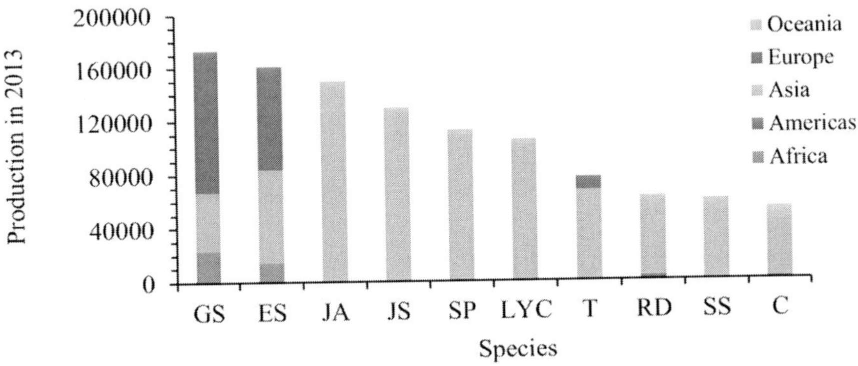

Figure 6. Aquaculture production of the top ten farmed marine species. Gilthead seabream (GS); European seabass (ES); Japanese amberjack (JA); Japanese seabass (JS), subnose pompano (SP); large yellow croaker (LYC); turbot (T); red drum (RD); silver seabream (SS) and cobia (C) (see Appendix for scientific names).

4. DISCUSSION

4.1. Evolution of Marine Fish Aquaculture Production

Aquaculture, the farming of aquatic organisms, including fish, shellfish and mollusks (excluding plants), is the fastest growing food production system globally, with an increase in production of about 9.3% per year since 1985 [14]. In 2012, aquaculture production has reached 66.7 million tons [2]. Despite a strong increase in the past two decades, the production of marine fish species is still small compared to global aquaculture, and particularly mariculture, which is dominated by mollusks and crustaceans [7]. In comparison, fish is totally dominant in freshwater aquaculture (>99%) [7].

When compared with the two other groups of fish listed in the FAO database, it appears that marine fish species represent a very small amount of global fish production (Figure 2). This is partly due to the fact that farming of most marine fish species is very recent, and thus the life cycle is controlled in captivity for only a handful of species (see below). In addition, several constraints have restricted the expansion of aquaculture of marine fish species, particularly in North America and Europe (Figure 3), among which limited areas sheltered from ocean swells, regulatory restrictions on sites, other competitive factors, such as tourism and port development, the relatively high costs (e.g., investments in infrastructure, maintenance, cost and transport of feed), and the high developmental costs and risks associated with off-shore aquaculture technologies [7,9,24,25,26,27,28].

Today, the marine environment contributes less than two percent of the human food supply [27,29]. It is largely because the development of controlled food production in the ocean lags several millennia behind that on land [12,27,29]. The space used for mariculture production is estimated at about 0.01 million km^2, or about 0.04% of the global shelf area [29]. Mariculture production is concentrated in a selected number of countries (e.g., China, Spain, Greece, Norway, Chile, and Scotland), particularly in sheltered bays and lagoons [29]. In future decades, it is anticipated that mariculture would increase significantly [9,27,29]. The FAO forecasts that mariculture will reach 54 million metric tons to 70 million metric tons by year 2020 [29]. However, further development of mariculture will run into major bottlenecks concerning the availability, suitability, and cost of feed; space availability; and adverse environmental impacts, which must be overcome if mariculture is to become a major component of global food production [28,29].

4.2. Evolution of the Number of Farmed and Domesticated Marine Species

The number of farmed marine species has strongly increased in the past decades, as observed for other fish groups [5,13,28]. However, only 10 percent of the marine species listed in the FAO in 2013 accounted for nearly 90% of global production (Figure 6). Eight of the top ten farmed species have reached the domestication level 4 ($n = 5$) or 5 ($n = 3$). This implies that the entire life cycle of these species has been closed in captivity without the need to use wild inputs, and for three of them, breeding programs have been developed: five for the gilthead seabream (*Sparus aurata*), three for the European sea bass (*Dicentrarchus labrax*) and two for the turbot (*Psetta maxima*) [25]. However, reaching level 4 and 5 does not necessarily imply that the entire aquaculture production of the 19 marine fish species classified at these two domestication levels (Figure 5) is based on domesticated or genetically improved stocks only [11,20]. For instance, even though sea bass has reached the level 5, most farms still rely today on wild broodstock for reproduction or, to lesser extent, from first-generation (F1) individuals and rarely from selected F2 or F3 fish [11].

Achieving full life cycle in captivity over several generations, which could then be called domesticated fish, thus appears an important progress in mastering the sustainability and the increase of production [17]. However, domestication (and notably selection) of a new species is a long and difficult process [30] that requires, among others, broodstock management (production of high quality broodstock, gonad and gamete development, ovulation/oviposition in females and ejaculation of milt in males), incubation of eggs, and rearing of larvae and juveniles [20,25,31]. Therefore, domestication needs access to specific skills, knowledge, and technology, and both long-term public and private funding [2,8,9,13]. This explains why it is has been primarily carried out in developed countries, notably in Europe [25,32].

For all the top ten farmed species, except the large yellow croaker (*Larimichthys croceus*), the aquaculture production now strongly exceeds capture fisheries (Figure 7). This may be caused by competition between these two sources, because a wild caught fish can commonly be sold at a relatively low price, but cannot be cultured at this low price for a profit. As capture fisheries decline because of overharvesting, the prices of target species often increase dramatically. Under these conditions, aquaculture can thrive, thereby further reducing the value of that capture fishery [14].

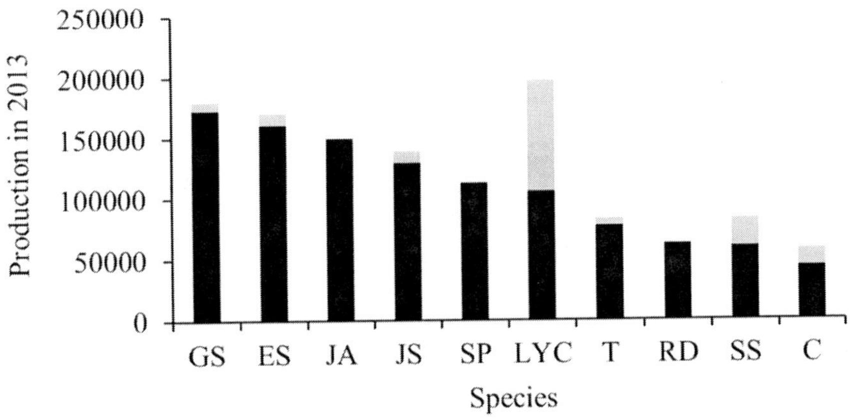

Figure 7. Global production of the top ten farmed marine species, aquaculture (**black**) and fisheries (**grey**). Gilthead seabream (GS); European seabass (ES); Japanese amberjack (JA); Japanese seabass (JS); subnose pompano (SP); large yellow croaker (LYC); turbot (T); red drum (RD); silver seabream (SS) and cobia (C). The fisheries production of JA, SP and RD are 0, 79, and 204 tons respectively (see Appendix for scientific names).

Two among the top ten farmed species, the Japanese amberjack (*Seriola quinqueradiata*) and the snubnose pompano (*Trachinotus blochii*), have only reached level 2, which implies that the entire aquaculture production is based on wild input. This method of production, known as capture-based aquaculture, consists of growing and fattening individuals removed from wild populations [17]. Tuna fattening and much of the marine cage culture in Asia, relies directly on wild-caught small pelagic fish with relatively low market price [19,28,31]. The aquaculture process transforms fish protein from low to high value for human consumption [28]. However, such systems can only function as long as survival and sustainable utilization of the affected wild stocks are warranted [17,19,31]. Therefore, capture-based aquaculture can only be seen as a transitory form of fish production, viable only as long as the wild resource is still available for seed withdrawal [17,31]. Efforts have to be made to domesticate species (*i.e.*, the closing of the life cycle in captivity) to allow a reliable production, independent of wild inputs, and then improve desirable traits through selective breeding [2,25,31].

One-third of the marine species listed in FAO in 2013 are no longer produced, and 50 percent more have a production less than 1000 tons. Nearly all species with a production inferior to 1000 tons have a domestication level between 1 and 3 (Appendix). This highlights that for most species, farming corresponds to one or a few years of aquaculture trial before being abandoned [15,20]. The main reason why numerous attempts with new species fail is that premature attempts to develop industrial enterprises were based on overly optimistic speculation about market demand, rather than on biological and technical knowledge and adequate information about economic feasibility [20,27].

5. CONCLUSIONS

Compared to the domestication of land animals, the domestication of aquatic animals, and particularly marine fish species, is a recent phenomenon. Mariculture of fish has only started a few decades ago, and today only a handful of species can be considered domesticated. In contrast, for numerous species, farming was only performed for a few years before being stopped. The future growth and sustainability of mariculture will depend partly on our ability to domesticate (*i.e.*, control the life cycle in captivity) of both currently farmed and new species.

ACKNOWLEDGMENTS

The author would like to thank two anonymous reviewers for their comments on a previous draft of the manuscript.

APPENDIX

Table A1. Domestication level for marine fish species listed in the 2013 FAO report (*n*= 100 species). Group of species were excluded. Species on the list for the first time are in bold (*n* = 17). Aquaculture production is in tons. When no reliable scientific information was found, species were arbitrarily considered to belong to level 1 when their production was less than five continuous years (*n* = 15) to the level 2 when their production was between five and up to 10 continuous years (*n* = 3), and to the level 3 when their production was greater than 10 years (*n* = 3).

Scientific Name	Common Name	Production in 2013	Domestication Level	Main Reference
Anarhichas lupus	Atlantic wolffish	0	1	Gunnarsson *et al.*, 2009, [33]
Atherina boyeri	**Big-scale sand smelt**	**0**	**1**	**Dulcic *et al.*, 2008, [34]**
Bolbometopon muricatum	**Green humphead parrotfish**	**1**	**1**	
Carangoides malabaricus	**Malabar trevally**	**387**	**1**	
Caranx hippos	Crevalle jack	0	1	
Caranx sexfasciatus	**Bigeye trevally**	**1**	**1**	
Centropristis striata	Black seabass	0	1	Rezek *et al.*, 2010, [35]
Chaetodipterus faber	**Atlantic spadefish**	**0**	**1**	**Gaspar 2005, [36]**
Dentex tumifrons	Yellowback seabream	0	1	

Dicentrarchus punctatus	**Spotted seabass**	**2**	**1**	**Ly *et al.*, 2012, [37]**
Labrus bergylta	Ballan wrasse	25	1	Muncaster *et al.*, 2010, [38]
Lethrinus miniatus	Trumpet emperor	45	1	
Lutjanus bohar	Two-spotted red snapper	0	1	
Megalops atlanticus	**Tarpon**	**0**	**1**	
Micropogonias furnieri	Whitemouth croaker	5	1	Velloso and Pereira Jr. 2010, [39]
Mugil liza	Lebranche mullet	7	1	
Muraenesox cinereus	Daggertooth pike conger	0	1	
Mycteroperca bonaci	**Black grouper**	**2**	**1**	
Pagrus major	**Japanese seabream**	0	**1**	
Platichthys flesus	European flounder	0	1	Engel-Sorensen *et al.*, 2004, [40]
Pleurogrammus azonus	Okhotsk atka mackerel	0	1	
Pomatomus saltatrix	Bluefish	0	1	
Scophthalmus rhombus	Brill	0	1	Cruzado *et al.*, 2004, [41]
Siganus canaliculatus	White-spotted spinefoot	1	1	Xu *et al.*, 2011, [42]
Siganus javus	Streaked spinefoot	0	1	
Siganus rivulatus	Marbled spinefoot	0	1	El Dakar *et al.*, 2011, [43]
Valamugil seheli	Bluespot mullet	0	1	Belal 2004, [44]
Acanthopagrus berda	Goldsilk seabream	0	2	Liao *et al.*, 2001, [45]
Argyrosomus japonicus	**Japanese meagre**	**130**	**2**	**Mirimin and Roodt-Wilding, 2015, [46]**
Boleophthalmus pectinirostris	**Great blue spotted mudskipper**	**0**	**2**	**Zhang *et al.*, 1989, [47]**
Centropomus undecimalis	Common snook	0	2	Carter *et al.*, 2010a, [48]
Eleutheronema tetradactylum	Fourfinger threadfin	4173	2	Liao *et al.*, 2001, [45]
Epinephelus areolatus	Areolate grouper	47	2	Ottolenghi *et al.*, 2004, [49]

Epinephelus coioides	Orange-spotted grouper	492	2	Ottolenghi *et al.* 2004, [49]
Epinephelus fuscoguttatus	Brown-marbled grouper	86	2	Ottolenghi *et al.*, 2004, [49]
Epinephelus lanceolatus	**Giant grouper**	**36**	**2**	**Peng *et al.*, 2015, [50]**
Epinephelus malabaricus	Malabar grouper	68	2	Ottolenghi *et al.*, 2004, [49]
Gnathanodon speciosus	Golden trevally	58	2	Liao *et al.*, 2001, [45]
Liza ramada	Thinlip grey mullet	0	2	Marino *et al.*, 1999, [51]
Lutjanus goldiei	Papuan black snapper	0	2	
Lutjanus guttatus	Spotted rose snapper	2	2	García-Ortega 2009, [52]
Lutjanus johnii	John's snapper	278	2	Liao *et al.*, 2001, [45]
Miichthys miiuy	**Mi-iuy (brown) croaker**	**0**	**2**	**An *et al.*, 2012, [53]**
Mugil soiuy	So-iuy mullet	905	2	
Plectropomus maculatus	Spotted coralgrouper	7	2	
Polydactylus sexfilis	Sixfinger threadfin	0	2	Deng *et al.*, 2011, [54]
Psammoperca waigiensis	Waigieu seaperch	5,704	2	Pham *et al.*, 2010, [55]
Seriola quinqueradiata	Japanese amberjack	149,766	2	Bilio 2007b, [56]
Seriola rivoliana	**Longfin yellowtail**	**400**	**2**	**Roo *et al.*, 2014, [57]**
Thunnus albacares	Yellowfin tuna	171	2	Wexler *et al.*, 2011, [58]
Thunnus maccoyii	Southern bluefin tuna	3,482	2	Carter *et al.*, 2010b, [59]
Thunnus thynnus	Atlantic bluefin tuna	3,445	2	Carter *et al.*, 2010b, [59]
Trachinotus blochii	Snubnose pompano	112,499	2	Liao *et al.*, 2001, [45]
Trachinotus carolinus	Florida pompano	350	2	Pfeiffer and Riche 2001, [60]
Cromileptes altivelis	Humpback grouper	2	3	Hong and Zhang 2003, [61]
Dentex dentex	Common dentex	54	3	Suquet *et al.*, 2009, [62]
Diplodus sargus	White seabream	24	3	Suquet *et al.*, 2009, [62]

Diplodus vulgaris	Common two-banded seabream	0	3	Suquet *et al.*, 2009, [62]
Epinephelus akaara	Hong Kong grouper	0	3	Hong and Zhang 2003, [61]
Epinephelus tauvina	Greasy grouper	5,354	3	Hong and Zhang 2003, [61]
Evynnis japonica	Crimson seabream	0	3	
Liza vaigiensis	Squaretail mullet	0	3	
Lutjanus argentimaculatus	Mangrove red snapper	5,357	3	Hong and Zhang 2003, [61]
Lutjanus russelli	Russell's snapper	13	3	Hong and Zhang 2003, [61]
Melanogrammus aeglefinus	Haddock	0	3	Roselund and Skretting 2006, [63]
Pagellus bogaraveo	Blackspot seabream	2	3	Suquet *et al.*, 2009, [62]
Pagellus erythrinus	Common pandora	0	3	Suquet *et al.*, 2009, [62]
Platax orbicularis	**Orbicular batfish**	**8**	**3**	**Coeurdacier and Gasset 2013, [64]**
Pollachius pollachius	Pollack	0	3	Roselund and Skretting 2006, [63]
Pseudocaranx dentex	White trevally	3,155	3	
Rhabdosargus sarba	Goldlined seabream	3	3	Hong and Zhang 2003, [61]
Sciaena umbra	Brown meagre	0	3	Bilio 2007a, [65]
Seriola dumerili	Greater amberjack	0	3	Hong and Zhang 2003, [61]
Solea senegalensis	Senegalense sole	640	3	Imsland 2010, [66]
Solea solea	Common sole	45	3	Imsland 2010, [66]
Sparidentex hasta	Sobaity seabream	551	3	Teng *et al.*, 1999, [67]
Takifugu obscurus	**Obscure pufferfish**	**4,860**	**3**	**Kim *et al.*, 2010, [68]**
Takifugu rubripes	**Tiger pufferfish**	**19,359**	**3**	**Wu *et al.*, 2015, [69]**
Thunnus orientalis	Pacific bluefin tuna	16,624	3	Carter *et al.*, 2010b, [59]
Trachurus japonicas	Japanese jack mackerel	958	3	Masuda 2006, [70]
Umbrina cirrosa	Shi drum	1,070	3	Suquet *et al.* 2009, [62]
Acanthopagrus latus	Yellowfin seabream	0	4	Hong and Zhang 2003, [61]

Acanthopagrus schlegeli	Blackhead seabream	1,161	4	Hong and Zhang 2003, [61]
Anarhichas minor	Spotted wolfish	0	4	Le François *et al.* 2010, [71]
Argyrosomus regius	Meagre	6,659	4	Lazo *et al.*, 2010, [72]
Diplodus puntazzo	Sharpsnout seabream	250	4	Suquet *et al.*, 2009, [62]
Hippoglossus hippoglossus	Atlantic halibut	1,485	4	Imsland 2010, [66]
Larimichthys croceus	Large yellow croaker	105,230	4	Bilio 2007b, [56]
Lateolabrax japonicus	Japanese seabass	129,334	4	Hong and Zhang 2003, [61]
Mugil cephalus	Flathead grey mullet	12,245	4	Hong and Zhang 2003, [61]
Pagrus auratus	Silver seabream	59,616	4	Suquet *et al.*, 2009, [62]
Pagrus pagrus	Red porgy	350	4	Suquet *et al.*, 2009, [62]
Rachycentron canadum	Cobia	43,395	4	Bilio 2007a,b, [56,65]
Sciaenops ocellatus	Red drum	62,197	4	Lazo *et al.*, 2010, [72]
Sebastes schlegeli	Korean rockfish	23,757	4	Bilio 2007b, [56]
Dicentrarchus labrax	European seabass	161,059	5	Jobling *et al.*, 2010, [73]
Gadus morhua	Atlantic cod	4,252	5	Björnsson *et al.*, 2010, [74]
Paralichthys olivaceus	Bastard halibut	39,445	5	Bilio 2007b, [56]
Psetta maxima	Turbot	76,998	5	Hulata 2001, [75]
Sparus aurata	Gilthead seabream	173,062	5	Jobling and Perruzi 2010, [76]

REFERENCES

1. Price, E.O. Behavioral development in animals undergoing domestication. *Appl. Anim. Behav. Sci.* **1999**, *65*, 245–271.
2. Olesen, I.; Bentsen, H.B.; Phillips, M.; Ponzoni, R.W. Can the global adoption of genetically improved farmed fish increase beyond 10%, and how? *J. Mar. Sci. Eng.* **2015**, *3*, 240–266.
3. Zeder, M.A. Core questions in domestication research. *Proc. Natl. Acad. Sci. USA* **2015**, *112*, 3191–3198.
4. Diamond, J. Evolution, consequences and future of plant and animal domestication. *Nature* **2002**, *418*, 700–707.
5. Balon, E.K. About the oldest domesticates among fishes. *J. Fish Biol.* **2004**, *65*, 1–27.

6. Mirkena, T.; Duguma, G.; Haile, A.; Tibbo, M.; Okeyo, A.M.; Wurzinger, M.; Sölkner, J. Genetics of adaptation in domestic farm animals: A review. *Livest. Sci.* **2010**, *132*, 1–12.
7. Olsen, Y. How can mariculture better help feed humanity? *Front. Mar. Sci.* **2015**, *2*.
8. Gjedrem, T. The first family-based breeding program in aquaculture. *Rev. Aquac.* **2010**, *2*, 2–15.
9. Gjedrem, T.; Robinson, N.; Morten, R. The importance of selective breeding in aquaculture to meet future demands for animal proteins: A review. *Aquaculture* **2012**, *350–353*, 117–129.
10. Lind, C.E; Ponzoni, R.W.; Nguyen, N.H.; Khaw, H.L. Selective breeding in fish and conservation of genetic resources for aquaculture. *Reprod. Domest. Anim.* **2012**, *47*, 255–263.
11. Teletchea, F. Domestication and genetics: What a comparison between land and aquatic species can bring. In *Evolutionary Biology. Biodiversification from Genotype to Phenotype*; Pontarotti, P., Ed.; Springer: Cham, Switzerland, 2015; pp. 389–402.
12. Duarte, C.M.; Marbá, N.; Holmer, M. Rapid domestication of marine species. *Science* **2007**, *316*, 382–383.
13. Bartley, D.M.; Nguyen, T.T.T.; Halwart, M.; de Silva, S.S. Use and exchange of aquatic genetic resources in aquaculture: Information relevant to access and benefit sharing. *Rev. Aquac.* **2009**, *1*, 157–162.
14. Diana, J.S. Aquaculture production and biodiversity conservation. *BioScience* **2009**, *59*, 27–37.
15. Hedgecock, D. Aquaculture, the next wave of domestication. In *Biodiversity in Agriculture, Domestication, Evolution, and Sustainability*; Gepts, P., Famula, T.R., Bettinger, R.L., Brush, S.B., Damania, A.B., McGuire, P.E., Qualset, C.O., Eds.; Cambridge University Press: Cambridge, UK, 2012; pp. 538–548.
16. Gjedrem, T. Genetic improvement for the development of efficient global aquaculture: A personal opinion review. *Aquaculture* **2012**, *344–349*, 12–22.
17. Bilio, M. Plenary lecture: the future of capture and culture fisheries. In Proceedings of the 7th International Symposium, Keeping and Creating Diversity in Fish Production—Innovative Marine Life Science for Three Es, Edibles Environment and Education, in 21st century, Hokkaido, Japan, 17–19 November 2008; pp. 5–24.
18. Teletchea, F. Qu'est-ce qu'un poisson domestique? Implications pour le développement futur de l'aquaculture. *Ethnozootechnie* **2012**, *90*, 7–12.
19. Klinger, D.H.; Turnipseed, M.; Anderson, J.L.; Asche, F.; Crowder, L.B.; Guttormsen, A.G.; Halpern, B.S.; O'Connor, M.I.; Sagarin, R.; Selkoe, K.A.; *et al.* Moving beyond the fished or farmed dichotomy. *Mar. Policy* **2013**, *68*, 369–374.
20. Teletchea, F.; Fontaine, P. Levels of domestication in fish: Implications for the sustainable future of aquaculture. *Fish Fish.* **2014**, *15*, 181–195.
21. Watson, R.; Pauly, D. Systematic distortions in world fisheries catch trends. *Nature* **2001**, *414*, 534–536.
22. Campbell, B.; Pauly, D. Mariculture: A global analysis of production trends since 1950. *Mar. Pol.* **2013**, *39*, 94–100.
23. Available online: http://www.fao.org/fishery/statistics/global-aquaculture-production/ query/en (accessed on 10 July 2015).
24. De Silva, S.S. Aquaculture: A newly emergent food production sector and perspectives of its impacts on biodiversity and conservation. *Biodivers. Conserv.* **2012**, *21*, 3187–3220.
25. Migaud, H.; Bell, G; Cabrita, E.; McAndrew, B.; Davie, A.; Bobe, J.; Herráez, M.P.; Carillo, M. Gamete quality and broodstock management in temperate fish. *Rev. Aquac.* **2013**, *5*, S194–S223.
26. Natale, F.; Hofherr, J.; Fiore, G.; Virtanen, J. Interactions between aquaculture and fisheries. *Mar. Policy* **2013**, *38*, 205–113.

27. Sorgeloos, P. Aquaculture: The blue biotechnology of the future. *World Aquac.* **2013**, *44*, 16–25.
28. Bostock, J.; McAndrew, B.; Richards, R.; Jauncey, K.; Telfer, T.; Lorenzen, K.; Little, D.; Ross, L.; Handisyde, N.; Gatward, I.; *et al.* Aquaculture: Global status and trends. *Philos. Trans. R. Soc. B* **2010**, *365*, 2897–2912.
29. Duarte, C.M.; Holmer, M.; Olsen, Y.; Soto, D.; Marbà, N.; Guiu, J.; Black, K.; Karakassis, I. Will the oceans help feed humanity? *BioScience* **2009**, *59*, 967–976.
30. Larson, G.; Fuller, D.Q. The evolution of animal domestication. *Annu. Rev. Ecol. Evol. Syst.* **2014**, *66*, 115–136.
31. Mylonas, C.C.; de la Gándara, F.; Corriero, A.; Belmonde Ríos, A. Atlantic Bluefin tuna (*Thunnus thynnus*) farming and fattening in the Mediterranean sea. *Rev. Fish. Sci.* **2010**, *18*, 266–280.
32. Trujillo, P.; Piroddi, C.; Jacquet, J. Fish farms at sea: The ground truth from google Earth. *PLoS ONE* **2012**, *7*, e30546.
33. Gunnarsson, S.; Sigurdsson, S.; Thorarensen, H.; Imsland, A.K. Cryopreservation of sperm from spotted wolffish. *Aquac. Int.* **2009**, *17*, 385–389.
34. Dulcic, J.; Grubisic, L.; Pallaoro, A.; Glamuzina, B. Embryonic and larval development of big-scale sand smelt *Atherina boyeri* (Atherinidae). *Cybium* **2008**, *32*, 27–32.
35. Rezek, T.C.; Watanabe, W.O.; Harel, M.; Seaton, P.J. Effects of dietary docosahexaenoic acid (22:6*n*-3) and arachidonic acid (20:4*n*-6) on the growth, survival, stress resistance and fatty acid composition in black sea bass *Centropristis striata* (Linnaeus 1758) larvae. *Aquac. Res.* **2010**, *41*, 1302–1314.
36. Gaspar, A.G. Induced spawning and rear larvae of spadefish *Chaetodipterus faber* in Margarita Island, Venezuala. In Proceedings of the 47th Gulf and Caribbean Fisheries Institute; Gulf and Caribbean Fisheries Institute: Margarita Island, Venezuala, 1995; Volume 48, pp. 15–24.
37. Ly, C.-L.; Vergnet, A.; Molinari, N.; Fauvel, C.; Bonhomme, F. Fitness of early life stages in F1 interspecific hybrids between *Dicentrarchus labrax* and *D. punctatus*. *Aquat. Living Resour.* **2012**, *25*, 67–75.
38. Muncaster, S.; Andersson, E.; Kjesbu, O.S.; Taranger, G.L.; Skiftesvik, A.B.; Norberg, B. The reproductive cycle of female Ballan wrasse *Labrus bergylta* in high latitude, temperate waters. *J. Fish Biol.* **2010**, *77*, 494–511.
39. Velloso, A.L.; Pereira, J., Jr. Influence of ectoparasitism on the welfare of *Micropogonias furnieri*. *Aquaculture* **2010**, *310*, 43–46.
40. Engell-Sørensen, K.; Støttrup, J.G.; Holmstrup, M. Rearing of flounder (*Platichthys flesus*) juveniles in semiextensive systems. *Aquaculture* **2004**, *230*, 475–491.
41. Cruzado, I.H.; Herrera, M.; Quintana, D.; Rodiles, A.; Navas, J.I.; Lorenzo, A.; Almansa, E. Total lipid and fatty acid composition of brill eggs *Scophthalmus rhombus* L. relationship between lipid composition and egg quality. *Aquac. Res.* **2004**, *42*, 1011–1025.
42. Xu, S.; Zhang, L.; Wu, Q.; Liu, X.; Wang, S.; You, C.; Li, Y. Evaluation of dried seaweed *Gracilaria lemaneiformis* as an ingredient in diets for teleost fish *Siganus canaliculatus*. *Aquacult. Int.* **2011**, *19*, 1007–1018.
43. El-Dakar, A.Y.; Shalaby, S.M.; Saoud, I.P. Dietary protein requirement of juvenile marbled spinefoot rabbitfish *Siganus rivulatus*. *Aquac. Res.* **2011**, *42*, 1050–1055.
44. Belal, I.E.H. Replacement of corn with mangrove seeds in bluespot mullet *Valamugil seheli* diets. *Aquac. Nutr.* **2004**, *10*, 25–30.
45. Liao, I.C.; Su, H.M.; Chang, E.Y. Techniques in finfish larviculture in Taiwan. *Aquaculture* **2001**, *200*, 1–31.
46. Mirimin, L.; Roodt-Wilding, R. Testing and validating a modified CTAB DNA extraction method to enable molecular parentage analysis of fertilized eggs and larvae of an emerging South African aquaculture species, the dusky kob *Argyrosomus japonicas*. *J. Fish Biol.* **2015**, *86*, 1218–1223.

47. Zhang, Q.-Z.; Hong, W.S.; Dai, Q.N.; Zhang, J.; Cai, Y.-Y.; Huang, J.L. Studies on induced ovulation, embryonic development and larval rearing of the mudskipper (*Boleophthalmus pectinirostris*). *Aquaculture* **1989**, *83*, 375–385.
48. Carter, C.; Glencross, B.; Katersky, R.S.; Bermudes, M. Chapter 14: The Snooks (Family: Centropomidae). In *Finfish Aquaculture Diversification*; Le François, N., Jobling, M., Carter, C., Blier, P., Eds.; CABI: Oxfordshire, UK, 2010; pp. 323–336.
49. Ottolenghi, F.; Silvestri, C.; Giodano, P.; Lovatelli, A.; New, M.B. *Capture-Based Aquaculture, the Fattening of Eels, Groupers, Tuna and Yellowtails*; FAO: Rome, Italy, 2004.
50. Peng, C.; Ma, H.; Su, Y.; Wen, W.; Feng, J.; Guo, Z.; Qiu, L. Susceptibility of farmed juvenile giant grouper *Epinephelus lanceolatus* to a newly isolated grouper iridovirus (genus *Ranavirus*). *Vet. Microbiol.* **2015**, *177*, 270–279.
51. Marino, G.; Ingle, E.; Cataudella, S. Status of aquaculture in Italy. *Cah. Opt. Méditerr.* **1999**, *43*, 117–126.
52. García-Ortega, A. Nutrition and feeding research in the spotted rose snapper (*Lutjanus guttatus*) and bullseye puffer (*Sphoeroides annulatus*), new species for marine aquaculture. *Fish Physiol. Biochem.* **2009**, *35*, 69–80.
53. An, H.S.; Kim, E.M.; Lee, J.W.; Kim, D.J.; Kim, Y.C. New polymorphic microsatellite markers in the Korean mi-iuy croaker, *Miichthys miiuy*, and their application to the genetic characterization of wild and farmed populations. *Anim. Cells Syst.* **2012**, *16*, 41–49.
54. Deng, D.-F.; Ju, Z.Y.; Dominy, W.; Murashige, R.; Wilson, R.P. Optimal dietary protein levels for juvenile Pacific threadfin (*Polydactylus sexfilis*) fed diets with two levels of lipid. *Aquaculture* **2011**, *316*, 25–30.
55. Pham, H.Q.; Nguyen, A.T.; Nguyen, M.D.; Arukwe, A. Sex steroid levels, oocyte maturation and spawning performance in Waigieu seaperch (*Psammoperca waigiensis*) exposed to thyroxin, human chorionic gonadotropin, luteinizing hormone releasing hormone and carp pituitary extract. *Comp. Biochem. Phys. A* **2010**, *155*, 223–230.
56. Bilio, M. Controlled reproduction and domestication in aquaculture—The current state of the art, Part II. *Aquac. Eur.***2007**, *32*, 5–23.
57. Roo, J.; Fernandez-Palacios, H.; Hernandez-Cruz, C.M.; Mesa-Rodriguez, A.; Schuchardt, D.; Izquierdo, M. First results of spawning and larval rearing of longfin yellowtail *Seriola rivoliana* as a fast-growing candidate for European marine finfish aquaculture diversification. *Aquac. Res.* **2010**, *45*, 689–700.
58. Wexler, J.B.; Margulies, D.; Scholey, V.P. Temperature and dissolved oxygen requirements for survival of yellowfin tuna, *Thunnus albacares*, larvae. *J. Exp. Mar. Biol. Ecol.* **2011**, *404*, 63–72.
59. Carter, C.; Nowak, B.; Clarke, S. Chapter 20: The Tunas (Family: Scombridae). In *Finfish Aquaculture Diversification*; Le François, N., Jobling, M., Carter, C., Blier, P., Eds.; CABI: Oxfordshire, UK, 2010; pp. 432–449.
60. Pfeiffer, T.J.; Riche, M.A. Evaluation of a low-head recirculating aquaculture system used for rearing Florida pompano to market size. *J. World Aquac. Soc.* **2011**, *42*, 198–208.
61. Hong, W.; Zhang, Q. Review of captive bred species and fry production of marine fish in China. *Aquaculture* **2003**, *227*, 305–318.
62. Suquet, M.; Divanach, P.; Hussenot, J.; Coves, D.; Fauvel, C. Pisciculture marine de «nouvelles espèces» d'élevage pour l'Europe. *Cah. Agric.* **2009**, *2–3*, 148–156.
63. Roselund, G.; Skretting, M. Worldwide status and perspective on gadoid culture. *ICES J. Mar. Sci.* **2006**, *63*, 194–197.
64. Coeurdacier, J.L.; Gasset, E. Dossier récapitulatif du développement de l'élevage du *Platax orbicularis* en Polynésie Française. Avaliable online: http://archimer.ifremer.fr/doc/00134/24490/ (accessed on 15 July 2013).
65. Bilio, M. Controlled reproduction and domestication in aquaculture—The current state of the art, Part I. *Aquac. Eur.***2007**, *32*, 5–14.

66. Imsland, A.K. Chapter 21: The Flatfishes (Order: Pleuronectiformes). In *Finfish Aquaculture Diversification*; Le François, N., Jobling, M., Carter, C., Blier, P., Eds.; CABI: Oxfordshire, UK, 2010; pp. 450–496.
67. Teng, S.-K.; El-Zahr, C.; Al-Abdul-Elah, K.; Almatar, S. Pilot-scale spawning and fry production of blue-fin porgy, *Sparidentex hasta* (Valenciennes), in Kuwait. *Aquaculture* **1999**, *178*, 27–41.
68. Kim, J.H.; Rhee, J.-S.; Lee, J.S.; Dahms, H.U.; Lee, J.; Han, K.N.; Lee, J.S. Effect of cadmium exposure on expression of antioxidant gene transcripts in the river pufferfish, *Takifugu obscurus* (Tetraodontiformes). *Comp. Biochem. Phys. C* **2010**, *152*, 473–479.
69. Wu, F.; Tang, K.; Yuan, M.; Shi, X.; Shakeela, Q.; Zhang, X.H. Studies on bacterial pathogens isolated from diseased torafugu (*Takifugu rubripes*) cultured in marine industrial recirculation aquaculture system in Shandong Province, China. *Aquac. Res.* **2015**, *46*, 736–744.
70. Masuda, R. Ontogeny of anti-predator behavior in hatchery-reared jack mackerel *Trachurus japonicus* larvae and juveniles: Patchiness formation, swimming capability, and interaction with jellyfish. *Fish. Sci.* **2006**, *72*, 1225–1235.
71. Le François, N.; Tveiten, H.; Halfyard, L.C.; Foss, A. Chapter 19. The Wolffishes (Family: Anarhichadidae). In *Finfish Aquaculture Diversification*; Le François, N., Jobling, M., Carter, C., Blier, P., Eds.; CABI: Oxfordshire, UK, 2010; pp. 417–431.
72. Lazo, J.P.; Holt, J.G.; Fauvel, C.; Suquet, M.; Quéméner, L. Chapter 18. Drum-Fish or Croakers (Family: Sciaenidae). In *Finfish Aquaculture Diversification*; Le François, N., Jobling, M., Carter, C., Blier, P., Eds.; CABI: Oxfordshire, UK, 2010; pp. 397–416.
73. Jobling, M.; Peruzzi, S.; Woods, C. Chapter 15. The Temperate Basses (Family: Moronidae). In *Finfish Aquaculture Diversification*; Le François, N., Jobling, M., Carter, C., Blier, P., Eds.; CABI: Oxfordshire, UK, 2010; pp. 337–360.
74. Björnsson, B.; Litvak, M.; Trippel, E.A.; Suquet, M. Chapter 13. The Codfishes (Family: Gadidae). In *Finfish Aquaculture Diversification*; Le François, N., Jobling, M., Carter, C., Blier, P., Eds.; CABI: Oxfordshire, UK, 2010; pp. 290–322.
75. Hulata, G. Genetic manipulations in aquaculture: A review of stock improvement by classical and modern technologies. *Genetica* **2001**, *111*, 155–173.
76. Jobling, M.; Peruzzi, S. Chapter 16. Seabreams and Porgies (Family: Sparidae). In *Finfish Aquaculture Diversification*; Le François, N., Jobling, M., Carter, C., Blier, P., Eds.; CABI: Oxfordshire, UK, 2010; pp. 361–373.

CHAPTER 2

Breeding Success of Four Male Life History Types of Spring Chinook Salmon Spawning in an Artificial Stream

Steve L. Schroder; Curtis M. Knudsen; Todd N. Pearsons; Todd W. Kassler; Edward P. Beall; Sewall F. Young; David E. Fast

Washington Department of Fish and Wildlife, 600 Capitol Way North, Olympia, WA 98501–1091, USA

Oncorh Consulting, 2623 Galloway SE, Olympia, WA 98501, USA

Grant County Public Utility District, P.O. Box 878, Ephrata, WA 98823, USA

Institut National de la Recherche Agronomique, 64 310 Saint Pée Sur Nivelle, France

Yakama Nation, P.O. Box 151, Toppenish, WA 98948, USA

ABSTRACT

In 1997 the Cle Elum Supplementation Research Facility was established to enhance spring Chinook salmon returning to the upper Yakima River, Washington State. This effort increased spring Chinook abundance, yet conditions at the hatchery also significantly elevated the occurrence of jacks and yearling precocious males. The potential genetic effect that a large influx of early maturing males might have on the upper Yakima River spring Chinook population was examined in an artificial stream. Seven independent groups of fish were placed into the stream from 2001 through 2005. Males with four different life history strategies, large anadromous, jacks, yearling precocious, and sub-yearling precocious were used. Their breeding success or ability to produce offspring was estimated by performing DNA-based pedigree assessments. Large anadromous males spawned with the most females and produced the greatest number of offspring per mate. Jacks and yearling precocious males spawned with more females than sub-yearling precocious males. However, jacks, yearling and sub-yearling precocious males obtained similar numbers of fry per mate. In the test groups, large anadromous males produced 89%, jacks 3%, yearling precocious 7%, and sub-yearling precocious 1% of the fry. These percentages remained stable even though the proportion of large anadromous males in the test groups ranged from 48% to 88% and tertiary sex ratios varied from 1.4 to 2.4 males per female. Our data suggest that large anadromous males generate most of the fry in natural settings when half or more of the males present on a spawning ground use this life history strategy.

Keywords: Spring Chinook salmon, Jacks, Precocious males, Breeding success, Relative breeding success

INTRODUCTION

Life history strategies in fishes are defined by rates of juvenile development, age of sexual maturation, offspring production, levels of parental investment, and the onset of senescence and death. They represent pathways designed to maximize reproduction under diverse environmental conditions and their occurrence is controlled by interactions between an individual's genotype and its environment (Thorpe 1994). Salmonid fishes exhibit a large array of life history types that are quite malleable (Caswell et al. 1984; O'Connell and Ash 1993). Thus, environmental variation, among other factors, will change the prevalence of different life history types within a salmonid population from one year to the next.

In the upper Yakima River, Washington State, spring Chinook salmon (*Oncorhynchus tshawytscha*) males exhibit four different life history strategies. Each type matures at a different age; two are anadromous while two others become mature without going to sea. Anadromous males that mature at the same modal age as females are referred to as large anadromous males. Most of them reach maturation at age-4 after spending two years at sea, although a few are age-5 at maturity and have resided in marine waters for three years (Knudsen et al. 2006). Anadromous males that mature a year earlier than the modal female age are referred to as jacks (Healey 1991). In this population they are age-3 and have spent one year at sea. Males that mature after rearing 18 to 20 months in freshwater are called yearling precocious males while non-anadromous males maturing after just 7 to 8 months post emergence are referred to as sub-yearling precocious males.

The occurrence of such a diverse array of male life history types in upper Yakima River spring Chinook is not unusual. Precocious spring Chinook males have been recognized for over 100 years (Rutter 1902 as cited by Pearsons et al. 2009) and their presence in spawning populations can be quite variable, ranging from 1 to 12% or more (Rich 1920 as cited by Larsen et al 2004; Gebhards 1960; Mullan et al. 1992). In the Upper Yakima River, yearling precocious males of hatchery- and wild-origin and wild sub-yearling males have been observed (Pearsons et al. 2009) and numerous jacks and large anadromous males may also be present (Knudsen et al. 2006; Dittman et al. 2010).

In 1997 a spring Chinook supplementation program began in the upper Yakima River with the establishment of the Cle Elum Supplementation Research Facility (CESRF). This program has significantly enhanced the number of spring Chinook salmon returning to the upper Yakima River (Sampson et al. 2009). However, hatchery conditions have inadvertently increased the number of yearling precocious males in the upper Yakima River. Analyses done by Larsen et al. (2004) indicated that 37–49% of the males produced by the CESRF underwent precocious development and became sexually mature in the second year of life. Many of these fish emigrated downstream or perished after release but a portion of them remained or migrated back upstream and became part of the upper Yakima River spring Chinook spawning population (Larsen et al. 2004; Pearsons et al. 2009). The CESRF also significantly augmented the number of jacks returning to the upper Yakima River (Knudsen et al. 2006; 2010). Increases in the abundance of precocious males and jacks due to artificial culture has been observed in other spring Chinook populations (Gebhards 1960; Mullan et al. 1992; Unwin and Glova 1997) and is likely caused by the growth patterns and energy stores salmon experience under artificial culture (Thorpe 1994; Shearer and Swanson 2000; Larsen et al. 2010).

The enhanced occurrence of jacks and yearling precocious males due to supplementation raised several questions. First, even though large numbers of hatchery-origin yearling precocious males and jacks were produced by the CESRF how many of them are present on upper Yakima River spawning grounds? Second, if they are present, how successful are they in producing offspring? Pearsons et al. (2009) partially addressed the first question with their comprehensive assessment of the occurrence of hatchery-origin yearling precocious males in spring Chinook spawning areas in the upper Yakima River. Their annual surveys of upper Yakima River spring Chinook spawning grounds showed that on average just 22 hatchery precocious males (range 0–78) were found over 60 km of spawning habitat.

Our objective in this paper is to estimate the potential genetic contributions that early maturing hatchery-origin males may make to the Yakima River spring Chinook salmon population, and by extension, the effect that such males may have on other natural populations undergoing supplementation. Concerns about how supplementation may change the genetic composition of salmonid populations prompted Mobrand et al. (2005) to propose that the proportion of natural-origin fish on spawning grounds and in hatchery broodstocks should always be greater than the corresponding proportion of hatchery fish. This protocol dilutes potential domestication because fifty percent or more of the genes transmitted from one generation to the next by hatchery fish are subjected to natural selection pressures (Mobrand et al. 2005).

In some cases, it is possible to monitor and control the occurrence of hatchery fish on spawning grounds. However, this becomes problematic when dealing with precocious males that reside and mature in freshwater and sometimes with jacks that may be able to pass through weirs or other obstructions designed to stop larger fish. Additionally, early maturing males produced from hatcheries potentially represent a greater source of domestication than hatchery fish maturing at older ages. This happens because they are subject to natural selection pressures for shorter periods of time (Pearsons et al. 2009). And if produced by hatchery-origin parents they may be the most domesticated fish in a population since they have been exposed to an additional generation of artificial culture.

In two previous papers we compared the breeding success of first-generation hatchery and wild female (Schroder et al. 2008) and male (Schroder et al. 2010) spring Chinook salmon spawning in an artificial stream. Because of the ubiquitous presence of precocious males and jacks in the upper Yakima River, and the desire to mimic natural social conditions, we included precocious males and jacks along with large anadromous males and females in seven test groups placed into the artificial stream. Their inclusion in our test groups gave us an opportunity to evaluate offspring production by males having different male life history types when they occurred simultaneously in the same spawning community. Our earlier work (Schroder et al. 2010) indicated no detectable differences occurred in the behavior and breeding success of large anadromous males of hatchery and wild origin. Consequently, here we combine data from hatchery and wild males assuming that both types within the same male life history strategy had similar breeding success values. This increased statistical power and provided an overall estimate that each male life history type had on offspring production in our test groups.

In making this evaluation we examined the importance of four factors, relative body size, spawning ground longevity, sex ratios, and the prevalence of each male life history type on the percentage of offspring they produced within a spawning population. These factors were chosen because they are linked to male breeding success. Male body weight, for example, is known to be positively correlated with breeding success in large anadromous males (Schroder et al. 2010). Similarly, males that reside on spawning grounds for relatively long periods of time should have more breeding opportunities than those with shorter life times. Additionally, the operational sex ratio or number of males competing for each female, influences intra-sexual competition and the expression of behavioral tactics within and across male life history types (Emlen and Oring 1977; Quinn et al 1996). Finally, the prevalence of males representing different life history types may also determine how many offspring each type produces and the behavioral tactics that are employed on spawning grounds (Berejikian et al. 2010).

We used DNA-based pedigree assessments to explore the potential importance of each of these factors on the breeding success of males placed into our artificial stream. Results were used to evaluate the effects that increased numbers of early maturing males may have on spawning populations. The insights gained also provided new information about the relative reproductive success of alternative male life history strategies in spring Chinook salmon.

METHODS

Origin and collection of spring Chinook salmon

Adult female, large anadromous males, jacks, and yearling precocious males were collected from the upper Yakima River from April through August at the Roza Adult Monitoring Facility. This facility is located at rkm 206 (as measured from the confluence with the Columbia River) and upper Yakima River spring Chinook must pass through it before reaching their spawning grounds. Methods described by Knudsen et al. (2006) were used to collect a representative sample of fish. Collected fish were transported 81 km to the CESRF and females, large anadromous males, and jacks were held in a 30.5 m long by 4.6 m wide by 3 m deep pond while yearling precocious males were placed into a 30.5 m long by 3.5 m wide by 1.1 m deep hatchery raceway. Sub-yearling precocious males were collected off natural redds in the upper Yakima River by using hook and line or electrofishing gear. After capture they were transported to the CESRF and held in small circular tanks.

Fish selection and introduction into the artificial stream followed the protocols previously described in Schroder et al. (2008, 2010). Briefly, beginning in early September, fish were examined weekly to determine their maturation status. Only ripe fish were selected for use in the artificial stream. Prior to being placed into the stream they were anesthetized in a 1:19,000 part solution of tricaine methanesulfonate (Bell 1964). Once docile, large anadromous males, females, and jacks were weighed to the nearest gram while yearling and sub-yearling precocious males were weighed to the nearest tenth of a gram on electronic balances. Fork lengths to the nearest mm were taken on all fish. The large anadromous fish and jacks were tagged with Petersen disks but no external tags were applied to precocious males due to their small size. A tissue sample from the posterior edge of the dorsal fin was removed from each fish and placed into 100% ethanol for later DNA extraction and genetic analysis. The fish were then transported 200 m and released into the artificial stream. The origin, number, fork lengths, tertiary sex ratios, and the percentage of the males in each test group that represented different life history types are shown in Table 1.

Artificial stream

The artificial stream is located on the grounds of the CESRF. It is 127 m long by 7.9 m wide and has six straight sections that are 15.2 m long by 7.9 m wide plus one curved section that is 21 m long by 7.9 m wide. The sections are separated from one another by concrete cross weirs. Each section has a level gradient and possesses gravel consisting of material ranging from 0.7 to 10.0 cm in diameter with a Fredle Index (Lotspeich and Everest 1981) of seven. A series of 30-cm falls separate one section from another. Water velocity (0.1–2.0 m/s), discharge (0.37 m^3/s) and depth (mean 0.4 m) mimicked conditions that spawning Chinook salmon typically prefer (Bjornn and Reiser 1991; Healey 1991). A 2.1-m high wall of camouflage netting was installed on both banks with observation openings at eye level every m along its length (Schroder et al. 2008).

Seven test groups of Chinook salmon were placed into the artificial stream from 2001 through 2005. In 2001 and 2002, the artificial stream was subdivided into an up and downstream part, each having three straight sections (total length equaled 45.6 m). In these two years, test groups were placed into each part of the stream. Single test groups were placed into the entire stream in 2003, 2004, and 2005. In all years, picket barriers were used to confine the fish to parts of the stream set aside for their test group.

Fish were introduced into a test group on the same day over a 2 to 3 h period. They were always released into the uppermost section of the artificial stream allotted to their test group and could move freely within the part of the stream that had been sequestered for their population. In those years where two test groups were placed into the stream, seven days elapsed between when fish were placed into the upper and lower parts of the artificial stream (Schroder et al. 2010).

Table 1 The number, fork length, tertiary sex ratio, and percentage of males using different life history strategies in the test groups placed into the artificial stream from 2001–2005

Test Group	Sex	Type	No.	Fork Length-Range (mm)	Mean Fork Length	Tertiary Sex Ratio[e]	% Males[f]
2001 A	Female	Large Anadromous[a]	16	589-774	729	1.5	
	Male	Large Anadromous	21	591-908	743		87.5%
		Jack[b]	3	526-531	528		12.5%
		Yr-Precocious[c]	0	—	—		0.0%
		Sub-Yr Precocious[d]	0	—	—		0.0%
2001 B	Female	Large Anadromous	17	668–770	715	1.8	
	Male	Large Anadromous	18	560–821	716		60.0%
		Jack	2	498–519	509		6.7%
		Yr-Precocious	5	177–227	203		16.7%
		Sub-Yr Precocious	5	77–87	81		16.7%
2002 A	Female	Large Anadromous	22	662–832	756	1.4	
	Male	Large Anadromous	22	660–841	733		73.3%
		Jack	2	442–458	450		6.7%
		Yr-Precocious	6	142–190	166		20.0%
		Sub-Yr Precocious	0	—	—		0.0%
2002 B	Female	Large Anadromous	17	660–760	718	1.4	
	Male	Large Anadromous	16	650–880	749		66.7%
		Jack	2	492–520	506		8.3%
		Yr-Precocious	6	145–206	184		25.0%
		Sub-Yr Precocious	0	—	—		0.0%
2003	Female	Large Anadromous	24	655–900	757	2.1	
	Male	Large Anadromous	24	560–890	758		48.0%
		Jack	6	470–580	545		12.0%
		Yr-Precocious	7	122–203	163		26.0%
		Sub-Yr Precocious	13	85–117	102		14.0%
2004	Female	Large Anadromous	20	644–800	721	2.4	
	Male	Large Anadromous	29	560–900	728		61.7%
		Jack	4	402–521	469		8.5%
		Yr-Precocious	12	122–221	177		25.5%
		Sub-Yr Precocious	2	114–115	115		4.3%
2005	Female	Large Anadromous	23	610–767	703	2.0	
	Male	Large Anadromous	25	680–920	772		55.6%
		Jack	4	494–530	514		8.9%
		Yr-Precocious	8	170–215	191		17.8%
		Sub-Yr Precocious	8	61–110	93		17.8%

[a] Anadromous fish maturing at ages 4 or 5

[b] Anadromous males maturing at age 3

[c] Non-anadromous males maturing after 18 to 20 months of freshwater residency

[d] Non-anadromous males maturing after 7 to 8 months of freshwater residency

[e] Number of males per female

[f] Percentage of males in a test group representing a life history type

Estimating male breeding success

In 2001 and 2002 modified fyke nets with floating live boxes were placed just below each part of the stream that contained a test group. In 2003, 2004, and 2005 the traps were installed immediately below the last section of the artificial stream. The traps captured newly emerged fry and parr (Allan and Ritter 1977) hereafter referred to collectively as fry, as they emigrated from the stream. Traps were installed prior to emergence and were in continuous operation until fry migration ceased, then electrofishing gear and seines were used to capture any remaining juveniles. Ten percent of each day's catch was preserved in 100% ethanol and used in microsatellite DNA pedigree assessments. The pedigree analyses indicated the number of fry each male had contributed to the ten percent sample collected from his test group. The breeding success of each male was estimated by dividing the number of fry he produced in the pedigree sample by the total number of fry in the sample.

Pedigree analyses

Methods used to perform the pedigree analyses have been previously described in Schroder et al. (2008, 2010). For test groups 2001A and 2001B, genomic DNA was extracted from the fry and adult samples by digesting their tissues in a 5% chelex solution containing 0.4 mg proteinase K. Following digestion the samples were heated to denature proteins and DNA extracts were stored at 5°C until all analyses were completed. Spin-column extraction kits from Machery-Nagel were used to purify genomic DNA from fish in test groups 2002A–2005. Adults and fry were genotyped at 10 to 13 loci. The number of alleles per locus ranged from 5 in Ots-1 to 40 at Ots-100. Microsatellite DNA loci were amplified via the polymerase chain reaction (PCR) using fluorescent-labeled primers obtained from Applied Biosystems or Integrated DNA Technologies. Data were collected using an ABI-3100 Genetic Analyzer. Applied Biosystems Genemapper 3.0 software was used to collect, analyze and determine genotypes at each locus. Allele identification on sampled fry was attempted on all loci and fry had to be genotyped at six or more loci before they were assigned to a parent fish. A maximum likelihood procedure in Cervus 2.0 (Marshall et al. 1998) was used to infer parent-offspring relationships. The procedure uses allele frequency data to assign likelihoods to parent-offspring combinations and allows mismatching genotypic data to be evaluated concurrently with matching genotype data.

Data analysis

Statistical analyses were performed using SYSTAT 12 software (SYSTAT 2007). All tests were two-sided and an alpha level of 0.05 was used.

Body lengths and longevity Body length comparisons were performed using a two-step process. First, Model I, 2-way ANOVAs were performed on the fork length data collected on males within a life history type. One fixed factor was test group origin while the other fixed factor was male origin, i.e. were the fish of wild or hatchery-origin. These analyses were performed to determine if all the length data obtained from males having the same life history could be pooled. A one-way ANOVA was then performed to assess whether males with different life histories had different mean fork lengths at maturity. A *post hoc* Tukey's Honestly-Significiant-Difference Test was used to determine which male life history types had different mean fork lengths.

Longevity or the number of hours a male resided in the artificial stream was calculated by subtracting the median time and date his test group entered the artificial stream from the time and date he was first observed dead. Mean longevity values for males representing each life history type within a test group were calculated by summing all the longevity data for a given male type and then dividing this value by the number of males of that type that were present. The resulting mean longevity values were used in a non-parametric (Friedman's Test) randomized block analysis of variance and subsequent Student-Newman Multiple range test to determine if longevity differences existed among males having different life histories. The longevity of sub-yearling precocious males was not compared to the other male life history types because too few were recovered.

The effect of body length on breeding success in all four types of males was examined by using regression analyses. Data from each test group were analyzed separately. Seven regressions, one for each test group, were performed on large anadromous males. In these regressions fork length (log 10) was the independent variable and breeding success, after it had been normalized by the arcsine transformation, was the dependent variable. Six similar regressions were conducted on data obtained from yearling precocious males, and three each were run on jacks and sub-yearling precocious males. The *p* values obtained from the regressions performed on each male type were used in meta-analyses referred to as combined probability tests (Sokal and Rohlf 1995). The combined probability tests allowed us to discover if fork length affected breeding success within a male life history type across multiple test groups.

The effect of longevity on breeding success within each male type was evaluated by using non-parametric Kendall tau rank correlations. In this instance, individual breeding success values were correlated with longevity values. Seven such correlations were performed on data collected from large anadromous males. Four were performed on yearling precocious males while three were conducted on jacks. Not enough longevity data were collected on sub-yearling precocious males to assess the importance of this trait on their breeding success.

Tertiary sex ratios Each test group had a different tertiary sex ratio or absolute number of males per female (Table 1). Because all the fish placed into the test groups were ripe at the time of entry and no new recruits were allowed to enter any of the test groups, the tertiary sex ratio was used to estimate male intra-sexual competition for females in each of the test groups. Regression analyses were performed to determine how much variation in male breeding success (the dependent variable) could be explained by the tertiary sex ratios in the test groups (independent variable). Three regressions were performed; one for large anadromous males, another for jacks, and one for yearling precocious males.

The effect of male intra-sexual competition for females on the proportion of fry produced by each male life history type in the test groups was also examined by regression. Four separate regressions were performed, one for each male life history type. The independent variable in these analyses was the tertiary sex ratio that was present in each test group. The dependent variable was the percentage of the fry population in a test group that a male life history type had produced. The arcsine transformation was used to normalize the percentage of the fry population in a test group that had been fathered by a male type. No data transformation was used on the tertiary sex ratios.

Prevalence of male life history types Four regression analyses, one for each male life history type, were performed to assess how much variation in fry paternity could be explained by how prevalent a male type was in a spawning population. In these analyses the percentage of the male spawning population representing a life history type was the independent variable. The dependent variable was the percentage of the fry population that had been fathered by that life history. Prior to performing the regressions both percentages were normalized by using the arcsine transformation.

Breeding success Average breeding success values for each male life history type were determined in the 2001B, 2003, 2004, and 2005 test groups. These test groups were chosen because males with all four male life history types were present. Mean values were calculated by summing individual breeding success values of all the males within the same life history type and then dividing that sum by the number of males having that strategy. These average values estimated the mean percentage of fry produced from a test group that had been fathered by males in each life history type. They were normalized by using the arcsine transformation. A randomized block analysis of variance and a *post hoc* Student-Newman Kuels multiple range test were used to determine if life history type affected male breeding success. Differences in breeding success were observed. To help explain why there were differences we also evaluated whether male history type affected mate number and progeny produced per mate.

Number of mates The pedigree analyses revealed the number of different females each male in a test group had spawned with. These data were sorted by male type, summed, and divided by the number of males in each life history type. The resulting quotients were used to estimate the mean number of mates males with different life history types had spawned with. Average mate numbers were determined for each male life history type in the 2001B, 2003, 2004, and 2005 test groups. A randomized block analysis of variance and *post hoc* Student-Newman Kuels multiple range test were used to see if male life history affected mean mate numbers.

Number of fry per mate The pedigree analyses also indicated the number of fry produced by each female a male spawned with. These values were sorted by male life history type, summed and divided by the number of male x female pairs that occurred within a male type. The quotients equaled the average number of progeny produced per mate in the fry used for the pedigree analysis. Estimates of the number of fry produced per mate were determined by dividing the above average values by the percentage of the fry from a test group that had been used in a pedigree assessment. For example, in the 2004 test group, 86 different large anadromous male x female pairs produced 2699 fry. The average number of fry produced by each pair was 31 (2,699/86). This test group produced 47 140 fry and 2892 or 6.1% of them were assigned to parental fish. Consequently, large anadromous males produced an average of 511 fry per mate. Similar calculations were done for each male life history type in the 2001B, 2003, 2004, and 2005 test groups. The resulting values were used in a randomized block analysis of variance and a *post hoc* Student-Newman Kuels multiple range test to compare the mean number of fry produced per mate among the four male life history types.

Relative breeding success To estimate the relative breeding success of jacks and both types of precocious males compared to large anadromous males, we determined the mean number of fry each type of male produced in every test group. This was done by dividing the total number of fry a male type had contributed to the fry used in our pedigree assessments by the number of males having that life history type. Relative breeding success was then calculated by dividing the mean fry values of jacks, and the two types of precocious males by the mean value obtained from the large anadromous males that had been in their test group.

RESULTS

Production of juveniles from the test groups varied from 18 960 (test group 2001 A) to 64 494 (test group 2003). Altogether 30 683 fry were sampled from the test groups for DNA analysis. A total of 13 779 were analyzed and 96% of those were assigned to parental fish. The percentage of the fry population produced from a test group that were successfully assigned to parental fish ranged from 1.8% (test group 2001B) to 6.1% (test group 2004) and averaged 4.3% (Schroder et al. 2010).

Fork length

The fixed factor two-way ANOVAs that examined whether male origin (first-generation hatchery or wild) or test group affected fork lengths in large anadromous males and jacks were non-significant ($p = 0.169$ for origin and $p = 0.116$ for test group in large anadromous males; $p = 0.915$ for origin and $p = 0.133$ for test group in jacks) and therefore length data within these two male types were pooled. Test group origin also had no effect on the fork lengths of yearling or sub-yearling precocious males ($p = 0.394$ for yearling precocious; $p = 0.093$ for sub-yearling precocious males). However, hatchery-origin yearling precocious males were significantly longer than wild yearling precocious males ($p < 0.001$). The one-way ANOVA and *post hoc*Tukey's Honestly Significant Difference tests that compared the fork lengths among males with different life histories indicated that there were dissimilarities in the mean fork lengths among all the life history types, with large anadromous males being the largest followed by jacks, hatchery yearling precocious, wild yearling precocious, and wild sub-yearling precocious males (Table 2).

Longevity

Male life history type affected longevity or the number of hours males with different life histories resided in the artificial stream ($p = 0.027$). On average, yearling precocious males lived for longer periods of time than both large anadromous males ($p = 0.017$) and jacks ($p = 0.035$). No difference, however, was observed between the longevity of large anadromous males and jacks ($p = 0.416$). Large anadromous males had a mean longevity of 152 h; jacks had an overall longevity value of 173 h while yearling precocious males lived for an average of 216 h (Table 2).

Table 2 Results of the randomized block and *post hoc* tests used to compare the fork lengths and spawning ground longevity of spring Chinook males having different life history strategies that were placed into the artificial stream

Trait	Male life history types compared	Mean values	Difference	*p* value	Conclusion
Fork Length (mm)	Large Anadromous vs. Jack	744 vs. 510	234	<0.001	Reject H_o
	Large Anadromous vs. Hatchery Yearling Precocious	744 vs. 194	550	<0.001	Reject H_o
	Large Anadromous vs. Wild Yearling Precocious	744 vs. 141	603	<0.001	Reject H_o
	Large Anadromous vs. Sub-Yearling Precocious	744 vs. 93	651	<0.001	Reject H_o
	Jack vs. Hatchery Yearling Precocious	510 vs. 194	316	<0.001	Reject H_o
	Jack vs. Wild Yearling Precocious	510 vs. 141	369	<0.001	Reject H_o
	Jack vs. Sub-Yearling Precocious	510 vs. 93	417	<0.001	Reject H_o
	Hatchery Yr Precocious vs. Wild Yearling Precocious	194 vs. 141	53	<0.001	Reject H_o
	Hatchery Yr Precocious vs. Sub-Yearling Precocious	194 vs. 93	101	<0.001	Reject H_o
	Wild Yr Precocious vs. Sub-Yearling Precocious	141 vs. 93	48	<0.001	Reject H_o
Spawning Ground Longevity (hrs)	Yearling Precocious vs. Large Anadromous	216 vs. 152	64	0.017	Reject H_o
	Yearling Precocious vs. Jack	216 vs. 173	43	0.035	Reject H_o
	Jack vs. Large Anadromous	173 vs. 152	21	0.416	Fail to reject H_o

Fork length and longevity effects on breeding success

The effect of fork length on male breeding success was examined by using regression analyses. For large anadromous males, regression coefficients ranged from 15 to 47% and averaged 25%. The combined probability test that examined the importance of this trait on breeding success in large anadromous males showed this trait was positively associated with breeding success ($p < 0.001$). The combined probability tests that examined the importance of length on male breeding success in the other male life history types were all non-significant ($p = 0.429$ for jacks, $p = 0.942$ for yearling precocious, and $p = 0.254$ for sub-yearling precocious males). Figure 1 illustrates how fork length and breeding success were associated with one another in all four types of males placed in the 2003 test group.

In two of the test groups, significant negative Kendall tau rank correlations between spawning ground longevity and breeding success in large anadromous males were found. The combined probability test that examined the overall effect of longevity on breeding success showed that longevity in large anadromous males was negatively correlated to breeding success ($p = 0.046$). Conversely, both positive and negative relationships between these two variables were seen in jacks and yearling precocious males and thus no consistent trends occurred in these two life history types.

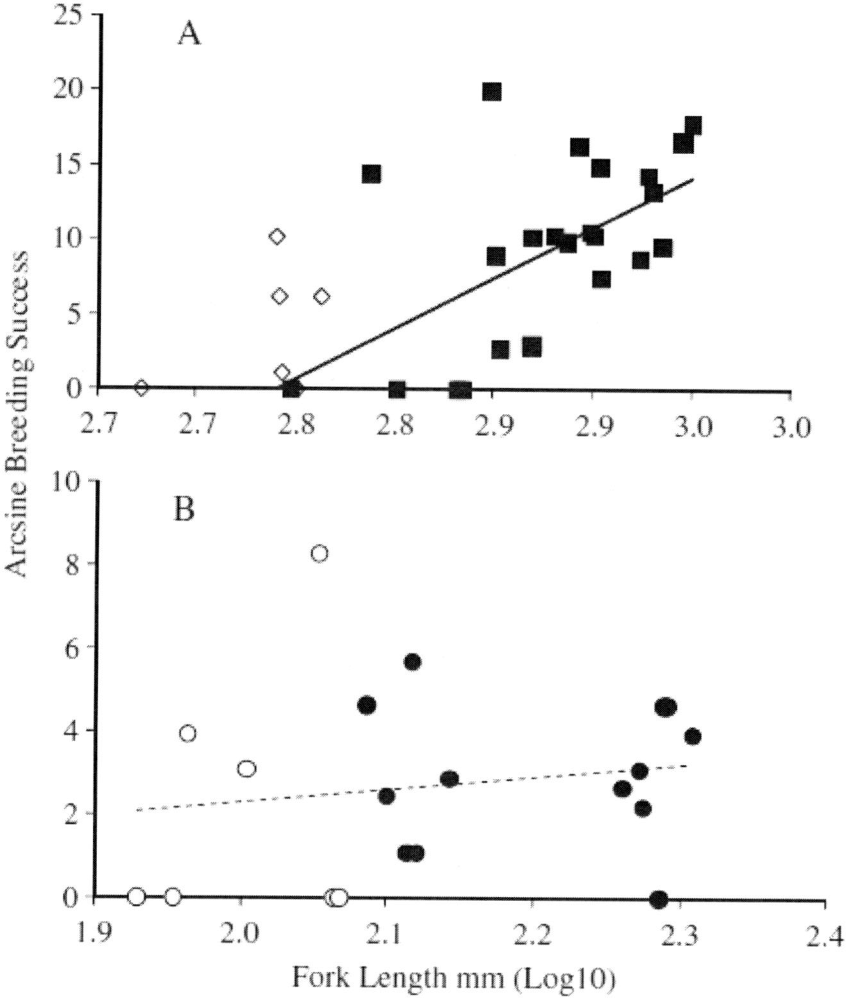

Figure 1 The effect of fork length on the breeding success of large anadromous males (*black squares*), jacks (*white diamonds*), yearling (*black circles*) and sub-yearling precocious males (white circles) spawning in the 2003 test group. Part A illustrates the significant relationship ($r^2 = 0.314$; $p = 0.004$) between these two variables (solid line) in large anadromous males and the non-significant association between them in jacks ($r^2 = 0.163$; $p = 0.428$). Part B shows the non significant relationship

($r^2 = 0.035$; $p = 0.431$) between fork length and breeding success (*dashed line*) in yearling and sub-yearling precocious males

Tertiary sex ratio

As the absolute number of males per female or tertiary sex ratios in the test groups increased, mean breeding success in large anadromous males decreased (Fig. 2). Similar, but non-significant decreases in mean breeding values occurred in jacks ($r^2 = 0.411, p = 0.121$) and yearling precocious males ($r^2 = 0.432, p = 0.156$). The effect that tertiary sex ratios had on the proportion of fry each male life history type produced was also examined by regression. These analyses showed that tertiary sex ratios in our test group did not influence the percentage of fry each male life history type produced (Fig. 3). Regression coefficients equaled 0.002 ($p = 0.919$) for large anadromous males, 0.063 ($p = 0.587$) for jacks, 0.228 ($p = 0.338$) for yearling precocious, and 0.003 ($p = 0.946$) for sub-yearling precocious males. Thus, under the array of tertiary sex ratios present in our test groups, large anadromous males consistently produced around 89% of the offspring and on average jacks produced 3%, yearling precocious 7% and sub-yearling males 1%.

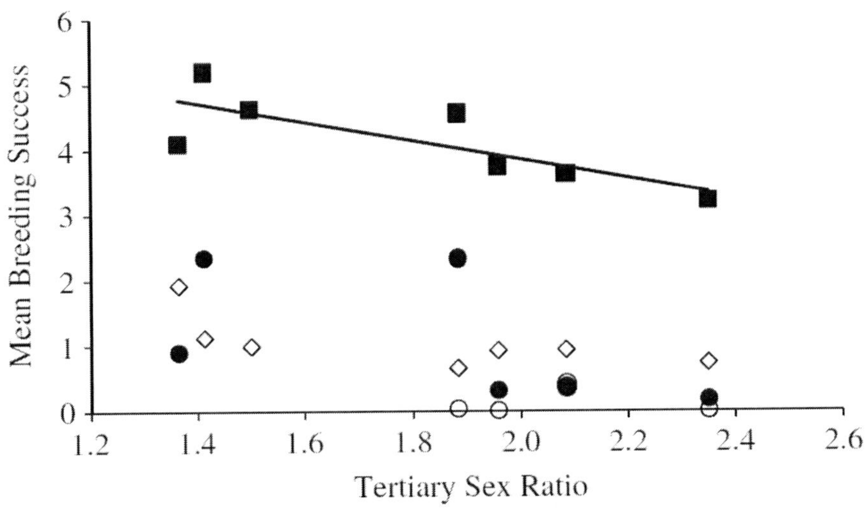

Figure 2 The influence of tertiary sex ratios on mean breeding success in large anadromous males (*black squares*), jacks (*white diamonds*), yearling (*black circles*) and sub-yearling precocious males (*white circles*). A significant negative relationship between these two variables occurred in large anadromous males ($r^2 = 0.643$; $p = 0.030$). Negative, but non-

significant relationships occurred in jacks ($r^2 = 0.411$; $p = 0.121$) and yearling precocious males ($r^2 = 0.432$; $p = 0.156$)

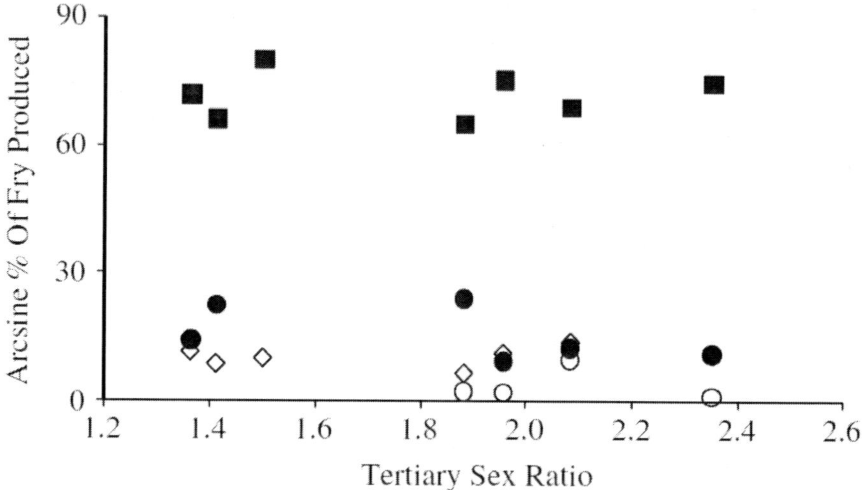

Figure 3 The proportion of fry fathered by large anadromous males (*black squares*), jacks (*white diamonds*), yearling (*black circles*) and sub-yearling precocious males (*white circles*) under different tertiary sex ratios in the test groups

Prevalence of male life history types

If each male life history type were equally successful at producing offspring, then there would be a one-to-one relationship between the percentage of a male type present on a spawning ground and the percentage of the fry population originating from that type of male. The relationship between the prevalence of male life history types spawning in our test groups and the paternal origin of fry produced from those groups is shown in Fig. 4. In every test group, large anadromous males produced about 89% of the fry regardless of their relative occurrence. No relationship was found between the percentage of large males in the test groups and the percentage of fry that life history type produced ($r^2 = 0.334$, $p = 0.174$). Similarly, the percentage of jacks ($r^2 = 0.142$; $p = 0.405$), yearling ($r^2 = 0.048$; $p = 0.676$) and sub-yearling precocious males ($r^2 = 0.068$; $p = 0.740$) in our test groups did not influence the percentage of fry they were able to produce.

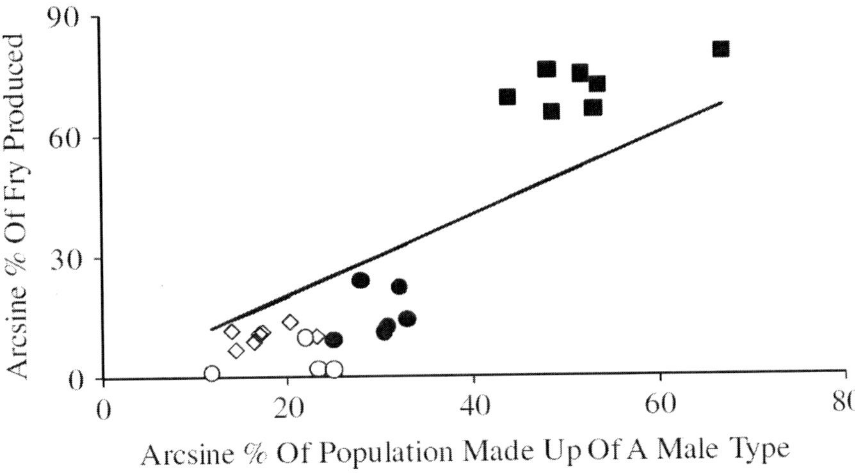

Figure 4 The occurrence of male life history types in the test groups versus the proportion of juveniles they fathered. The *black diagonal line* represents the percentage of fry each male life history type would be expected to produce if individuals within each type had an equal ability to produce offspring. Large anadromous males are represented by black squares, jacks by white diamonds, yearling precocious males by *black circles* and sub-yearling precocious males by *white circles*

Mean breeding success comparisons

Mean breeding success values among the four male life history types were different ($p < 0.001$, Table 3). Large anadromous males had greater mean breeding values than jacks ($p = 0.001$), yearling precocious ($p < 0.001$) and sub-yearling precocious males ($p < 0.001$). No difference was found in the breeding success values of jacks and yearling precocious males ($p = 0.555$). However, both had higher mean breeding success values than sub-yearling precocious males ($p = 0.016$ for jacks and $p = 0.020$ for yearling precocious males).

To understand why differences in breeding success had occurred we evaluated whether male life history type affected the average number of mates individuals in each life history type spawned with. Male life history type did influence mate number ($p = 0.002$). On average large anadromous males mated with more females than sub-yearling precocious males ($p < 0.001$), yearling precocious males ($p = 0.034$), and jacks ($p = 0.029$). Jacks and yearling precocious males had similar mean mate numbers ($p = 0.686$) and both had higher mean mate numbers than sub-yearling

Table 3 Results of the randomized block and *post hoc* tests used to compare the breeding success, number of mates, and number of fry produced per mate of spring Chinook salmon males possessing different life history strategies that were used in the test groups

Trait	Male life history types compared	Mean values	Difference	*p* value	Conclusion
Breeding Success[a]	Large Anadromous vs. Sub-Yearling Precocious	9.3 vs. 0.8	8.5	<0.001	Reject H_o
	Large Anadromous vs. Yearling Precocious	9.3 vs. 3.8	5.5	0.001	Reject H_o
	Large Anadromous vs. Jack	9.3 vs. 4.5	4.8	0.001	Reject H_o
	Jack vs. Sub-Yearling Precocious	4.5 vs. 0.8	3.7	0.016	Reject H_o
	Jack vs. Yearling Precocious	4.5 vs. 3.8	0.7	0.555	Fail to reject H_o
	Yearling Precocious vs. Sub Yearling Precocious	3.8 vs. 0.8	3.0	0.020	Reject H_o
Mate Number[b]	Large Anadromous vs. Sub-Yearling Precocious	3.4 vs. 0.5	2.9	0.001	Reject H_o
	Large Anadromous vs. Yearling Precocious	3.4 vs. 1.9	1.5	0.034	Reject H_o
	Large Anadromous vs. Jack	3.4 vs. 2.1	1.3	0.029	Reject H_o
	Jack vs. Sub-Yearling Precocious	2.1 vs. 0.5	1.6	0.020	Reject H_o
	Jack vs. Yearling Precocious	2.1 vs. 1.9	0.2	0.686	Fail to reject H_o
	Yearling Precocious vs. Sub Yearling Precocious	1.9 vs. 0.5	1.4	0.017	Reject H_o
Progeny Per Mate[c]	Large Anadromous vs. Sub-Yearling Precocious	580 vs. 131	449	0.002	Reject H_o
	Large Anadromous vs. Yearling Precocious	580 vs. 175	405	0.004	Reject H_o
	Large Anadromous vs. Jack	580 vs. 212	368	0.006	Reject H_o
	Jack vs. Sub-Yearling Precocious	212 vs. 131	81	0.332	Fail to reject H_o
	Jack vs. Yearling Precocious	212 vs. 175	37	0.506	Fail to reject H_o
	Yearling Precocious vs. Sub-Yearling Precocious	175 vs. 131	44	0.369	Fail to reject H_o

[a] Breeding success is the arcsine transformed average percentage of fry in a pedigree analysis produced by males in each life history strategy

[b] Mate number equals the mean number of different females males in each life history strategy spawned with as indicated by the pedigree analyses

[c] Progeny per mate is the mean number of fry males in each life history strategy produced per female they spawned with

precocious males ($p = 0.020$ for jacks and $p = 0.017$ for yearling precocious males) (Table 3). Male life history type also influenced the average number of progeny produced per mate ($p < 0.001$). Large anadromous males produced an average of 580 fry per mate which was greater than the 131 fry per mate for sub-yearling precocious males ($p = 0.002$), 175 fry per mate in yearling precocious ($p = 0.004$) and 212 fry per mate for jacks ($p = 0.006$). No differences, however, were seen in fry production per mate among jacks and the two types of precocious males ($p = 0.506$ for jacks and yearling precocious males; $p = 0.332$ for jacks and sub-yearling precocious males; and $p = 0.369$ for yearling and sub-yearling precocious males, Table 3).

The biological consequences of differences in average mate number, fry per mate, and overall breeding success in the four male life history types can be examined by comparing their relative breeding success values (Table 4). On average, jacks and yearling precocious males had breeding success values that were 25.0% and 22.0% respectively of those obtained by large anadromous males. Overall sub-yearling precocious males had breeding success values that were 3.2% of that realized by large anadromous males. These values were variable however. For example, within our test groups, breeding success of jacks relative to large anadromous males ranged from 16.6% to 46.8%. Comparable variation was found in yearling precocious males whose breeding success relative to large anadromous males varied from 8.5% to 51.0%. Similarly, breeding success in sub-yearling precocious males relative to large anadromous males fluctuated from 0.3% to 11.2%.

Table 4 The breeding success of jacks, yearling precocious and sub-yearling precocious males in each test group relative to that achieved by large anadromous males

Test group	Breeding success relative to large anadromous males[a]		
	Jacks	Yearling precocious	Sub-yearling precocious
2001 A	16.6%	—	—
2001 B	14.1%	51.0%	0.6%
2002 A	46.8%	14.3%	—
2002 B	21.3%	39.2%	—
2003	25.3%	9.7%	11.2%
2004	26.4%	9.4%	0.5%
2005	24.2%	8.5%	0.3%
Over All Average	25.0%	22.0%	3.2%

[a] Relative breeding success was estimated by dividing the mean number of offspring produced in a test group by either jacks, yearling, or sub yearling precocious males by the mean number of offspring produced by large anadromous males

DISCUSSION

Spring or stream-type Chinook salmon (Healey 1991) populations probably possess the most diverse array of male life history strategies in semelparous Oncorhynchids. Some males reach maturation after just 7 to 8 or 18 to 20 months of freshwater growth (sub-yearling and yearling precocious males), while others utilize an anadromous life history strategy and reach maturation after spending one or more years in the ocean (jacks and large anadromous males) before returning to freshwater to spawn. In our experimental setting, large anadromous males achieved significantly higher mean individual breeding success values than the other three male life history phenotypes examined even when they were outnumbered by males using different life history strategies. Jacks and yearling precocious males had comparable mean breeding success values, while sub-yearling precocious males had the lowest breeding success values.

Significant differences existed among the male life history types in their body sizes, spawning ground longevity, and their ability to fertilize eggs from multiple females. Of these factors, body size was probably the proximate factor responsible for the detected differences in breeding success. Large anadromous males were on average over 200 mm longer than jacks, the next largest male life history type, and obviously had even greater size advantages over the precocious males (Table 2). Their superior size allowed them to be socially dominant over the other three male life history types and provided them with a greater number of opportunities to spawn with females. For example, on average, large anadromous males in our test groups mated with 3.4 females, jacks with 2.1, yearling precocious with 1.9 and sub-yearling precocious with 0.5 females (Table 3).

Even though there was a large disparity in their body sizes, jacks and yearling precocious males spawned with a similar number of females. Both located themselves in close proximity to spawning pairs but used slightly different tactics to participate in spawnings. Jacks typically positioned themselves just downstream of a dominant pair and persistently remained in this location even though they were often subject to repeated attacks by both members of the dominant pair. When able to detect a spawning event, they rushed upstream and participated by lying adjacent to a spawning pair or in a few cases by inserting themselves between the dominant male and the spawning female. Yearling precocious males, on the other hand, were often seen in the nest depression or lying closely adjacent to a dominant male or female. Initially these small fish were subject to repeated chases and attacks by both members of a courting pair. Their ability to use shallow water refugia, great maneuverability, tenacious fidelity to a developing nest, and the eventual habituation of their presence by the dominant pair, allowed them to be in close proximity to the larger fish during egg deposition. This same tactic was employed by sub-yearling precocious males (Schroder and Knudsen unpubl. data). The capacity of sub-yearling precocious males to mate with different females, however, may have been limited by the amount of milt they were able to produce.

Although important, the ability to spawn with numerous females only partially explained why differences existed among the male life history types in their mean individual breeding success values. The number of fry produced per mate was another critical factor. In our test groups large anadromous males produced three to four times as many fry per mate as jacks, yearling precocious and sub-yearling precocious males (Table 3). A variety of factors may have contributed to this difference. The aggressive behavior of large anadromous males, for instance, often delayed sperm release by competitors. This gave their sperm precedence over any released by rivals and therefore increased the likelihood they would fertilize most of the eggs in a multiple-male spawning (Berejikian et al. 2010). Large salmonid males also release significantly greater volumes of milt (Taborsky 1998) at higher pressures (Mjolnerod et al. 1998) than smaller competitors. Together these two factors may enhance their ability to fertilize eggs as sperm released by rivals could be flushed out or numerically overwhelmed (Mjolnerod et al. 1998). A proclivity for salmonid females to prefer relatively large mates (Taborsky 1998; Blanchfield and Ridgway 1999) might have further contributed to their higher breeding success values. The number of offspring produced per mate in jacks and the two types of precocious males was similar. Thus, even though precocious males and jacks used alternative tactics to obtain fertilizations, they produced comparable numbers of progeny per mate.

The effects that body size, longevity, and different tertiary sex ratios had on individual breeding success values helped expose the selection pressures males in each life history type experienced while spawning. Large anadromous males mainly competed among themselves for access to females and in this life history type; body size was positively associated with social dominance and breeding success (Fig. 1). In the present study, body size did not influence the breeding success of jacks and the two precocious male life history types. As male densities around a courting pair increased it becomes more difficult for any male to exclude all his potential rivals. In brook trout, for example, per capita aggressive interactions among males surrounding an active female decreased as male numbers increased (Blanchfield and Ridgway 1999). The inability to exclude multiple rivals leads to scramble competition among peripheral males and under this circumstance body size is not as important as being in close proximity when spawning events occur (Jones and Hutchings 2001).

The negative relationship between longevity and male breeding success in large anadromous males was probably due to energy expenditures. Large anadromous males expend a considerable amount of effort on agonistic interactions and courtship. In our test groups, large anadromous males with high breeding success values engaged in more attacks and courting activities than those that had lower breeding success values (Schroder et al. 2010). Conversely, spring Chinook jacks and precocious males engage in less courtship and agonistic behavior (Schroder unpubl. data); instead they spend time and energy avoiding attacks and on finding and remaining in close proximity to spawning pairs. Because their breeding success is often determined by the outcomes of scramble competition (but *see* Grimardias et al. 2010a, 2010b), we speculate that energy expenditures directed toward excluding rivals does not affect their breeding success as strongly as it does in large anadromous males. Support for this speculation can be found in the relationship between tertiary sex ratios in the test groups and how this factor influenced individual breeding success values in the various male life history types. A negative relationship occurred between these two variables within each male life history type (Fig. 2). However, it was only significant in large anadromous males, suggesting that intra-sexual competition among males possessing the same life history type is probably more important for large anadromous males than it is for the other male life history types.

Individuals typically mating at a disadvantage, such as jacks and precocious males, are theoretically expected to employ behavioral, physiological or morphological tactics designed to offset this disadvantage (Andersson 1994; Gage et al. 1995; Taborsky 1998). Precocious males and jacks did have longer spawning ground longevities than large anadromous males. Relatively long spawning ground residency may be a reproductive adaptation by these life history types to increase their likelihood of participating in spawning events. Additionally, Mjolnerod et al. (1998) found that the recent spawning history of large anadromous Atlantic salmon (*Salmo salar*) had an effect on their ability to fertilize eggs. Individuals that had spawned extensively prior to a spawning event released fewer sperm. This meant that secondary males associated with such a spawning were able to achieve high breeding success values even though they had been socially dominated. Relatively long spawning ground residency may increase the probability of jacks and precocious males participating in a spawning of this type. The possible reproductive advantage associated with increased longevity was not fully evaluated in our experiment because additional males and females were not introduced into our test groups over time. Consequently, females were available to all the males over a shorter time period than they would have been under natural conditions.

Other reproductive adaptations observed in precocious males include higher gonadosomatic indices (GSI) than anadromous males (Gage et al 1995; Vladic and Jarvi 2001; Koseki and Maekawa 2002). And when adjusted for body size, precocious males also produce greater quantities of sperm that are more motile and live longer (but *see* Mjolnerod et al. 1998) than those produced by anadromous males (Gage et al. 1995; Vladic and Jarvi 2001). Alternatively, when Hoysak et al. (2004) examined sperm competition between sockeye jacks (3-yr-old fish) and large anadromous males (5 yr-old fish) they found no differences in sperm quality. Instead they hypothesized that the relatively small size of jack sockeye allowed them to gain closer access to females during spawning and thus enhanced their fertilization success. Precocious spring Chinook salmon have significantly higher GSI values than those found in large anadromous males (Knudsen and Schroder unpubl. data). No evaluations, however, have been made on gamete quality in jack and precocious male spring Chinook to determine whether their sperm and milt have traits similar to those found in other salmonid species with early maturing males.

Even though precocious males and jacks may have made greater investments in relative gonad size and possessed more motile and longer lived sperm; large anadromous males consistently produced about 89% of the fry generated from each of our test groups. Similar results in other salmonid species have been found. In an 8 km long Scottish river containing 34 to 66 anadromous Atlantic salmon females, 38 to 43 large anadromous males competed with an estimated two to three thousand precocious males for fertilization opportunities. In this population, large anadromous males represented just 1 to 2% of the male population yet they still fertilized about 50% of the eggs that were deposited (Taggart et al. 2001). Additionally, Hutchings and Myers (1988) established populations in simulated streams where large anadromous male Atlantic salmon represented 5% of the spawning males, yet these individuals fertilized approximately 75% of the deposited eggs. In another study using Atlantic salmon spawning in an artificial stream, 16% of the males were large anadromous individuals while the remaining 84% were precocious parr; in this setting anadromous males fertilized between 63 and 77% of the deposited eggs (Jones and Hutchings 2002). In a natural spawning population of brook trout, Blanchfield and Ridgway (1999) and Blanchfield et al. (2003) found that male size and a female preference for large males meant that over 90% of the fry produced originated from large males, even though there were numerous smaller "peripheral" males on the spawning grounds. In aggregate, our study and those cited above indicate that the largest males in a breeding population often produce a disproportionate number of fry when salmonids are allowed to spawn naturally.

The reproductive tactics used by each male life history type are conditional and may change with the relative abundance of the other types. For example, jacks will court females and fight among themselves for primary male status if large anadromous males become scarce (Allen et al. 2007). Even precocious males may assume this role if other male life history types are absent (Hutchings and Myers 1985; Beall et al. 1997; Martinez et al. 2001). Thus, the potential genetic contribution rates of different male life history types is dynamic; it may increase or decrease depending upon its prevalence in a spawning population. Similarly, the amount of intra-sexual competition occurring among males will alter how genes are exchanged. When there are just a few males competing among themselves for numerous sexually active females, each male, regardless of life history type, is expected to be successful. As intra-sexual competition among males escalates, average breeding success values for individuals within each life history type will decline. This decline is probably asymmetric as breeding success values in males using life history strategies that reach maturation at relatively small sizes should decline more sharply than those belonging to the male life history type that has the largest size at maturation. This occurs because when intra-sexual competition is high, large males that would normally assume primary courting roles will be forced by competitors to employ a peripheral or satellite tactic. Their utilization of this tactic further diminishes fertilization opportunities for males that matured at earlier ages and smaller body sizes because they must now compete with larger fish for satellite positions. Habitat complexity is another factor that could affect the relative breeding success of males adopting different life history strategies (Beall and de Gaudemar 1999). Grimardias et al. (2010a) for example, found a positive relationship between substrate complexity and the number of precocious males that fertilized eggs in Atlantic salmon redds. They speculated that substrate diversity provided precocious males with shelters that protected them from male aggression and allowed them close access to spawning females.

The primary purpose of our investigation was to estimate the potential genetic impact that a large influx of early maturing hatchery males might have on the upper Yakima River spring Chinook salmon population. Under the conditions in the artificial stream, jacks and precocious males fathered about 10% of the fry produced from each test group. The applicability of this result depends upon how closely conditions in the test groups resembled those occurring in the upper Yakima River. Currently, multiple thousands of anadromous spring Chinook spawn in the upper Yakima River (Dittman et al. 2010). Data from Pearsons et al. (2009) and Knudsen et al. (2010) were used to estimate the percentage of males representing each life history type present on upper Yakima River spawning grounds. From 1999 through 2007, 50 to 89% (x¯x¯ = 76%) of the males were large anadromous fish, 2 to 47% (x¯x¯ = 19%) were jacks, and 1 to 8% (x¯x¯ = 4%) were precocious males (sub-yearling and yearlings combined). In our test groups 48 to 87% (x¯x¯ = 65%) were large anadromous males, 7 to 13% (x¯x¯ = 9%) were jacks and 20 to 40% (x¯x¯ = 26%) were precocious males (Table 1). The proportion of large anadromous males in the test groups was similar to that found in the upper Yakima River, although the test groups did have fewer jacks and a greater percentage of both types of precocious males. Data from the artificial stream, however, suggest that when 50% or more of the males on a spawning ground are large anadromous individuals they typically produce 90% of the offspring. Since the majority of males spawning in the upper Yakima used the large anadromous strategy it is likely that they accounted for most of the progeny production. Thus, in the upper Yakima River, intra-sexual competition among male life history types has likely limited gene flow from early maturing hatchery fish to the naturally reproducing population.

Under different circumstances (e.g., low abundance of large anadromous males) early maturing hatchery males could make significant genetic contributions to natural populations, becoming potential agents of genetic change and domestication (Garant et al. 2003). Nevertheless, their occurrence on natural spawning grounds in the upper Yakima River appears to be quite low (Pearsons et al. 2009). Beckman and Larsen (2005) suggest two reasons for this; first, mortality during the downstream migration period may be high. And second, many of these fish appear to migrate downstream to the nearest large reservoir they encounter where they stay until they mature and attempt to ascend back upstream. Their upstream migration is several months later than that used by the large anadromous fish and consequently they may encounter thermal blocks which prevent them from completing a migration to a spawning area (Beckman and Larsen 2005). Thus, the principal impacts of hatchery precocious males in the upper Yakima River appear to be linked to the potential loss of adult salmon for harvest or broodstock (Zimmerman et al. 2003; Beckman and Larsen 2005) and to interactions they may have with native fishes in the natural environment (Beckman and Larsen 2005).

Conversely, the production of early maturing males from a local or native population may be beneficial when hatcheries are used as agents of conservation. The infusion of genes from these males would increase the effective size of a recipient population and therefore help retain its genetic diversity and variation (Saura et al. 2008). They could also be used as hatchery broodstock. Flain (1971) and more recently Unwin et al. (1999) report that precocious male Chinook salmon can survive from one spawning season to the next if held under artificial culture. Consequently, when fish abundance is extremely low precocious males could be used as broodstock over several spawning seasons.

Our investigation examined a small part of a larger question, what is the genetic legacy of supplementation? Recent reviews by Araki (2008) and Araki and Schmid (2010) suggest that rapid domestication occurs when salmonids are under culture and that domestication will often lead to reduced performance under natural conditions (also see Blanchet et al. 2008; Roberge et al. 2008). Alternatively, significant benefits from salmonid supplementation programs have been reported (e.g. Sharma et al. 2006; Berejikian et al. 2008; Small et al. 2009). Because supplementation is a human enterprise, social and economic values along with biological considerations will determine whether it is instituted, maintained or stopped. Importantly, every supplementation project has the potential to provide managers with information if its effects are monitored and evaluated. Consequently careful assessments and planned evaluation programs should be incorporated into every supplementation program since data derived from such efforts will help refine how supplementation is used in the future.

ACKNOWLEDGEMENTS

We thank State of the Salmon and the organizers of the "Ecological Interactions between Wild and Hatchery Salmon" symposium for giving us an opportunity to present this information. Our field work would have been impossible without considerable assistance from Yakama Nation staff Charles Strom, Vernon Bogar, DJ Brownlee, Greg Strom, Simon Goudy, Quinn Jones, Annie Jo Parrish, Jason Rau, and Dan Barrett all located at the CESRF. Mark Johnston, Joe Hoptowit, Gerry Lewis, Ray Decoteau, and Antoine Marek collected and transported all the fish used in our experiments from the Roza Adult Monitoring Facility to the CESRF. Other Yakama Nation staff, Bill Bosch, David Lind, and Paul Huffman also supported us in numerous ways. Jen Scott, Mike Hamlin, Anthony Fritts, Gene Sanborn, Jordan Vandal, Kurt Fresh, Eric Volk, Gabriel Temple, Timothy Webster, Charity Davidson, Chris Johnson and Molly Kelly all from the Washington Department of Fish and Wildlife helped make observations, assisted us while we cleaned and repaired the artificial stream, or were involved in the capture and sampling of fry. Personnel in WDFW's Molecular Genetics Laboratory, Jim Shaklee, Alice Frye, Jennifer Von Bargen, Norm Switzler, Cherril Bowman, Mo Small, Janet Loxterman, and Denise Hawkins helped with pedigree analyses. Two anonymous reviewers and Barry Berejikian considerably improved the manuscript by their helpful comments and suggestions. We also extend our appreciation to the Bonneville Power Administration (BPA) for funding this work as part of the Yakima/Klickitat Fisheries Project (project 1995-063-25). David Byrnes and Patty Smith of BPA provided administrative support for the project.

REFERENCES

1. Allan IRH, Ritter JA (1977) Salmonid terminology. J Conseil Int 'Exploration La Mer 37:293–299
2. Allen CS, Rich HB Jr, Quinn TP (2007) Condition-dependent reproductive tactics by large and small anadromous male sockeye salmon *Oncorhynchus nerka*. J Fish Biol 70:1302–1307
3. Andersson M (1994) Sexual selection. Princeton University Press, Princeton New Jersey
4. Araki H (2008) Hatchery stocking for restoring wild populations: A genetic evaluation of the reproductive success of hatchery fish vs. wild fish. In: Tsukamoto K, Kawamura T, Takeuchi T, Beard TD Jr., Kaiser MJ (ed.s) Fisheries for global welfare and environment, 5th world fisheries congress 2008, pp 153–167
5. Araki H, Schmid C (2010) Is hatchery stocking a help or harm? Evidence, limitations and future direction in ecological and genetic surveys. Aquaculture 308:S2–S11

6. Beall E, de Gaudemar B (1999) Plasticity of reproductive behavior in Atlantic salmon (*Salmo salar)* (Salmonidae) in relation to environmental factors. Cybium 23:9–28
7. Beall E, Moran P, Pendas A, Izquierdo J, Garcia-Vazquez E (1997) L'hybridation dans les populations naturelles de salmonidés dans le sud-ouest de l'Europe et en milieu experimental. Bull Fr Pêche Pisciculture 344(345):271–285
8. Beckman BR, Larsen DA (2005) Upstream migration of minijack (age-2) Chinook salmon in the Columbia River: behavior, abundance, distribution, and origin. Trans Am Fish Soc 134:1520–1541
9. Bell GR (1964) A guide to the properties, characteristics, and uses of some general anesthetics for fish. Fisheries Research Board Canada Bulletin 148. 4 pp
10. Berejikian BA, Johnson T, Endicott RS, Lee-Waltermire J (2008) Increases in steelhead (*Oncorhynchus mykiss*) redd abundance resulting from two conservation hatchery strategies in the Hamma Hamma River, Washington. Can J Fish Aquat Sci 65:754–764
11. Berejikian BA, Van Doornik DM, Endicott RC, Hoffnagle TL, Tezak EP, Moore ME, Atkins J (2010) Mating success of alternative male phenotypes and evidence for frequency-dependent selection in Chinook salmon, *Oncorhynchus tshawytscha*. Can J Fish Aquat Sci 67:1933–1941
12. Bjornn TC, Reiser DW (1991) Habitat requirements of salmonids in streams. In: Meehan WR (ed.) Influences of forest and rangeland management on salmonid fishes and their habitats. American Fisheries Society Special Publication 19, Bethesda, Maryland, pp 83–138
13. Blanchet S, Páez DJ, Bernatchez L, Dodson JJ (2008) An integrated comparison of captive-bred and wild Atlantic salmon (*Salmo salar*): implications for supportive breeding programs. Biol Conserv 141:1989–1999
14. Blanchfield PJ, Ridgway MS (1999) The cost of peripheral males in a brook trout mating system. Anim Behav 57:537–544
15. Blanchfield PJ, Ridgway MS, Wilson CC (2003) Breeding success of male brook trout (*Salvelinus fontinalis*) in the wild. Mol Ecol 12:2417–2428
16. Caswell H, Naiman RJ, Morin R (1984) Evaluating the consequences of reproduction in complex salmonid life-cycles. Aquaculture 43:123–134
17. Dittman AH, May D, Larsen DA, Moser ML, Johnston M, Fast D (2010) Homing and spawning site selection by supplemented hatchery- and natural-origin Yakima River spring Chinook salmon. Trans Am Fish Soc 139:1014–1028
18. Emlen ST, Oring LW (1977) Ecology, sexual selection, and the evolution of mating systems. Science 173:215–223
19. Flain M (1971) Repeated spawning and survival of quinnat salmon in New Zealand (note). NZ J Mar Freshw Res 5:519–521
20. Gage MJG, Stockley P, Parker GA (1995) Effects of alternative male mating strategies on characteristics of sperm production in the Atlantic salmon (*Salmo salar*): theoretical and empirical investigations. Philos Trans R Soc Lond 350:391–399
21. Garant D, Fleming IA, Einum S, Bernatchez L (2003) Alternative male life-history tactics as potential vehicles of speeding introgression of farm salmon traits into wild populations. Ecol Lett 6:541–549
22. Gebhards SV (1960) Biological notes on precocious male chinook salmon parr in the Salmon River drainage, Idaho. Prog Fish Cult 22:121–123
23. Grimardias D, Merchermek N, Manicki A, Chebaux J, Gaudin P, Jarry M, Beall E (2010a) Effects of habitat on individual reproductive success of mature parr of Atlantic salmon *Salmo salar*. J Fish Biol 77:2460–2466
24. Grimardias D, Merchermek N, Manicki A, Garnier J, Gaudin P, Jarry M, Beall E (2010b) Reproductive success of Atlantic salmon (*Salmo salar*) mature parr in a small river, the Nivelle: influence of shelters. Ecol Freshw Fish 2010:1–10
25. Healey MC (1991) Life history of Chinook salmon. In: Groot C, Margolis L (eds) Pacific salmon life histories. UBC Press, Vancouver, Canada, pp 313–393

26. Hoysak DJ, Liley NR, Taylor EB (2004) Raffles, roles, and outcome of sperm competition in sockeye salmon. Can J Zool 82:1017–1026
27. Hutchings JA, Myers RA (1985) Mating between anadromous and nonanadromous Atlantic salmon, *Salmo salar*. Can J Zool 63:2219–2221
28. Hutchings JA, Myers RA (1988) Mating success of alternative maturation phenotypes in male Atlantic salmon, *Salmo salar*. Oecologia 75:169–174
29. Jones MW, Hutchings JA (2001) The influence of male parr size and mate competition on fertilization success and effective population size in Atlantic salmon. Heredity 86:675–684
30. Jones MW, Hutchings JA (2002) Individual variation in Atlantic salmon fertilization success: implications for effective population size. Ecol Appl 12:184–193
31. Koseki Y, Maekawa K (2002) Differential energy allocation of alternative male tactics in masu salmon (*Oncorhynchus masou*). Can J Fish Aquat Sci 59:1717–1723
32. Knudsen CM, Schroder SL, Busack CA, Johnston MV, Pearsons TN, Bosch WJ, Fast DE (2006) Comparison of life-history traits between first-generation hatchery and wild upper Yakima River spring Chinook salmon. Trans Am Fish Soc 135:1130–1144
33. Knudsen CM, Schroder SL, Johnston MV, Fast DE, Fritts A (2010) Comparison of hatchery- and natural-origin upper Yakima River spring Chinook salmon life history and demographic traits. In: Knudsen CM (ed.) Reproductive Ecology of Yakima River Hatchery and Wild Spring Chinook, Yakima/Klickitat Fisheries Project Monitoring and Evaluation, Annual Report 2009
34. Larsen DA, Beckman BR, Cooper KA (2010) Examining the conflict between smolting and precocious male maturation in spring (stream-type) Chinook salmon. Trans Am Fish Soc 139:564–578
35. Larsen DA, Beckman BR, Cooper KA, Barrett D, Johnston M, Swanson P, Dickhoff WW (2004) Assessment of high rates of precocious male maturation in a spring Chinook salmon supplementation hatchery program. Trans Am Fish Soc 133:98–120
36. Lotspeich FB, Everest FH (1981) A new method for reporting and interpreting textural composition of spawning gravel. U.S. Forest Service, Pacific Northwest Research Note 369, Portland, Oregon
37. Marshall TC, Slate J, Kruuk L, Pemberton JM (1998) Statistical confidence for likelihood-based paternity inference in natural populations. Mol Ecol 7:639–655
38. Martinez JL, Dumas J, Beall E, Garcia-Vazquez E (2001) Assessing introgression of foreign strains in wild Atlantic salmon populations: variation in microsatellites assessed in historic scale collections. Freshw Biol 46:835–844
39. Mjolnerod IB, Fleming IA, Refseth UH, Hindar K (1998) Mate and sperm competition during multiple-male spawning of Atlantic salmon. Can J Zool 76:70–75
40. Mobrand LE, Barr J, Blankenship L, Campton DE, Evelyn TTP, Flagg TA, Mahnken CVW, Seeb LW, Seidel PR, Smoker WW (2005) Hatchery reform in Washington State: Principles and emerging issues. Fisheries 30:11–23
41. Mullan JW, Rockhold A, Chrisman CR (1992) Life histories and precocity of Chinook salmon in the mid-Columbia River. Prog Fish Cult 54:25–28
42. O'Connell MF, Ash EGM (1993) Size and growth of smolts in relation to age at maturity of Atlantic salmon (*Salmo salar* L.): the role of lacustrine habitat. J Fish Biol 42:551–569
43. Pearsons TN, Johnson CL, James BB, Temple GM (2009) Abundance and distribution of precociously mature male spring Chinook salmon of hatchery and wild origin in the Yakima River. N Am J Fish Manage 29:778–790
44. Quinn TP, Adkison MD, Ward MB (1996) Behavioral tactics of male sockeye salmon (*Oncorhynchus nerka*) under varying operational sex ratios. Ethology 102:304–322
45. Rich WH (1920) Early history and seaward migration of chinook salmon in the Columbia and Sacramento rivers. U.S. Bur Fish Bull 37:1–73

46. Roberge C, Normandeau E, Einum S, Guderley H, Bernatchez L (2008) Genetic consequences of interbreeding between farmed and wild Atlantic salmon: insights from the transcriptome. Mol Ecol 17:314–324
47. Rutter C (1902) Natural history of the quinnat salmon. U.S. Fish Comm Bull 22:65–141
48. Sampson M, Fast D, Bosch B (2009) Yakima/Klickitat Fisheries Project; Monitoring and Evaluation, Final Report for the Performance Period May 1, 2008 through April 30, 2009, Project No. 199506325, 304 electronic pages, (BPA Report DOE/BP-00037822-1)
49. Saura M, Caballero A, Caballero P, Moran P (2008) Impact of precocious male parr on the effective size of a wild population of Atlantic salmon. Freshw Biol 53:2375–2384
50. Schroder SL, Knudsen CM, Pearsons TN, Kassler TW, Young SF, Beall EP, Fast DE (2010) Behavior and breeding success of wild and first generation hatchery male spring Chinook salmon spawning in an artificial stream. Trans Am Fish Soc 139:989–1003
51. Schroder SL, Knudsen CM, Pearsons TN, Kassler TW, Young SF, Busack CA, Fast DE (2008) Breeding success of wild and first generation hatchery female spring Chinook salmon spawning in an artificial stream. Trans Am Fish Soc 137:1475–1489
52. Sharma R, Morishima G, Wang S, Talbot A, Gilbertson L (2006) An evaluation of the Clearwater River supplementation program in western Washington. Can J Fish Aquat Sci 63:423–437
53. Shearer KD, Swanson P (2000) The effect of whole body lipid on early sexual maturation of 1+ age male chinook salmon (*Oncorhynchus tshawytscha*). Aquaculture 190:343–367
54. Small MP, Currens K, Johnson TH, Frye AE, Von Bargen JF (2009) Impacts of supplementation: genetic diversity in supplemented and unsupplemented populations of summer chum salmon (*Oncorhynchus keta*) in Puget Sound (Washington, USA). Can J Fish Aquat Sci 66:1216–1229
55. Sokal RR, Rohlf FJ (1995) Biometry, 3rd edn. W.H. Freeman and Company, New York
56. SYSTAT (2007) SYSTAT for Windows, Version 12.0. SYSTAT Software Inc., Richmond, California
57. Taborsky M (1998) Sperm competition in fish: 'bourgeois' males and parasitic spawning. Trends Ecol Evol 13:222–227
58. Taggart JB, McLaren IS, Hay DW, Webb WH, Youngson AF (2001) Spawning success in Atlantic salmon (*Salmo salar* L.). Mol Ecol 10:1047–1060
59. Thorpe JE (1994) Reproductive strategies in Atlantic Salmon, *Salmo salar*. Aquac Fish Manage 25:77–87
60. Unwin MJ, Glova GJ (1997) Changes in life history parameters in a naturally spawning population of chinook salmon (*Oncorhynchus tshawytscha*) associated with releases of hatchery-reared fish. Can J Fish Aquat Sci 54:1235–1245
61. Unwin MJ, Kinnison MT, Quinn TP (1999) Exceptions to semelparity: postmaturation survival, morphology, and energetics of male Chinook salmon (*Oncorhynchus tshawytscha*). Can J Fish Aquat Sci 56:1172–1181
62. Vladic TV, Jarvi T (2001) Sperm quality in the alternative reproductive tactics of Atlantic salmon: the importance of the loaded raffle mechanism. Proc R Soc Lond B 268:2375–2381
63. Zimmerman CE, Stonecypher RW Jr, Hayes MC (2003) Migration of precocious male hatchery Chinook salmon in the Umatilla River, Oregon. N Am J Fish Manage 23:1006–1014

CHAPTER 3

Genetic and Otolith Isotopic Markers Identify Salmon Populations in The Columbia River at Broad and Fine Geographic Scales

Rachel Barnett-Johnson & David J Teel & Edmundo Casillas R. Barnett-Johnson

Institute of Marine Sciences, University of California Santa Cruz, Santa Cruz, CA, USA

Northwest Fisheries Science Center, NOAA Fisheries Service, Manchester, WA, USA

Northwest Fisheries Science Center, NOAA Fisheries Service, East Seattle, WA, USA

Division of Environmental Affairs, US Bureau of Reclamation, Sacramento, CA, USA

ABSTRACT

Processes occurring in freshwater, estuarine, and marine habitats strongly influence the growth, survival and reproductive success of salmonids. Nonetheless, implementing an ecosystem model explicitly linking these important habitats has been hindered by the inability to track the source identity of individuals where they co-occur. Here we explore the development and integration of natural markers- molecular and isotopic to characterize the natal sources of Chinook salmon (*Oncorhynchus tshawytscha*) in the Mid and Upper Columbia River summer/fall-run (UCR

Su/F) population. Microsatellite DNA markers identified the majority of juveniles collected in rivers and hatcheries in the Mid and Upper Columbia River watershed to the Summer/Fall-run population in this watershed with 90% posterior probabilities of group membership. Strontium isotopes ($^{87}Sr/^{86}Sr$) measured in the natal rearing portion of the otolith showed significant geographic variation among natal rivers and hatcheries. Natal sites exhibited a wide dynamic range in $^{87}Sr/^{86}Sr$ source signatures (0.7043–0.7142), such that on average 61% of individuals were correctly classified to the location from which they were collected. We found that multilocus genotypes and otolith $^{87}Sr/^{86}Sr$ ratios collected on the same individuals were complementary markers when applied in a hierarchy. Microsatellites successfully assigned individuals to the broader UCR Su/F genetic group and $^{87}Sr/^{86}Sr$ provided finer-scale geographic assignments to five natal river and hatchery groups nested within the UCR Su/F population. The temporal stability of both genetic and $^{87}Sr/^{86}Sr$ markers, together with the coast-wide microsatellite baseline currently being used for mixed-stock fisheries management supports the further development and integration of $^{87}Sr/^{86}Sr$ markers to potentially achieve finer levels of stock resolution. Stock identification at the scales of individual rivers and hatcheries would help elucidate the abundance, distribution, and the relative contributions of natal sources important for the recovery and spatial management of Chinook salmon.

Keywords: Conservation Ecology Freshwater Habitat Management Marine

INTRODUCTION

A successful life history depends on an organism's ability to find suitable habitat in all phases of their life cycle. This can be particularly challenging for the many fish species that rely on different habitats between juvenile and adult stages that are spatially distant from one another. Thus, one of the outstanding issues in fisheries ecology is identifying habitats critical to the growth, survival and reproductive success of species that are endangered and targeted by fisheries. Yet, there are relatively few methods to estimate movement patterns of migratory fish and their habitat linkages among freshwater, estuarine and marine environments.

Many fisheries applications would benefit from identifying connectivity of populations over short and long temporal scales as well as across broad geographic (e.g., species' distribution) and local spatial scales (Cadrin et al. 2005; Botsford et al. 2009; Cowen and Sponaugle 2009). Several biogeography studies in marine fishes have found no genetic differentiation across species' distributions, although only a few migrants per generation could account for this lack of genetic structure (Palumbi 2004; reviewed in Selkoe et al. 2008). Similarly, recent studies using otolith microchemistry have found significant variation of elements at surprisingly small spatial (10 s of km) and temporal scales, limiting its utility for many marine applications (Gillanders 2002; Ruttenberg and Warner 2006; Miller 2007; Standish et al. 2008). There is a growing appreciation that genetic and otolith elemental signatures generally test for connectivity among populations at different scales (Thresher 1999; Barnett-Johnson 2007; Selkoe et al. 2008).

Accurate homing behavior in space and time for salmonids results in reproductively isolated populations and allows adaptive divergence to occur among populations in the freshwater (e.g., reviewed in Taylor 1991). Indeed, salmon populations have diverged in a number of traits due to spatially varying selection pressures (e.g., Carlson and Quinn 2007; Carlson et al. 2009) combined with the fact that many traits are heritable (reviewed in Carlson and Seamons 2008). Information for several traits that differ among Chinook salmon (*Oncorhynchus tshawytscha*) populations has been synthesized with genetic and environmental data to identify broad conservation units in the United States that generally correspond geographically to regional watersheds (Evolutionarily Significant Units (ESUs); Myers et al. 1998). The genetic divergence among freshwater salmon populations results in genetic markers that can be used to track the abundance and distribution of salmon populations during their non-reproductive life history stage in estuarine and marine systems (Utter et al. 1987; Seeb et al. 2007).

Microsatellites have been the genetic marker of choice in ecological and conservation genetics studies in the last decade due to their high variability and power to resolve population structure and to assign individual fish to populations or parents (O'Connell and Wright 1997; Hansen et al. 2001; Planes et al. 2009). For Pacific salmon, efforts to standardize genetic data among labs have permitted the development of geographically broad datasets (Moran et al. 2006) including a coast wide microsatellite baseline for Chinook salmon stock identification (Seeb et al. 2007). More recently, differentiation among Chinook salmon populations has also been examined using single nucleotide polymorphisms (SNPs; Narum et al. 2008). However, these genetic approaches are only effective for differentiating among genetically distinct groups of fish and many spawning populations of Chinook salmon are connected by recent or ongoing interbreeding. While populations in proximate rivers are often linked by natural exchanges of spawners, many populations are currently connected due to alterations in migration and spawning habitats and by transfers of hatchery broodstocks across watersheds. Therefore, the need remains to identify the specific natal river or hatchery of origin of fish in population mixtures to meet conservation and management goals.

The Columbia River was historically among the most productive rivers for anadromous salmonids in the world (Chapman 1986). Summer and fall run (time of adult return to freshwater) Chinook salmon populations in the upper river are major contributors to the commercial and recreational fishery off the coasts of Washington, British Columbia, and southeast Alaska (Utter et al. 1987; Weitkamp 2010). These populations comprise the Upper Columbia Summer and Fall Run ESU (UCR Su/F; Myers et al. 1998) and are genetically distinct from Mid and Upper Columbia River spring-run populations and from other summer and fall-run groups in Columbia River watersheds (Utter et al. 1995; Waples et al. 2004). However, overall genetic differentiation among natal rivers within the UCR Su/F ESU is small, likely a result of blockage of spawning habitat upstream of Grand Coulee Dam and subsequent fishery and hatchery management actions (Utter et al. 1995). In addition, several current hatcheries in different rivers rear and release fish derived from common UCR Su/F stocks, making fine-scale genetic identification of individuals very difficult. Otolith-based techniques have provided finer scale identification of sources in high gene flow systems where environmental heterogeneity exists (Secor 1999; Campana and Thorrold 2001; Gillanders 2002).

Strontium isotope ($^{87}Sr/^{86}Sr$) ratios in otoliths and rivers largely reflect the age and mineralogy of the surrounding surficial geology (Kennedy et al. 1997; Barnett-Johnson et al. 2008). Since Sr isotopes, like other trace elements, are permanently incorporated in the otolith's carbonate matrix, $^{87}Sr/^{86}Sr$ can be measured from discrete daily growth increments deposited throughout the life of a fish (Kennedy et al. 1997, Ingram and Weber 1999, Barnett-Johnson et al. 2005). Barnett-Johnson et al. (2008) found significant variation in natal otolith $^{87}Sr/^{86}Sr$ values among the major salmon rivers and hatcheries in the California Central Valley, which permitted Chinook salmon to be linked to their natal sources within the fall-run ESU with high classification accuracy (98%). Rivers providing salmon rearing habitat in the Columbia River watershed system drain areas of contrasting lithologies, thus showing similar promise for the development of Sr isotopic markers within the watershed (Fig. 1).

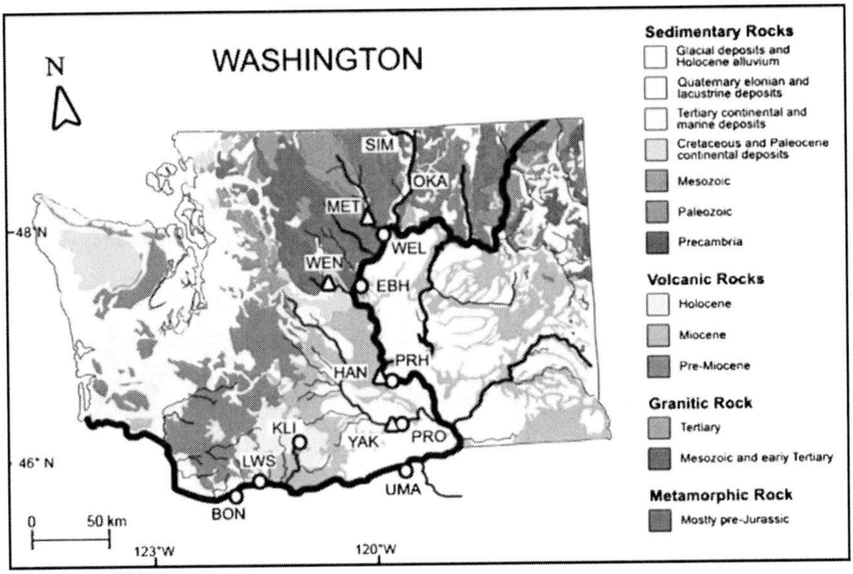

Figure. 1 Map of geology, and major salmon rivers (*open triangles*) and hatcheries (*open circles*) for the Mid and Upper Columbia River summer/fall Chinook salmon (*Oncorhynchus tshawytscha*) population (modified from Weissenborn 1969; river abbreviations referenced in Table 1). Okanagon (OKA) and Similkameen (SIM) rivers have summer-run Chinook salmon populations not sampled in this study. Surficial geology in the upper Columbia River watershed varies in age and lithology ranging from young (Cenozoic) volcanic rocks (*light pink*, *orange* and *brown*) in the Columbia Basin and Southern Cascades, older granitic rocks (*red*) and metamorphic rocks (*purple*) in the Northern Cascades, to varying ages of sedimentary rocks with the oldest rocks in the state found in the East Okanogan Highlands (*blue* and *dark green*)

Integrating intrinsic markers is a powerful approach to tracking fish movement at different scales, as no single marker has yet achieved this goal (Wood et al. 1989; Hobson et al. 2009; Planes et al. 2009). Molecular techniques can detect restricted gene flow and the resultant genetic divergence among groups over evolutionary time scales, whereas some chemical signatures in otoliths reflect only those environmental differences experienced by individuals in their lifetime (i.e., ecological time scales). To date, only work by Milton and Chenery (2001) on a Tropical Shad (*Tenualosa ilisha*), Miller et al. (2005) on Black Rockfish (*Sebastes melanops*), Barnett-Johnson (2007) on Chinook salmon, Feyrer et al. (2007) on Splittail (*Pogonichthys macrolepidotus*), Bradbury et al. (2008) on Rainbow Smelt (*Osmerus mordax*), and Planes et al. (2009) on Clown

Fish (*Amphiprion percula*) have combined genetic and otolith elemental analyses on the same individuals. Chinook salmon in the Mid and Upper Columbia River watershed are an ideal species and system to develop an integrated genetic and otolith microchemistry approach because (1) limited genetic differentiation occurs among rivers within the UCR Su/F population, yet restoration and management actions occur at finer-scales of natal sources (Waples et al. 2001), (2) high diversity in watershed geology exists, establishing a mechanistic foundation for variation in $^{87}Sr/^{86}Sr$ markers at small spatial scales, and (3) populations of Chinook salmon are of economic value and conservation concern.

An integrated baseline of isotopic and genetic information may act synergistically to improve assignment probabilities where stocks co-occur in estuarine and marine systems. Several Chinook salmon ESUs show gene flow at the scale of natal sources while $^{87}Sr/^{86}Sr$ are likely to be limited as broad spatial scale markers due to natal sources having similar $^{87}Sr/^{86}Sr$ in other watersheds along the species' distribution. In this study, we characterize for the first time the $^{87}Sr/^{86}Sr$ values of natal sources of summer and fall run Chinook salmon from the Mid and Upper Columbia River population and examine the utility of using genetic and isotopic information as natural population markers on the same individuals to identify broad (e.g., regional genetic stock groups/ ESUs) and fine-scale (natal source) populations.

METHODS

Our approach to exploring the use of Sr isotopes and genetic information to identify the natal sources of Chinook salmon was to first establish a baseline of $^{87}Sr/^{86}Sr$ ratios in the otoliths collected from each natal source. Then we characterized the variation in microsatellite DNA from the same individuals and natal sources. Finally, to assess the utility of integrating genetic and isotopic markers at different scales of population structure, we used a step-wise hierarchy approach by: (1) testing the accuracy of a pre-existing coast wide microsatellite DNA baseline to identify individuals to the UCR Su/F genetic group (Seeb et al. 2007), (2) testing the accuracy of our $^{87}Sr/^{86}Sr$ baseline to identify fish to natal sources within the UCR Su/F genetic group, and (3) compared the accuracy of the coast wide genetic baseline to identify individuals to natal sources present in the genetic baseline, with the results using $^{87}Sr/^{86}Sr$.

Study system

The native populations of summer and fall run Chinook salmon in the Columbia River above the confluence with the Snake River comprise the UCR Su/F ESU (Myers et al. 1998). This is one of eight ESUs of Chinook salmon that have been identified within the Columbia River Basin based on a synthesis of genetic, life-history, biogeographic, geologic, and environmental information (Myers et al. 1998). The boundaries of the ESU are the Grand Coulee Dam upstream and down-river at the crest of the Cascade Range. Most of the current natural production of fish in the UCR are summer run populations in the Wenatchee, Methow, Okanagon, and Similkameen rivers and fall run populations in the Yakima and Klickitat rivers and in the mainstem Columbia River in the Hanford Reach area (Waknitz et al. 1995, Myers et al. 1998; Fig. 1). Large hatchery programs associated with these populations also release summer and fall run fish in the Upper and Mid Columbia River (Table 1).

Table 1 Sample information, $^{87}Sr/^{86}Sr$ values, standard deviation (SD), and results of the linear discriminant function analysis to classify Chinook salmon (*Oncorhynchus tshawytscha*) to natal population based on $^{87}Sr/^{86}Sr$ values

Natal collection site; abbreviation (code)	River	Run[1]	Collection year	Rearing type	$^{87}Sr/^{86}Sr$	SD	N	Classification accuracies % correct
1. Methow River (MET)	Methow	SU	2007	Natural	0.70432	0.00011	8	100
2. Klickitat Hatchery (KLI)	Klickitat	F	2008	Hatchery	0.70568	0.00018	10	50
3. Wenatchee River (WEN)	Wenatchee	SU	2007	Natural	0.70572	0.00016	10	30
4. Yakima River (YAK)	Yakima	F	2008	Natural	0.70577	0.00024	10	20

5. Little White Salmon Hatchery (LWS)	L. White Salmon	F	2007	Hatchery	0.70606	0.00022	10	0
6. Prosser Hatchery (PRO)	Yakima	F	2008	Hatchery	0.70604	0.00039	8	25
7. Bonneville Hatchery (BON)	mainstem	F	2007	Hatchery	0.70818	0.00013	10	100
8. Umatilla Hatchery (UMA)	mainstem	F	2008	Hatchery	0.71058	0.00012	11	82
9. Wells Hatchery (WEL)	mainstem	SU	2007	Hatchery	0.71080	0.00015	11	55
10. Eastbank Hatchery (EBH)	mainstem	SU	2007	Hatchery	0.71113	0.00023	10	50
11. Priest Rapids Hatchery (PRH)	mainstem	F	2007	Hatchery	0.71154	0.00023	10	70
12. Hanford Reach (HAN)	mainstem	F	2007	Natural	0.71415	0.00012	9	100

Jack-knife classification accuracies and overall mean classification accuracies.[1] Progeny of fish returning to spawn in the summer (SU) and fall (F), except for adults sampled at Hanford Reach

The Columbia River watershed is characterized by complex and heterogeneous geology. There are several physiographic regions that influence the geochemistry of the Upper and Mid Columbia River including Okanogan Highlands, Northern and Southern Cascade Mountains, and the Columbia Basin. The geochemistry of the water where fish rear in the wild is derived from weathering of rocks within the entire upstream watershed. The Okanogan Highlands contain the oldest sedimentary and metamorphic rocks in the state, which influence the $^{87}Sr/^{86}Sr$ values in the mainstem waters of the Columbia River (Lasmanis 1991). The northwestern portion of the upper Columbia River watershed receives freshwater input from the

Northern Cascade Mountains composed primarily of Mesozoic crystalline and metamorphic rocks. The Southern Cascades are younger Tertiary and Holocene volcanic rocks (Lasmanis 1991; Fig. 1). Most of the salmon-bearing tributaries to the Columbia River drain the Southern Cascades or the Columbia Basin province comprised of basalt from the Miocene. This Columbia River Basalt group is characterized by several hundred individual flows that amount to ~234 000 km^3 of tholeiitic basalt, and andesite lava flows covering large areas including south central and eastern Washington along the Columbia River (Hooper 1997; Camp et al. 2004). This diverse surficial geology results in a wide range of $^{87}Sr/^{86}Sr$ values among Columbia River tributaries (Ramos et al. 2006; Singleton et al. 2006; Fig 1)

Natal source signatures

To develop a baseline of $^{87}Sr/^{86}Sr$ ratios across natal sources and acquire genotypes from individuals within the natal sources, otoliths and tissues were collected from sub-yearling juvenile Chinook salmon between May-August from seven hatcheries and five naturally spawning populations in 2007 and 2008 (Table 1). Wild juveniles (~55 mm fork length) were collected from streams with rotary screw traps operated by Yakama Nation Fisheries and Washington Department of Fish and Wildlife. Wild juveniles were assumed to originate from the streams where they were collected based on their small sizes and general downstream migration behavior out of tributaries. Hanford Reach, on the mainstem Columbia River, is known to be an area of active spawning and juvenile rearing. Yet juveniles collected from this location could represent out-migrants from several potential upstream rivers. Thus, to characterize the Hanford Reach natal site, we collected adults spawning at Hanford Reach in November of 2007 and analyzed the natal portion of adult otoliths. This approach assumes spawning fidelity to the site of origin and was later supported by the low variance in $^{87}Sr/^{86}Sr$ values characteristic of individuals at this site.

Genetic

Genomic DNA was extracted from fin tissue samples using Wizard genomic DNA purification kits (Promega Corp.) following the manufacturer's protocols. Isolated genomic DNA was amplified via the polymerase chain reactions (PCR) at 13 microsatellite loci using the standardized protocols discussed by Seeb et al. (2007). The resulting PCR products were analyzed with an Applied Biosystems 3100 capillary electrophoresis system. GeneScan and Genotyper software (Applied Biosystems) were used to determine the size and number of alleles observed at each locus.

Isotopic

Sagittal otoliths were extracted, washed, and stored in dry vials prior to cleaning and mounting. The left otolith was soaked for 6 h in 30% Suprapur® H_2O_2 to remove organic material and double rinsed in ultrafiltered water. The right otolith was used if the left otolith was compromised during preparation or composed of the vaterite form of calcium carbonate. Otoliths were mounted sulcus side up on microscope slides and polished using Al_2O_3 lapping paper until daily increments were revealed (Barnett-Johnson et al. 2005). Otoliths were transferred and grouped onto clean petrographic slides, randomized by source location and sampled using 60 × 500 × 80 μm (W × L × D) laser tracks in the dorsal region along the longest axis parallel to daily increments (5 μm increment^{-1}; ~12 days) where the plane of growth exhibits the least curvature (Barnett-Johnson et al. 2005). We measured $^{87}Sr/^{86}Sr$ in the region of the otolith accreted while in the natal tributary or hatchery, but after yolk absorption and prior to outmigration (~250 μm from primordia). This region was just distal to the dark band diagnostic of the onset of exogenous feeding (Barnett-Johnson 2007). Isolating this region is particularly important when characterizing natal $^{87}Sr/^{86}Sr$ values, because the prefeeding value reflects Sr derived from both marine (maternally inherited) and natal river sources (Bacon et al. 2004; Miller and Kent 2009).

Otolith $^{87}Sr/^{86}Sr$ data were collected using a NuPlasma HR multi-collector inductively coupled plasma mass spectrometer (MC-ICPMS) coupled to a Nd:YAG 213 nm laser (New Wave UP213; with a SuperCell™ sample chamber) with specific instrument, laser, and interference corrections described in Ramos et al. (2004) as applied to otoliths (Barnett-Johnson et al. 2005). All monitored isotopes were collected simultaneously and integrated for 2 s. Approximately thirty $^{87}Sr/^{86}Sr$ ratios were measured for each laser track and used to estimate within-run precision for each otolith (2 × standard error ± 0.00004). The laser ablated the otolith using a 20 Hz pulse-rate and moved at a rate of 10 μm·s^{-1}, removing material along a 60 μm wide transect 500 μm in length. While there are several potential interferences on Sr isotopes in carbonates, including Ca dimers, Ca argides and doubly-charged Er and Yb, only Rb and Kr affected the accuracy and precisions of $^{87}Sr/^{86}Sr$ (Barnett-Johnson et al. 2005). We accounted for the effects of ^{84}Kr and ^{86}Kr, present in the Ar gas by subtracting on-peak baselines (<1 mV ^{83}Kr) measured prior to ablation of otoliths from beam intensities during otolith ablation. To correct for ^{87}Rb, we used the measured $^{86}Sr/^{88}Sr$ and natural $^{85}Rb/^{87}Rb$ value to calculate a mass bias uncorrected $^{85}Rb/^{87}Rb$ ratio and the measured ^{85}Rb intensity to calculate and remove any ^{87}Rb contributions to measured ^{87}Sr intensities. $^{87}Sr/^{86}Sr$ ratios were then normalized to $^{88}Sr/^{86}Sr = 0.1194$ to account for any natural or machine induced fractionation. External precisions (reproducibility) were based on repeated measurements of our in-house gastropod shell standard. The mean (SD) value of $^{87}Sr/^{86}Sr$ measurements for the standard was 0.70922 (0.00004; $n = 417$), which is close to the global marine $^{87}Sr/^{86}Sr$ value of 0.70918.

Sr isotope ratios among natal sources were compared using analysis of variance (ANOVA) with Bonferroni post-hoc pairwise comparisons, to determine whether there were significant differences in the means in isotopic signatures among sources. Strontium isotope values met the assumptions of normality for ANOVA, but did not have equal variances among locations.

Genetic-isotope assignment hierarchy

We conducted our assignments using genetic and isotopic information in a hierarchical manner, similar to a method one may use in a realistic unknown mixture analysis for salmon. First, Chinook salmon genotyped in this study (Table 1) were assigned to genetic stock groups using microsatellite data compiled from a pre-existing coast wide genetic baseline representing 141 populations from Southeast Alaska to California described by Seeb et al. (2007). This baseline estimates the likely regional genetic stock group as well as population (e.g., natal origin) within the regional stock groups for each individual genotyped. We used the likelihood model of Rannala and Mountain (1997) as implemented by the genetic stock identification computer program ONCOR (Kalinowski et al. 2007). Posterior assignment probabilities for individual baseline sources were summed to regional genetic stock groups, the scale for which there is greater assignment accuracy (see Seeb et al. 2007). We used equal priors in our genetic assignments even though fish in this study did not have a random chance of being part of the broader baseline. In an unknown mixture analysis, one would assume equal priors and only individuals assigned to the UCR Su/F genetic group with high posterior probabilities would be evaluated using this $^{87}Sr/^{86}Sr$ baseline. However, since all samples collected for this study were known to originate from the UCR Su/F region, we retained all individuals in the isotope model.

Second, Chinook salmon were assigned to their natal origins using a single factor linear discriminant function analysis (DFA) with jackknife resampling to determine whether $^{87}Sr/^{86}Sr$ values alone could be used to correctly classify individual fish to natal origin (SAS version 8, SAS Institute Inc., Cary, New York). A linear function was used to estimate the variance-covariance matrix response and applied across all sites. Jackknife resampling used the same data set to generate and evaluate the discriminant function by removing an observation and reclassifying it to the source with the closest mean $^{87}Sr/^{86}Sr$ value. This procedure was repeated for all observations to determine correct assignment percentages for each natal source. Prior probability of group membership was assumed to be proportional.

To identify natal sources and groups with similar $^{87}Sr/^{86}Sr$ values, we conducted a cluster analysis (PRIMER version 6.1.5 software; Plymouth Routines in Multivariate Ecological Research; SIMPROF test). Cluster analysis used normalized $^{87}Sr/^{86}Sr$ for all fish in the study to generate a similarity matrix of Euclidean distances and implemented a hierarchical agglomerative clustering routine using the group average to determine the similarity of $^{87}Sr/^{86}Sr$ values among natal sources. The SIMPROF test was used to determine actual structure in the subset of data corresponding to each branch by calculating the distance between clusters of the dendrogram using 1000 permutations.

The pre-existing microsatellite baseline contains genotypes from approximately 150 individuals from each of four natal sources within the UCR Su/F regional genetic stock group also represented in this study- Methow River summer, Hanford Reach fall, Wells Hatchery summer, and upper Wenatchee summer/fall. Therefore, we compared the accuracies of fine-scale (natal source) assignments between the genetic and $^{87}Sr/^{86}Sr$ baselines for fish from these four sources.

RESULTS

The vast majority of Chinook salmon in our study were assigned to the UCR Su/F genetic stock group with strong group membership using the coast wide microsatellite baseline. The strength of assignment (posterior probability) for all individuals in this study averaged 92% (standard deviation (SD)=13). Of 86 individuals with assignment probabilities >0.90, 84 were correctly assigned to the UCR Su/F group. Of 32 fish with probabilities <0.90, 15 were correctly identified (Fig. 2). Individuals assigned to non-UCR Su/F groups had an average posterior probability of assigning to the UCR Su/F of 0.25 ($SD = 0.15$). Most of the incorrect assignments were to the genetically related Snake and Deschutes Fall-run. Five of the low probability fish were not assigned to the Columbia River (Fig. 2).

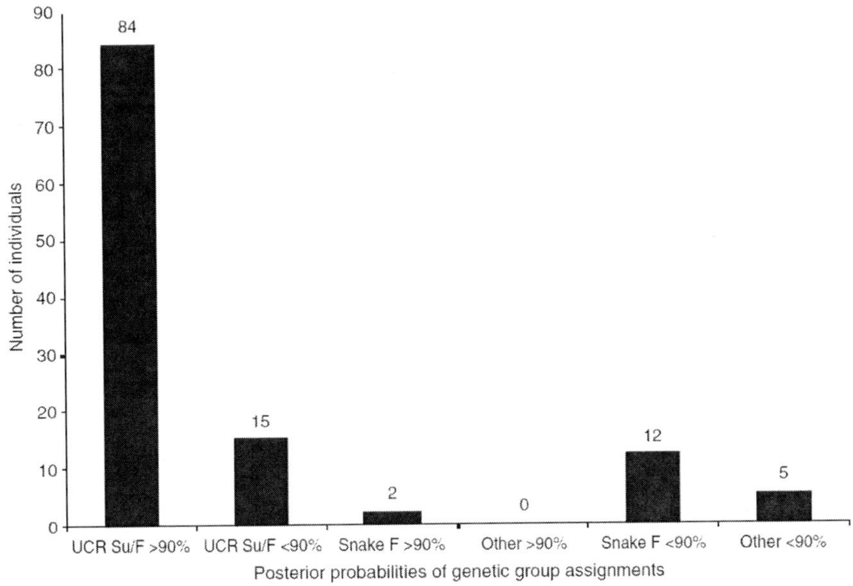

Figure. 2 Estimated genetic stock of origin and posterior probabilities of assignments for 118 upper Columbia River summer and fall Chinook Salmon using 13 standardized microsatellite DNA loci. Stock groups are Upper Columbia River Summer and fall (UCR Su/F), Snake River fall (Snake F), and 30 groups ranging from California to southern British Columbia (Other)

$^{87}Sr/^{86}Sr$ values were significantly different across natal sources and ranged in values from 0.7043–0.7142 (ANOVA, $F_{11,104}$= 2345, $p < 0.001$). The $^{87}Sr/^{86}Sr$ model showed that all 58 pairwise comparisons were significantly different (df= 105, $p < 0.05$) with the exception of eight comparisons (Klickitat Hatchery (KLI) with Wenatchee River (WEN) and Yakima River (YAK); Little White Salmon Hatchery (LWS) with Yakima River (YAK) and Prosser Hatchery (PRO); Umatilla Hatchery (UMA) and Wells Hatchery (WEL); PRO with WEN and YAK; and WEN and YAK; Table 1). Discriminant function analysis for the $^{87}Sr/^{86}Sr$ only model resulted in classifications that ranged from 0 to 100% (Wilks's Lambda, $F_{11,105}$= 2344, $p < 0.001$). All individuals from three natal sources were correctly assigned- Methow River (MET); Bonneville Hatchery (BON); and Hanford Reach (HAN), with only two individuals from UMA misclassified (82%; Fig. 3). Fish from KLI, WEN, YAK, LWS, PRO, WEL, EBH, and PRH had lower assignment success (Table 1). Some natal sources (e.g., Okanagon (OKA) and Similkameen (SIM) were not included in the $^{87}Sr/^{86}Sr$ baseline. This may influence the strength of assignment for some natal sources.

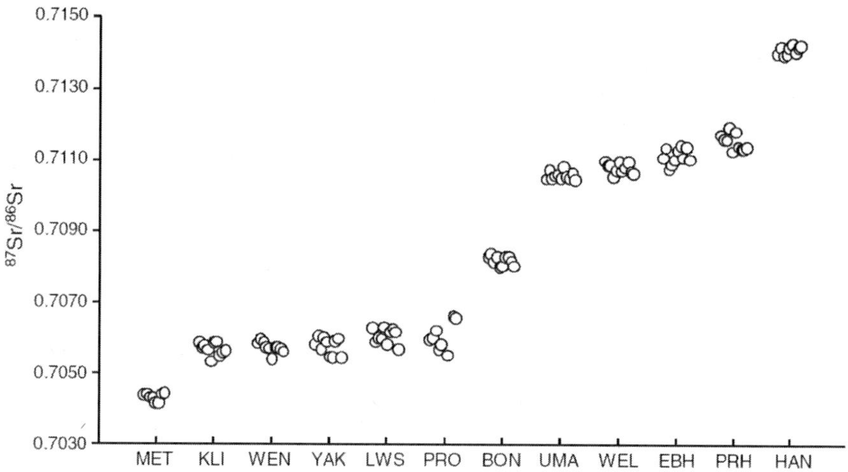

Figure. 3 $^{87}Sr/^{86}Sr$ values in the natal growth region of otoliths from Chinook Salmon (*Oncorhynchus tshawytscha*) collected in rivers and hatcheries in the upper Columbia River. Natal river codes correspond to populations listed in Table 1

The similarity dendrogram from the cluster analysis shows that $^{87}Sr/^{86}Sr$ resolved fine-scale assignments to three individual natal sources (MET, BON, and HAN), and two more regional groups with 100% correct classification success (Fig. 4). The two regional groups were comprised of mainstem hatchery sources (WEL, EBH, PRH, UMA) and sources located on tributaries to the Columbia River (WEN, YAK, PRO, KLI, LWS).

Of the 34 samples in our study from the four sources represented in the pre-existing microsatellite DNA baseline (MET, WEL, HAN, WEN) 31 (91%) were correctly assigned to the UCR Su/F genetic stock group. However, only about 1/3 of the fish from these four sources were assigned to the correct natal source using the entire coast-wide genetic baseline (Fig. 5). Individuals appeared randomly misassigned to non-natal sources. A significantly greater proportion of the 34 samples (approximately 2/3) were correctly assigned to their natal source using the $^{87}Sr/^{86}Sr$ baseline. The lower assignment success of individuals to their correct natal sources using the genetic baseline highlights its limitation at finer scales. In contrast, the $^{87}Sr/^{86}Sr$ baseline significantly improved classification success of individuals originating from three of the four natal sources. Assignments improved from (25 to 100%), (40 to 64%), (33 to 100%) for individuals from MET, WEL, HAN respectively (Fig. 5). Assignment success

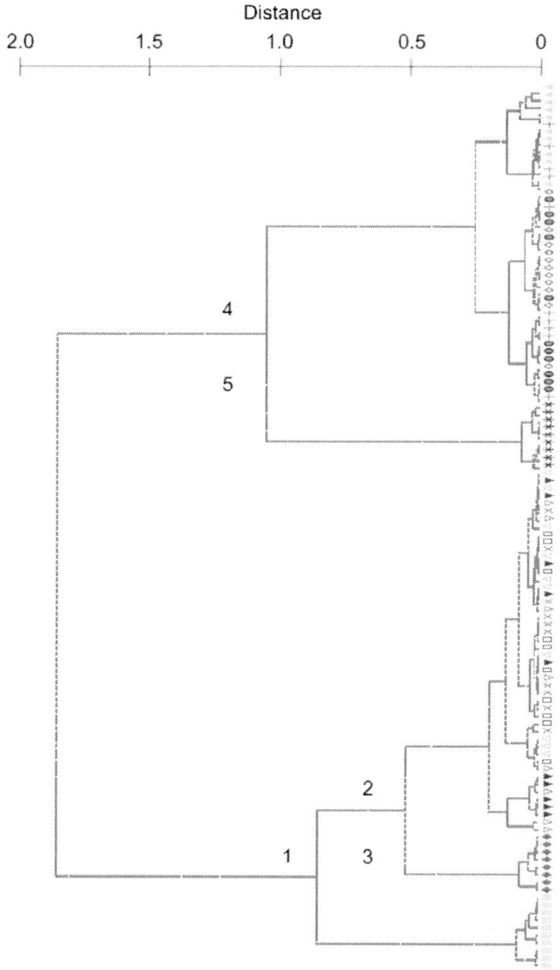

Figure. 4 Similarity dendrogram using normalized $^{87}Sr/^{86}Sr$ values for all individuals and a similarity matrix of Euclidean distances with a group average algorithm. Individuals originated from the summer/fall Chinook salmon from Mid and Upper Columbia River tributaries- MET (◆), WEN (✕), YAK (△), LWS (▼), KLI (□), PRO (▽), and mainstem Columbia River-HAN (✱), BON (■), UMA (◇), WEL (●), EBH (+), PRH (▲). Natal river and hatchery codes are referenced in Table 1. Five distinct groups (numbered) are identified with 100% correct classification success. Groups 1, 2, and 4 are resolved to individual natal sources whereas groups 3 and 5 reflect natal sources on tributaries and the mainstem Columbia River respectively

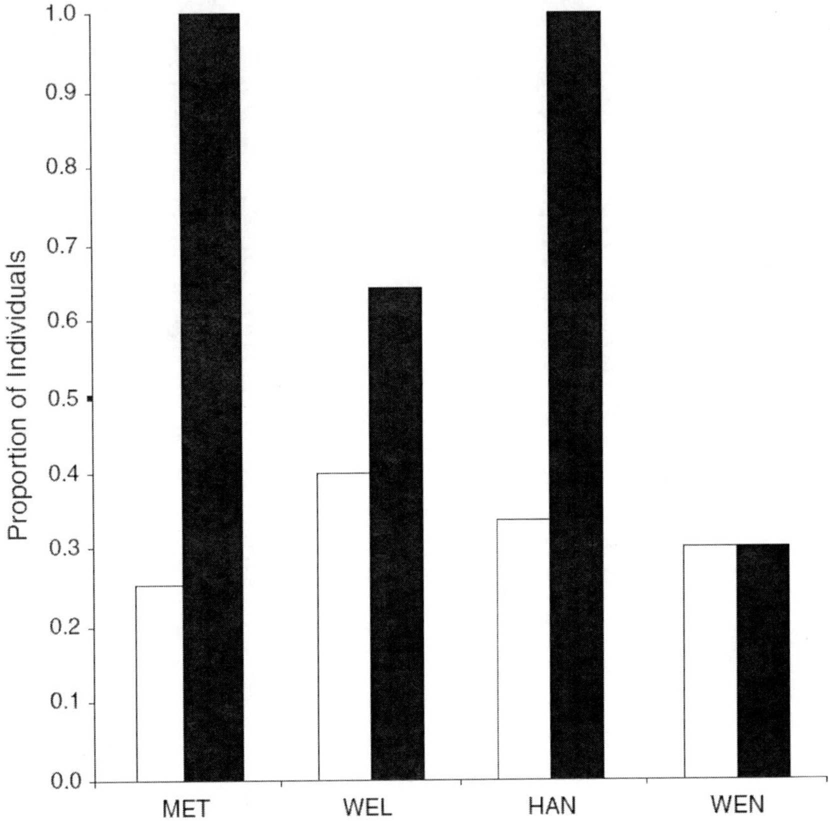

Figure. 5 Proportion of Chinook salmon (*Oncorhynchus tshawytscha*) identified as originating from their natal capture site using a coast wide microsatellite DNA baseline (*open bars*) and $^{87}Sr/^{86}Sr$ values (*filled bars*). Results for four populations represented in both this study and in the coast wide microsatellite DNA baseline are shown. Natal river codes correspond to populations listed in Table 1

for individuals from WEN remained the same (30%) with the inclusion of $^{87}Sr/^{86}Sr$ information.

DISCUSSION

This study provides support for the integration of otolith microchemistry and multilocus genotypes to provide broad (e.g., basin/ESU) and finer scale (e.g., natal source) identities for Chinook salmon. There is a long history and continued effort in the fisheries literature on documenting biological attributes that differ among stocks as intrinsic markers to minimize the need for expensive and laborious physical tags (Marques et al. 2006; Abaunza et al. 2008; MacKenzie 2009). For example, many salmonid stocks can be distinguished based on unique scale or otolith structure (Henry 1961; Marshall et al. 1987; Barnett-Johnson 2007), morphometric and meristic characters (Fukuhara et al. 1962; Winans 1984; MacCrimmon and Claytor 1985), parasites (Margolis 1963; Bailey and Margolis 1987), and genetic differences (Utter et al. 1987; Winans et al. 2001). Recent studies have explored integrating multiple biological markers in salmon with varying levels of success and resolution (Wood et al. 1989; Barnett-Johnson et al. 2008; Narum et al. 2008). Specifically, none have achieved a resolution that easily scales between local natal sources and regional populations applicable to mixed-stock groups at fine-scale resolution.

Molecular markers have previously been used to identify the genetic origins of Chinook salmon in mixed-stock ocean fisheries at broad scales with great success (Winans et al 2001; Beacham et al. 2006). The majority (84%) of Chinook salmon in this study were correctly identified to the UCR Su/F genetic stock group using a standardized microsatellite dataset. However, assignment success of natal sources nested within the UCR Su/F population was poor. This is likely due to naturally occurring gene flow among natal sources over short temporal and spatial scales as well as alterations in migration and spawning habitat, and among basin transfers of broodstocks for hatchery operations (see also Seeb et al. 2007; Narum et al. 2008). There is a growing need to identify specific natal sources to better understand salmon ocean ecology and to monitor recovery efforts that occur at the scale of individual rivers.

Previous studies have highlighted the utility of using $^{87}Sr/^{86}Sr$ to differentiate natal habitats at local scales (Kennedy et al. 1997; Ingram and Weber 1999; Hobbs et al. 2005). We found a wide range in $^{87}Sr/^{86}Sr$ values present among natal sites in the Mid and Upper Columbia River basin, which was 1.5 times greater than that observed among California's Central Valley rivers and hatcheries (Barnett-Johnson et al. 2008). The small errors around our population mean $^{87}Sr/^{86}Sr$ value suggests that individuals collected in our study were likely born or spent a significant amount of time rearing at the specific location of collection. Three sites (MET, BON, and HAN) had $^{87}Sr/^{86}Sr$ that were unique allowing for unequivocal assignment of individuals to correct natal locations.

We found general trends in $^{87}Sr/^{86}Sr$ values along the Columbia River and its tributaries that corresponded to surficial geology in the watershed. The mainstem Columbia River is influenced at its headwaters in Washington by old Paleozoic sedimentary rocks in the eastern Okanogan Highlands (Lasmanis 1991; Fig. 1). The Hanford Reach site (HAN) represents our only natural source of Chinook salmon on the mainstem and had the most radiogenic $^{87}Sr/^{86}Sr$ value (0.7142). This $^{87}Sr/^{86}Sr$ value is consistent with previous Columbia River water values and Columbia River Basalt Group values of 0.7034 in early-erupted basalts to 0.7145 in late-erupted flows (Ramos et al. 2006; Singleton et al. 2006). Four hatcheries (WEL, EBH, PRH, UMA) on the mainstem have similar $^{87}Sr/^{86}Sr$ ratios that are less radiogenic than HAN but show greater than ocean $^{87}Sr/^{86}Sr$ values (Table 1). This pattern is consistent with the influence of marine-derived feed on the $^{87}Sr/^{86}Sr$ in the otoliths of hatchery fish reflecting a mixture of the $^{87}Sr/^{86}Sr$ in source water on the mainstem Columbia River and less radiogenic $^{87}Sr/^{86}Sr$ value of hatchery feed (0.70918; Ingram and Weber 1999). Hatchery feed could affect otolith $^{87}Sr/^{86}Sr$ through dietary source of Sr or altered dissolved ambient Sr in the water (Barnett-Johnson et al. 2008).

Natal sources located along the tributaries to the west of the Columbia River and south of the Methow River (WEN, YAK, PRO, KLI, LWS) have similar $^{87}Sr/^{86}Sr$ values (~0.706). The $^{87}Sr/^{86}Sr$ values characterizing these tributary sources are influenced by the younger Miocene volcanic formations in the Columbia Basalt Group and the Southern Cascade Mountain range (Lasmanis 1991; Fig. 1). The Bonneville Hatchery isotopic signature (0.70818) is intermediate between the mainstem Columbia River and upstream tributary values. This result is consistent with the isotopic mixing of these waters.

For the few studies that have used both molecular genetic and otolith microchemical information on the same individuals, significant insight has been gained about the spatial scales of connectivity (Milton and Chenery 2001; Miller et al. 2005; Bradbury et al. 2008). The variation in otolith $^{87}Sr/^{86}Sr$ performs at its best when integrated with other data and approaches (Harrington et al. 1998; Barnett-Johnson et al. 2008). $^{87}Sr/^{86}Sr$ values are unlikely to be unique among natal sources across a species' distribution. Without a marker identifying individuals to broader geographic regions, the $^{87}Sr/^{86}Sr$ marker may be of limited utility for many fisheries applications where stocks co-occur. For example, Chinook salmon from the Yuba River in California's Central Valley and from the Bonneville Hatchery in the Columbia River basin have similar $^{87}Sr/^{86}Sr$ values (~0.70818; Barnett-Johnson et al. 2008). However, the California Central Valley and Columbia River stocks differ genetically (Seeb et al. 2007). Harrington et al. (1998) used variation in $\delta^{15}N$ among watersheds due to agricultural influences with $^{87}Sr/^{86}Sr$ to further delineate Atlantic Salmon (*Salmo salar*) rearing habitats. (Walther and Thorrold 2008) used oxygen isotopes that varied broadly along a latitudinal gradient of precipitation in conjunction with Sr/Ca, Ba/Ca, and $^{87}Sr/^{86}Sr$ in otoliths to characterize the freshwater natal rearing sites of American Shad (*Alosa sapidissima*) along the entire U.S. Atlantic coast.

Many Pacific salmon populations are demographically independent and exhibit stock-specific extinction risks. Identifying individual stocks in mixed-stock harvest management is fundamental for preventing overexploitation of vulnerable stocks while optimizing the harvest of abundant populations. The need for a coordinated coast-wide effort in stock identification for Pacific salmon has long been recognized, but the search for the ideal biological markers continues to evolve as new stock identification techniques and statistical approaches emerge (Wood et al. 1989; Seeb et al. 2007). Wood et al. (1989) evaluated the accuracy and precision of stock composition estimates of sockeye salmon using 51 principal stocks in British Columbia and Southeast Alaska with four types of biological markers- freshwater age, six scale pattern variables, the prevalence of a brain parasite, and five genetic (electrophoretic) traits. They concluded that in general, estimates for individual stocks are unreliable using these markers largely due to the annual variability in scale patterns, and lack of genetic or stable parasite markers to achieve desired levels of stock resolution.

Our integrated approach shows promise in addressing outstanding questions in salmon ecology and the potential for application at the coast-wide scale. The low annual variation in both the microsatellite and $^{87}Sr/^{86}Sr$ markers minimizes the frequency of new baseline data collection. An additional benefit to developing $^{87}Sr/^{86}Sr$ baselines is that multiple fish species occupying the same systems can be tracked using the same $^{87}Sr/^{86}Sr$ signatures of origin. For example, several rivers from Japan to California have multiple species of salmon-Chinook, Steelhead (*O. mykiss*) Coho (*O. kisutch*), and Sockeye (*O. nerka*). Developing a baseline of $^{87}Sr/^{86}Sr$ natal values within a river system could improve the management and conservation of several salmon species. Sophisticated genetic baselines (microsatellite and SNPs) are currently being considered by fisheries resource managers as integral tools to minimize mortality on protected stocks, and target healthy populations in the ocean fisheries (see Seeb et al. 2007). The development of $^{87}Sr/^{86}Sr$ baselines for individual rivers and hatcheries within key ESUs would allow for finer-scale identities in systems where molecular markers show limited resolution. Genetic tools alone could provide natal source identities or even finer-resolution (e.g., parentage) in some systems, but would require sampling each brood year, examination of more loci, and expensive and labor intensive efforts (Hauser et al. 2006; Planes et al. 2009). $^{87}Sr/^{86}Sr$ and microsatellite baselines have been developed for rivers and hatcheries within the UCR Su/F and California's Central Valley fall ESUs, two significant populations to west-coast salmon fisheries (Barnett-Johnson et al. 2008; Weitkamp 2010). Next steps would be to apply these tools for mixed collections of juveniles or adults in estuaries and coastal oceans to better understand source-specific distributions and individual growth and condition.

The need for reliable stock identification tools will continue to grow as recovery efforts for salmon stocks listed under the Endangered Species Act intensify. Our results suggest that integrating molecular genetic and otolith isotopic data in a Bayesian framework may be the next step in its application to coast-wide salmon conservation and harvest management for these stocks. Genetic stock identification programs can produce rapid turn-around times useful for real-time fisheries management decisions. To date, otolith $^{87}Sr/^{86}Sr$ measurements have provided retrospective insights, but could achieve comparable efficiencies with significant resources and advances in mass spectrometry. The development of additional baselines of $^{87}Sr/^{86}Sr$ variation within key genetic groups may provide fine-scale resolution in other salmon systems. In addition, other biological markers (e.g., otolith elemental ratios) may provide still further resolution for natal sources that have overlapping $^{87}Sr/^{86}Sr$ values, as is the case with two of our groupings within the UCR Su/F population. With continued research and the emergence of new technologies, the promise of understanding salmon ecology at broad and fine geographic scales will be realized.

ACKNOWLEDGEMENTS

We thank Melinda Davis, Paul Hoffarth, Andrew Murdoch, and Bill Sharp for providing samples; David Kuligowski for collecting genetic data; George Whitman for his technical assistance; anonymous reviewers for improving the paper; and the organizers for the 4thInternational Otolith Symposium.

REFERENCES

1. Abaunza P, Murta AG, Campbell N, Cimmaruta R, Comesana AS, Dahle G, Garcia Santamaria MT, Gordo LS, Iversen SA, MacKenzie K, Magoulas A, Mattiucci S, Molloy J, Nascetti G, Pinto AL, Quinta R, Ramos P, Sanjuan A, Santos AT, Stransky C, Zimmermann C (2008) Stock identity of horse mackerel (*Trachurus trachurus*) in the Northeast Atlantic and Mediterranean Sea: Integrating the results from different stock identification approaches. Fish Res 89:196–209
2. Bacon CR, Weber PK, Larson KA, Reisenbichler R, Fitzpatrick JA, Wooden JL (2004) Migration and rearing histories of Chinook salmon (*Oncorhynchus tshawytscha*) determined by ion microprobe Sr isotope and Sr:Ca transects of otoliths. Can J Fish Aquat Sci 61:2425–2439

3. Bailey RE, Margolis L (1987) Comparison of the parasite fauna of juvenile sockeye salmon (*Oncorhynchus nerka*) from southern British Columbia and Washington State lakes. Can J Zool 65:420–431
4. Barnett-Johnson R (2007) Spatial scales of mixing and natal source contributions of salmon populations in the coastal ocean detected by otolith and genetic signatures of origin. University of California, Dissertation
5. Barnett-Johnson R, Ramos FC, Grimes CB, MacFarlane RB (2005) Validation of Sr isotopes in otoliths by laser ablation multicollector inductively coupled plasma mass spectrometry (LA-MC-ICPMS): opening avenues in fisheries science applications. Can J Fish Aquat Sci 62:2425–2430
6. Barnett-Johnson R, Pearson TE, Ramos FC, Grimes CB, MacFarlane RB (2008) Tracking natal origins of salmon using isotopes, otoliths, and landscape geology. Limnol Oceanogr 53:1633–1642
7. Beacham TD, Candy JR, Jonsen KL, Supernault J, Wetklo M, Deng L, Miller KM, Withler RE, Varnavskaya N (2006) Estimation of stock composition and individual identification of Chinook salmon across the Pacific Rim using microsatellite variation. Trans Am Fish Soc 135:861–888
8. Botsford LW, Brumbaugh DR, Grimes C, Kellner JB, Largier J, O'Farrell MR, Ralston S, Soulanille E, Wespestad V (2009) Connectivity, sustainability, and yield: bridging the gap between conventional fisheries management and marine protected areas. Rev Fish Biol Fish 19:69–95
9. Bradbury IR, Campana SE, Bentzen P (2008) Estimating contemporary early life-history dispersal in an estuarine fish: integrating molecular and otolith elemental approaches. Mol Ecol 17:1438–1450
10. Cadrin SX, Friedland KD, Waldman JR (2005) Stock identification methods: applications in fisheries science. Elsevier Academic Press, San Francisco
11. Camp VE, Ross ME, Hanson WE (2004) Genesis of flood basalts and Basin and Rang volcanic rocks from the Stenns Mountain to the Malheur River Gorge, Oregon. Geol Soc Am Bull 115:105–128
12. Campana SE, Thorrold SR (2001) Otoliths, increments, and elements: keys to a comprehensive understanding of fish populations? Can J Fish Aquat Sci 58:30–38
13. Carlson SM, Seamons TR (2008) A review of quantitative genetic components of fitness in salmonids: implications for adaptation to future change. Evol Appl 1:222–238
14. Carlson SM, Quinn TP (2007) Ten years of varying lake level and selection in size-at-maturity in sockeye salmon. Ecology 88:2620–2629
15. Carlson SM, Rich HB, Quinn TP (2009) Does variation in selection imposed by bears drive divergence among populations in the size and shape of sockeye salmon? Evolution 63:1244–1261
16. Chapman DW (1986) Salmon and steelhead abundance in the Columbia River in the nineteenth century. Trans Am Fish Soc 115:662–670
17. Cowen RK, Sponaugle S (2009) Larval dispersal and marine population connectivity. Annu Rev Mar Sci 1:443–66
18. Feyrer F, Hobbs J, Baerwald M, Sommer T, Yin Q, Clark K, May B, Bennett W (2007) Otolith microchemistry provides information complementary to microsatellite DNA for a migratory fish. Trans Am Fish Soc 136:469–476
19. Fukuhara FM, Murai S, LaLanne JJ, Sribhibhadh A (1962) Continental origin of red salmon as determined from morphological characters. Int North Pacific Fish Comm Bull 8:15–109
20. Gillanders BM (2002) Temporal and spatial variability in elemental composition of otoliths: implications for determining stock identity and connectivity of populations. Can J Fish Aquat Sci 59:669–679
21. Henry K (1961) Radical identification of Fraser River sockeye salmon (*Oncorhynchus nerka*) by means of scales and its application to salmon management. Int North Pacific Fish Comm Bull 12

22. Hansen MM, Kenchington E, Nielsen EE (2001) Assigning individual fish to populations using microsatellite DNA markers. Fish Fish 2:93–112
23. Harrington RR, Kennedy BP, Chamberlain CP, Blum JD, Folt CL (1998) ^{15}N enrichment in agricultural catchments: field patterns and applications to tracking Atlantic salmon (*Salmo salar*). Chem Geol 147:281–294
24. Hauser L, Seamons R, Dauer M, Naish KA, Quinn TP (2006) An empirical verification of population assignment methods by marking and parentage data: hatchery and wild steelhead. Mol Ecol 15:3157
25. Hobbs JA, Yin Q, Burton J, Bennett WA (2005) Retrospective determination of natal habitats for an estuarine fish with otolith strontium isotope ratios. Mar Freshwat Res 56:655–660
26. Hobson K, Barnett-Johnson R, Cerling T (2009) Using isoscapes to track animal migration. In:West J, Bowen G, Dawson T, Tu K (eds.) Isoscapes: Understanding movement patterns and processes on earth through isotope mapping. Springer Press, Berlin, Germany, pp 273–298
27. Hooper PR (1997) The Columbia River flood basalt province: Current status. In: Mahoney JJ, Coffin MF (eds.) Large igneous provinces: Continental, oceanic, and planetary flood volcanism: Am Geophys Union Geophys Monogr 100:1–27
28. Ingram BL, Weber PK (1999) Salmon origin in California's Sacramento-San Joaquin river system as determined by otolith strontium isotopic composition. Geology 27:851–854
29. Kalinowski ST, Manlove KR, Taper ML (2007) ONCOR A computer program for genetic stock identification. Department of Ecology, Montana State University, Bozeman, MT. Available at http://www.montana.edu/kalinowski/ kalinowski_software.htm
30. Kennedy BP, Folt CL, Blum JD, Chamberlain CP (1997) Natural isotope markers in salmon. Nature 387:766–767
31. Lasmanis R (1991) Geology Wash Rocks Miner 66:262–277
32. MacKenzie K (2009) The use and abuse of parasites as stock markers for fish. Fish Res 97:1–2
33. MacCrimmon HR, Claytor RR (1985) Meristic and morphometric identify of Baltic stocks of Atlantic salmon (*Salmo salar*). Can J Zool 63:2032–2037
34. Margolis L (1963) Parasites as indicators of the geographic origin of sockeye salmon, *Oncorhynchus nerka* (Walbaum) occurring in the North Pacific Ocean and adjacent seas. Int North Pacific Fish Comm Bull 11:101–156
35. Marques JF, Teixeria CM, Cabral HN (2006) Differentiation of commercially important flatfish populations along the Portuguese coast: Evidence from morphology and parasitology. Fish Res 81:293–305
36. Marshall S, Bernard D, Conrad R, Cross B, McBride D, McGregor A, McPherson S, Oliver G, Sharr S, Ban Allen B (1987) Application of scale pattern analysis to the management of Alaska's sockeye salmon (*Oncorhynchus nerka*) fisheries. In: Smith HD, Margolis L, Wood CC (ed) Sockeye salmon (*Oncorhynchus nerka*) population biology and future management. Can Spec Publ Fish Aquat Sci 96:307–326
37. Miller JA (2007) Scales of variation in otolith elemental chemistry of juvenile staghorn sculpin (*Leptocottus armatus*) in three Pacific Northwest estuaries. Mar Biol 151:483–494
38. Miller JA, Kent AJR (2009) The determination of maternal run time in juvenile Chinook salmon (*Oncorhynchus tshawytscha*) based on Sr/Ca and $^{87}Sr/^{86}Sr$ within otolith cores. Fish Res 95:373–378
39. Miller JA, Banks MA, Gomez-Uchida D, Shanks AL (2005) A comparison of population structure in black rockfish (*Sebastes melanops*) as determined with otolith microchemistry and microsatellite DNA. Can J Fish Aquat Sci 62:2189–2198
40. Milton DA, Chenery SP (2001) Can otolith chemistry detect the population structure of the shad hilsa *Tenualosa ilisha*? Comparison with the results of genetic and morphometric studies. Mar Ecol Prog Ser 222:239–251

41. Moran P, Teel DJ, LaHood ES, Drake J, Kalinowski S (2006) Standardizing multi-laboratory microsatellite data in Pacific salmon: an historical view of the future. Ecol Freshw Fish 15:597–605
42. Myers JM, Kope RG, Bryant GJ, Teel D, Lierheimer LJ, Wainwright TC, Grant WS, Waknitz FW, Neely K, Lindley ST, Waples RS (1998) Status review of Chinook salmon from Washington, Idaho, Oregon, and California. U.S. Dept. Commer., NOAA Tech. Memo. NMFS–NWFSC–35, 443 p.
43. Narum SR, Banks M, Beacham TD, Bellinger MR, Campbell MR, Dekoning SJ, Eliz A, Guthrie CM III, Kozfkay C, Miller KM, Moran P, Phillips R, Seeb LW, Smith CT, Warheit K, Young SF, Garza JC (2008) Differentiating salmon populations at broad and fine geographic scales with microsatellite and single nucleotide polymorphisms. Mol Ecol 17:3464–3477
44. O' Connell M, Wright JM (1997) Microsatellite DNA in fishes. Rev Fish Biol Fisher 7:331–363
45. Palumbi SR (2004) Marine reserves and ocean neighborhoods: the spatial scale of marine populations and their management. Annu Rev Env Resour 29:31–68
46. Planes S, Jones GP, Thorrold SR (2009) Larval dispersal connects fish populations in a network of marine protected areas. P Natl Acad Sci USA 106:5693–5697
47. Ramos FC, Wolff JA, Tollstrup DL (2004) Measuring $^{87}Sr/^{86}Sr$ variations in minerals and groundmass from basalts using LA-MC-ICPMS. Chem Geol 211:135–158
48. Ramos FC, Wolff JA, Tollstrup DL (2006) Sr isotope disequilibrium in Columbia River flood basalts: Evidence for rapid shallow-level open-system processes. Geology 33:457–460
49. Rannala B, Mountain JL (1997) Detecting immigration by using multilocus genotypes. P Natl Acad Sci USA 94:9197–9201
50. Ruttenberg BI, Warner RR (2006) Spatial variation in the chemical composition of natal otoliths from a reef fish in the Galapagos Islands. Mar Ecol Prog Ser 328:225–236
51. Seeb LW, Antonovich A, Banks MA, Beacham TD, Bellinger MR, Blankenship SM, Campbell MR, Decovich NA, Garza JC, Guthrie CM III, Lundrigan TA, Moran P, Narum SR, Stephenson JJ, Supernault KJ, Teel DJ, Templin WD, Wenburg JK, Young SF, Smith CT (2007) Development of a standardized DNA database for Chinook Salmon. Fisheries 32:540–552
52. Secor DH (1999) Specifying divergent migrations in the concept of stock: the contingent hypothesis. Fish Res 43:13–34
53. Selkoe KA, Henzler CM, Gaines SD (2008) Seascape genetics and the spatial ecology of marine populations. Fish Fish 9:363–377
54. Singleton MJ, Maher K, DePaulo DJ, Conrad ME, Dresel PE (2006) Dissolution rates and vadose zone drainage from strontium isotope measurements of groundwater in Pasco Basin, WA unconfined aquifer. J Hydrol 321:39–58
55. Standish JD, Sheehy M, Warner RR (2008) Use of otolith natal elemental signatures as natural tags to evaluate connectivity among open-coast fish populations. Mar Ecol Prog Ser 356:259–268
56. Taylor EB (1991) A review of local adaptation in salmonidae, with particular reference to Pacific and Atlantic salmon. Aquaculture 98:185–207
57. Thresher RE (1999) Elemental composition of otoliths as a stock delineator in fishes. Fish Res 43:165–204
58. Utter F, Teel D, Milner G, McIsaac D (1987) Genetic estimates of stock compositions of 1983 Chinook salmon harvests off the Washington Coast and the Columbia River. Fish Bull 85:13–23
59. Utter FM, Chapman DW, Marshall AR (1995) Genetic population structure and history of Chinook salmon of the Upper Columbia River. Am Fish Soc Symp 17:149–165
60. Waknitz, FW, Mathews GM, Wainwright T, Winans GA (1995) Status review for Mid-Columbia River summer Chinook salmon. NOAA Tech. Mem. NMFS-NWFSC-22, 80 p. (Available from Natl. Mar. Fish. Serv., Northwest Fisheries Science Center, Coastal

Zone and Estuarine Studies Division, 2725 Montlake Blvd. E., Seattle, WA 98112–2097)
61. Walther BD, Thorrold SR (2008) Geochemical signatures in otoliths record natal origins in American shad. Trans Amer Fish Soc 137:57–69
62. Waples RS, Gustafson RG, Weitkamp LA, Myers JM, Johnson OW, Busby PJ, Hard JJ, Bryant GJ, Waknitz FW, Neely K, Teel D, Grant WS, Winans GA, Phelps S, Marshall A, Baker B (2001) Characterizing diversity in salmon from the Pacific Northwest Journal of Fish Biology 59 (Suppl. A):1–41
63. Waples RS, Teel DJ, Myers JM, Marshall AM (2004) Life history variation in Chinook salmon: historic contingency and parallel evolution. Evolution 58:386–403
64. Weitkamp LA (2010) Marine distributions of Chinook salmon (*Oncorhynchus tshawytscha*) from the West Coast of North America determined by coded wire tag recoveries. Trans Am Fish Soc 139:147–170
65. Weissenborn AE (1969) Geologic map of Washington: U.S. Geological Survey, Miscellaneous Geologic Investigations Map I-583, scale 1:2000000
66. Winans GA (1984) Multivariate morphometric variability in Pacific salmon: technical demonstration. Can J Fish Aquat Sci 41:1150–1159
67. Winans GA, Viele D, Grover A, Palmer-Zwahlen M, Teel D, Van Doornik D (2001) Update of genetic stock identification of Chinook salmon in the Pacific Northwest: Test fisheries in California. Rev Fish Sci 9:213–237
68. Wood CC, Rutherford DT, McKinnell S (1989) Identification of sockeye salmon (Oncorhynchus nerka) stocks in mixed-stock fisheries in British Columbia and southeast Alaska using biological markers. Can J Fish Aquat Sci 46:2108–2120

CHAPTER 4

Evaluation of Genetic Diversity in Fish Using Molecular Markers

Andreea Dudu[1], Sergiu Emil Georgescu[1] and Marieta Costache[1]

[1] University of Bucharest, Department of Biochemistry and Molecular Biology, Splaiul Independentei 91-95, Bucharest, Romania

1. INTRODUCTION

The variety of plants, animals and microorganisms on Earth, along with the tremendous diversity of genes in these species and the high variety of ecosystems on the globe are all constitutive parts of what is called "biodiversity". It is globally recognized that the conservation of nature and of its biological diversity should represent one of the priorities of the current human society. An important number of factors, sometimes interconnected, like climate change, pollution, anthropic activities destroying habitats for agriculture or logging, excessive exploitation of species with economic importance, hunting or poaching, represent serious treats for biodiversity.

The loss of biodiversity is a serious concern for multiple reasons, as a "healthy" biodiversity can provide various natural benefits including ecosystems services (contribution to climate stability, protection of water resources, nutrient storage and recycling, etc.), biological resources (food, medicinal and pharmaceutical resources, diversity in genes and species, breeding stocks and population reservoirs, etc.) and social benefits (research and education, recreation and tourism, etc.) [1].

Despite of well-known importance of biodiversity, human activity has been causing massive extinctions. In 2005, the report released by the Millennium Ecosystem Assessment pointed out a substantial and largely irreversible loss in the diversity of life on Earth, with some 10-50% of the well-studied higher taxonomic groups threatened with extinction, due to human actions. Over the past few hundred years, humans have increased species extinction rates by as much as 1000 times background rates that were typical over Earth's history [2]. Species are being lost at a rate that far exceeds the emergence of new species. The current extinction problem has been called the 'sixth extinction', as its magnitude compares with that of the other five mass extinctions revealed in geological records [3, 4].

The conservation of biodiversity at all levels from genes to ecosystems represents a global concern. Thus, the genetic diversity within species (between populations as well as among individuals within populations) as part of biodiversity is the result of the degree of variation at different levels (nucleotide, gene, chromosome, and genome). The presence of genetic variation plays an important role in species/populations survival and in their successful evolution in response to both short-term and long-term environmental changes [5]. To prevent the problems of genetic defects caused by inbreeding, species need a variety of genes to ensure successful survival. The decrease of this variety also translated as reduced genetic diversity is correlated with enhancing the chances of extinction.

The group of fish comprises over 32,000 species and an inestimable number of individuals inhabiting a large territory, from depths of 6000-7000 m on the bottom of oceans and seas to mountain waters found at an altitude of 2000 m. Thus, the fish group presents the most significant species variety among vertebrates [6]. The importance for humans derives mainly from economic significance, with aspects including consumption as food source, use in industry, aquaculture and fish farming. Overfishing and poaching along with habitat destruction (including water pollution, the building of dams, removal of water for use by humans, and the introduction of exotic species) represent the major threats to fish species and populations and currently, according to 2014 IUCN Red List, 2172 fish species are threatened with extinction [7].

Because of their life environment fish are more difficult to monitoring and study than terrestrial animal and to recover important information about fish populations is very challenging.

Sturgeons hold an important position in the category of the most threatened fish species in the world, mainly due to their particular scientific and commercial importance. Sturgeon species and the distantly related paddlefish as well as some extinct families are reunited in the Order Acipenseriformes and are generally regarded as "living fossils" and as the most primitive surviving bony fish. Sturgeons present some relict characters that differentiate them from other fish species and prove their ancient origin. The skeleton is primarily cartilaginous, with partial ossification only in cranium and maxilla, the posterior vertebrae continue far out into the dorsal lobe of the caudal fin (heterocercal condition), branchiostegal rays are absent or inconspicuous, undifferentiated vertebrae, ganoid scales only in caudal part of the body, five rows of bony scutes along the body. Sturgeons have a protrusible mouth, toothless in adults, a prominent snout with four barbells located about midway between the mouth and the snout tip [8].

Their extinct relatives (Infraclass Chondrostei) date to the Devonian (about 350 million years ago) and *Acipenser* fossils were found in the Western of North America dating from Upper Cretaceous. At present time within Acipenseriformes only two families exist, *Acipenseridae* and *Polyodontidae,* the first one counting 25 species inhabiting rivers, lakes, coastal waters and inner seas from the Northern hemisphere. The Acipenseriformes species are strictly Northern hemisphere fishes, some being resident to fresh water, while others are anadromous. Sturgeons are reunited in *Acipenseridae* family with four genera – *Acipenser, Huso, Scaphirhynchus* and *Pseudoscaphirhynchus.*

The area comprising the Black, Caspian and Aral Seas (Ponto-Caspian region) and the rivers that are tributaries to these seas, is the area with the greatest species diversity despite the fact the most of the populations in the region are facing extinction [9]. Currently in the Danube River three anadromous sturgeon species *Huso huso* (Beluga sturgeon), *Acipenser stellatus* (Stellate sturgeon) and *Acipenser gueldenstaedtii* (Russian sturgeon) and a potamodromous species *Acipenser ruthenus* (Sterlet sturgeon) are found [10]. The others two species that were present in the past in this area (*Acipenser sturio* – European sturgeon and *Acipenser nudiventris* – ship sturgeon) are considered to be extinct [11].

From ancient times, sturgeons had a great economic impact for the Danube region and they were an important part of the welfare of the local communities from this area. Due to human intervention, a decline of the sturgeon populations in this river has been observed starting with the 19th century, but much more accentuated in the 20th century. In Romania, in the first decade after the communist regime decline an overexploitation of sturgeon from Lower Danube has occurred mainly due to the absence of a legislation that protects these species [12, 13]. Thus, between 2002 and 2005 the sturgeon captures have severely diminished from 37 tons in 2002 to 11 tons in 2005. As a consequence of this evolution in 2006 the Romanian government adopted a law according to which the fishing of sturgeons for commercial purposes is prohibited for a period of 10 years. A solution to counter-balance the depletion of wild population was the one of sturgeon breeding in aquaculture. In the fish farms from Romania several species are preferred, especially Danube sturgeons as beluga sturgeon, stellate sturgeon and Russian sturgeon, considered to be ideal for obtaining caviar of superior quality and sterlet sturgeon that have the advantage of a very tasteful meat. Beside these species, the Siberian sturgeon (*Acipenser baerii*) and different inter-specific hybrids are also raised in aquaculture.

The causes for the dramatic decline of sturgeon populations all over the world are related mostly to the changes that are taking place in their habitats: pollution, dams and other constructions that are blocking the migration into the river for reproductive purposes, overfishing and poaching, mainly to provide caviar for the black market. Another factor that has a negative impact on the sturgeon populations is the practice of stocking the rivers with individuals from aquaculture, which becomes more and more popular with the years. Uncontrolled restocking presents as consequence the decrease of genetic diversity within and between populations, mainly due to the reduced number of adult individuals from the wild used as genitors for reproduction in aquaculture. Also, other practice with disastrous consequences for the populations is the one of restocking with inter-specific hybrids incorrectly labeled as individuals of native species. The inter-specific hybridization phenomenon is relatively frequent in sturgeons both in the wild and in aquaculture. Such hybrids were caught in the Danube River [14, 15] or are produced and raised in sturgeon farms. The species identifications based only on morphology might be misleading sometimes and molecular methods were proposed for species identification and caviar labeling [15-18].

Other fish species with a high ecological and economic importance are represented by salmonids. Romania is one of the European countries characterized by the presence of a significant number of wild salmonids populations and by an important potential for the development of intensive breeding of salmonids in aquaculture. The Salmoniformes order represents a heterogeneous group of fish reunited in one family *Salmonidae*, with three subfamilies (*Coregoninae* - whitefish and ciscos; *Thymallinae* – graylings; *Salmoninae* – trout, salmon and charr), nine genera and roughly 68 species [19]. The largest of salmoniform fishes are considered to be Chinook salmon (*Oncorhynchus tshawytscha*) and Danube huchen (*Hucho hucho*), that might reach at 1.5 m in length and a weight around 60 kg. Salmonids are characterized by body and fins that are streamlined and symmetrical, being covered with small and cycloid scales. All fins have soft rays. The representatives of this order possess several primitive anatomical features that are characteristics for an early stage in the evolution of modern bony fishes, like a small, fleshy adipose fin located between the dorsal fin and the powerful caudal fin [20].

They are native to the cooler climates of the Northern Hemisphere, but have been widely introduced around the world for angling and aquaculture.

In the Romanian fauna the following salmonid species are present: *Salmo trutta fario* (brown trout), *Salmo labrax* (Black Sea salmon), *Salvelinus fontinalis* (brook trout), *Hucho hucho* (huchen or Danube salmon), and *Thymallus thymallus* (grayling). The European whitefish (*Coregonus lavaretus maraenoides*) was introduced starting with 1957 in Lake Rosu, Bicaz and Tarcau rivers, the biological material originating from Poland and Russia. Unfortunately, there are no other studies regarding the adaptation of this species in the water systems where has been introduced. Apart the salmonids from the wild fauna previous mentioned the following species are bred for commercial purposes in aquaculture: brown trout, brook trout and rainbow trout (*Oncorhynchus mykiss*) and a series of hybrids of natural and fishery species.

The taxonomy of *Salmo* is still a matter of controversy [21] since more than 60 synonyms for varieties of brown trout and more than 20 for varieties of Atlantic salmon were described. The number of *Salmo*species recognized varies considerably not only because of highly phenotypic variation (body shape, colour, etc.), but also because the species inhabit and are adapted to very different habitats over large distribution areas. Due to taxonomic ambiguities, authors often refer to brown trout as *Salmo trutta*species complex [22, 23].

Salmo trutta comprises several distinct ecological and geographical morphs and with respect to this is still controversy as far as their classification as species or subspecies is concerned [24, 25]. Based on morphological and ecological variations, the existing populations of *Salmo trutta* from distinct areas are grouped into different taxa: i) Black Sea populations – *Salmo labrax*, ii) Caspian Sea populations – *Salmo caspius*, iii) Aral Sea populations – *Salmo oxianus* and iv) Mediterranean Sea populations – *Salmo macrostigma* [26].

The brown (common) trout (*Salmo trutta* morpha *fario* and *Salmo trutta* morpha *lacustris*) and the sea trout (*Salmo trutta* morpha *trutta*) are fish of the same species, considered by some taxonomists different subspecies in order to distinguish the anadromous *Salmo trutta trutta*, living in the sea and migrating in freshwater only to spawn, from *Salmo trutta fario*, residing in freshwater and the lake dwelling form *Salmo trutta lacustris*. Instead, other authors consider that these do not necessarily represent monophyletic groups [27].

In Romania, *Salmo trutta fario* (Linnaeus, 1758) is widely spread in a large number of water streams from the mountain area, along the Carpathian Arch, whereas the Black Sea trout, *Salmo labrax* (Pallas, 1814) is endemic to the Black Sea area and migrates for reproduction in the Danube River and its tributaries. Nowadays only few individuals are captured annually along the Black Sea coast and sporadically in the Lower Danube River. Due to the present situation of the species several measures were adopted to protect it. Thus, the fishing is completely prohibited and the Black Sea trout is included on the Red List of Danube Delta Biosphere Reserve [8].

Salvelinus fontinalis (Mitchill, 1815) is predominantly raised in fish farms for food consumption, but a low number of wild populations are still present in the Romanian mountain waters.

Thymallus thymallus (Linnaeus, 1758) is the only native salmonid species for which no imports of biological material and restocking programs were completed in Romania [26].

Until two or three decades ago, excellent habitat conditions for Danube huchen (*Hucho hucho*, Linnaeus, 1758) still existed in many rivers in Romania. The historical range of this species included the majority of the Carpathian river systems [28], but hydropower development, river pollution, and overfishing and poaching led to drastic declines in the area inhabited by huchen [29], and the species is now extinct in the Mures, Timis, Cerna, Olt, Arges, and Ialomita river systems. Data on the current area of huchen occurrence in Romania are fragmentary, and the species only occurs in a few rivers, including, among others, the Tisa River and its tributaries (Viseu, Ruscova, Crasna, Bistra, Vaser, Somes, and Cris) and in the Siret River.

In relation to the main rivers from Romania, there are 10 hydrographic basins in which significant salmonid populations are found. Nowadays is a well-known fact that the most prevalent salmonid species is the brown trout, while the most threatened are the Danube huchen and the Black Sea salmon.

Both sturgeons and salmonids are of crucial importance for Romanian fauna, having also a significant socio-economic value. In this context, well documented studies regarding the biology, taxonomic classification and ecology of the species were performed [10, 28, 30-32]. The development of technologies based on DNA markers has had a tremendous impact on animal genetics in generally, and changed the way in which studies were conducted inclusive in population genetics, phylogeny, phylogeography and conservation. Thus, in a progressive way, studies aiming at analyzing molecular aspects in sturgeons and salmonids populations from Romania founded their place in the research field, although the number of such studies is still relatively low.

2. MOLECULAR MARKERS FOR POPULATION GENETICS AND CONSERVATION

The definition of "marker" in wide sense is "something that serves to identify, predict, or characterize" [33]. In biology, the markers refer to any stable variation, which is heritable and can be measured or detected by an appropriate method. Such variations are produced at different levels for example, morphological, gene, chromosomal, biochemical or genomic. The markers that represent variations that appear at the DNA level are so-called *molecular markers*. The molecular markers occupy specific places in the genome and possess the role to "mark" the position of a specific gene or the inheritance of a particular character.

All organisms are subject to mutations that appear when genetic material fails to copy accurately or as a consequence of interactions with the environment, leading to genetic variation or polymorphism. Individuals of a species are distributed into more or less separate groups called populations, distributed over the species range. The genetic variation of a species is distributed both within and between populations. The diversity at the gene level is often referred to as intra-specific variability as it represents biological diversity within a single species. The researchers are very interested in assessing genetic variation within and between populations and detecting similarities as well as differences between individuals/ populations in order to establish optimum conservation strategies.

For this variation to be useful in different genetic studies it is necessary to be inherited and to be perceptible for the researchers, whether is recognizable as phenotypic variation or as a genetic mutation detectable by different molecular techniques. At the DNA level the genetic variation is represented by point mutations called also single nucleotide polymorphisms (SNPs), indels (insertions or deletions of nucleotide sequences), inversion of a segment of DNA within a locus, and rearrangement of DNA segments around a locus of interest. As the mutation rate is very low the evolution through mutation is extremely slow. The process of mutation is the only way in which genetic variation is created and in the lack of mutations there would be no biological diversity.

DNA marker technology can be applied to reveal these mutations. Large deletions and insertions determine shifts in the size of DNA fragments resulted consequently digestion by restriction enzymes, and are among the easiest type of mutations to detect by electrophoresis on agarose gel; smaller indels require DNA sequencing or more elaborate electrophoretic techniques, while the SNPs can be easily detected by DNA sequencing [34].

By using molecular markers is possible to observe and exploit genetic variation across the entire genome. Thus, the application of molecular markers in fish allowed recording rapid progress regarding the study of genetic variability and inbreeding, parentage determination, species identification, genetic linkage map construction for aquaculture species, identifying Quantitative Trait Loci (QTL) related to specific traits for marker assisted selection. One of the classifications of the molecular markers refers to markers of type I as markers associated with genes having a known function and markers of type IIas markers associated with unknown genomic segments [35]. The type I of DNA markers were not initially consider to be appropriate for genetic studies in fishes, but over time it became clear that these markers are very important both for the study of wild populations and aquaculture. Thus, this of markers has become very important for studying the phenomenon of linkage and for QTL mapping, being of great use in comparative genomic studies and for identification of candidate genes for quantitative traits in different fish species raised in aquaculture. An important number of studies were focused on the elucidation of the molecular basis of economically important traits in different species of salmonids. For example in the case of Atlantic salmon the mapping the QTLs was performed for the loci correlated with viral disease resistance [36], flesh color and growth traits [37], salinity tolerance [38], late sexual maturation [39], etc.

The type II of molecular markers (RAPD, AFLP, microsatellites) is considered to be non-coding. Such markers are used in population genetic studies aiming to characterize the genetic diversity. The type II of DNA markers proved useful in identifying species, populations and subpopulation, but also in identifying interspecific hybrids.

A second classification of markers is the one regarding their position in the cell. Thus we can distinguish between nuclear and mitochondrial DNA (mtDNA) markers, depending on their localization in the nuclear or mitochondrion genome.

The molecular markers might be highlighted by a variety of techniques that differ by difficulty, repeatability, cost and nature of polymorphism that is detected. The DNA markers detected by Polymerase Chain Reaction (PCR) are the most frequent used in assessing the genetic diversity and have a particular role in conservation. By PCR is possible to amplify DNA sequences up to several million times, so these markers present the advantage of nonlethal sampling. Small amount of tissue (fin clips or scales) are sufficient for analysis and is not necessary that the individuals to be sacrificed for sampling. This can be an important feature when evaluating genetic change in protected or declining populations and for providing access to DNA of ancient or archived tissue samples. It can provide information about genetic diversity over extensive temporal and spatial scales, especially for populations that no longer exist. Taken together, these characteristics suggest capability to monitoring populations that are small, exploited or declining.

Depending on the primers used for PCR amplification these markers can be divided in two groups: (i) PCR markers for target sequences – in this case the fragment of interest is amplified with two specific primers and (ii) PCR markers for arbitrary sequences – one primer with a arbitrary nucleotide sequence is used; the primer binds to randomly in the genome resulting unknown DNA fragments. The primer used in this type of techniques is usually short (of 8-10 nucleotides) and so the probability for it to bind at multiple sites in genome increase.

In the first category are found molecular markers like PCR-RFLP (PCR Restriction Fragment Length Polymorphism), PASA (PCR Amplification of Specific Alleles), SNP (Single Nucleotide Polymorphism), repetitive DNA sequences (minisatellites, microsatellites, etc.) [34]. The second category includes markers like RAPD (Randomly Amplified Polymorphic DNA), AP-PCR (Arbitrary Primed-PCR) and AFLP (Amplified Fragment Length Polymorphism), but these are less preferred comparing with PCR markers for target sequences due to the difficulty of analysis and lack of results accuracy and reproducibility.

PCR-RFLP

The analysis of these markers involves the amplification by PCR of a specific DNA region comprising one or more polymorphic sites for restriction enzymes (RE). With the increasing number of so-called "universal" primers available, can be targeted DNA regions that are relatively conserved among species. In addition, PCR products can be digested with restriction enzymes and visualized by ethidium bromide staining due to the increased amount of the resulting DNA amplification reaction. The ability of RFLP markers to highlight the genetic variation is relatively low compared with the one of other markers. Substitutions, insertions, deletions and rearrangements of the regions containing restriction sites are probably quite widespread in the genomes of many species, but the probability that they exist at the locus that we want to study is quite low. Because the difference in size between the restriction fragments is usually large these can be easily separated by agarose electrophoresis. The disadvantage of PCR-RFLP is that presents a relatively low level of polymorphism and requires knowing the sequence of the fragment amplified by PCR, fact that makes difficult to establish new markers.

In genetic studies regarding sturgeons the PCR-RFLP markers were mainly used in two directions: to assess the genetic diversity of sturgeon populations [40, 41] and for species identification and implicit for caviar traceability [42, 43]. Wolf et al. [42] has identified species specific restriction profiles in a fragment from the mitochondrial gene *cytochrome b* (*cyt b*) in 10 species from *Acipenser* and *Huso*genera. A similar study was performed by Ludwig et al [44] for 22 sturgeon species, an accurate identification being possible by analyzing the restriction profiles resulted consequently the action of five REs on a fragment of 1121 bp from *cyt b* gene. Panagiotopoulou et al. [43] proposed a molecular method based on PCR-RFLP to distinguish between Atlantic (*Acipenser oxyrinchus*) and European (*A. sturio*) sturgeon. The discrimination between the two species is difficult to be done exclusively by morphological traits, especially in their early life stages, while the application of two REs allowed the clear and unambiguous discrimination of 132 specimens of Atlantic and European sturgeon.

The identification of acipenserid species by PCR-RFLP presents the difficulty of correct diagnostic for species that are closely related from genetic and evolutionary point of view, like is the case of *A. gueldenstaedtii* / *A. persicus* or *Scaphirhynchus* genus species. Moreover, the analysis of length polymorphisms for mitochondrial DNA fragments put in evidence only the genetic variability originating from the maternal genitor, as the mtDNA is almost exclusively maternally inherited. This fact might lead to misinterpretations in case of inter-specific hybrids diagnostic and/ or an ancient introgression, both phenomena occurring in sturgeons [15, 45].

PCR-RFLP markers were used also in salmonid species for the genetic diversity and phylogeography analyses in different populations of brown trout [46, 47], charrs [48], grayling [49], etc.

Microsatellites also referred to as "simple sequence repeat" (SSR) [50] represent short repetitive sequences of 2-9 bp, wide spread in the genome and with a significant level of polymorphism. These markers are numerous in vertebrates, in fish appearing in every 10kb [51]. The majority of microsatellite loci are relatively small sizes, being amplified easily by PCR. Up to 70% of loci present dinucleotide repeats, the (AC) motif being the most common in the vertebrate genome [52]. The main features of microsatellites are co-dominant inheritance, high degree of polymorphism, hypervariability, higher mutation rate than standard. Through microsatellites analysis is possible to infer the genetic profile of an individual (genetic fingerprint) and to establish the relationships between individuals. Some microsatellites have a high number of alleles for a locus within population and are very suitable to identify the genitors and their progeny in hybrid populations. The higher level of allelic variation at microsatellite markers make them useful for addressing questions related to genetic structure, particularly where genetic differentiation may be limited. On the contrary other microsatellites present a low number of alleles or even a fixed allele and are more appropriate for phylogeny or species identification. The utility of microsatellite markers can be determined depending on the polymorphic information content (PIC), defined as the capacity of the marker to detect a polymorphism in the population [53]. PIC is direct correlated with the number of detected alleles and their frequency, a higher number of alleles determining a higher value for PIC. By comparing the values for PIC in different markers is possible to obtain valuable information about their power and efficiency in population genetic and conservation studies.

The primers designed for microsatellites amplification in a species give also cross-amplification for similar loci in related species, this being an important benefit in analyzing populations that are small or at the brink of extinction [54]. Microsatellite markers are widely used for population genetic and conservation studies in fishes.

The first studies based on microsatellite loci analysis were initiated at the North-American species, where it aimed at isolation and description of disomic loci [55, 56, 57].

The identification and characterization of new loci is complicated by the polyploidy of the sturgeon species. Lots of potential useful microsatellites were eliminated from analysis since these were polysomic and thus, they complicated the interpretation of the inheritance mode and of genetic variation within and between populations. Once a disomic set of microsatellites is established, this fact permits to analyze the genetic diversity and structure of wild populations and aquaculture stocks. Such disomic loci were isolated and characterized also for *A. naccarii* [58, 59] and *A. persicus* [60]. In the case of Ponto-Caspian sturgeons, a few studies based on microsatellites analysis in *A. stellatus* from the Caspian Sea [61] and, respectively, in *A. gueldenstaedtii* from the North-West of the Black Sea, North of the Caspian Sea and Azov Sea [62] were performed.

Beside their high applicability in for inferring the genetic diversity and structure of the wild populations, the microsatellites are appropriate for aquaculture stocks evaluation, selection of breeders and proper conservation. Different studies were conducted in several sturgeon species (*A. transmontanus, A. fulvescens, A. naccarii, A. sturio,* etc.) for *ex-situ* conservation purposes. The obtained data were used in the management program to adopt appropriate conservation methods.

In salmonids, the microsatellites were successfully applied in phylogeography studies [22], in determination of genetic variation in wild and farmed fish populations [63], in inferring the genetic diversity within population, fine-scale genetic differentiation and relationship of populations [64], in assessment of stocking impact on wild populations [65].

The mitochondrial genome possesses certain characteristics (compact organization, maternal inheritance, hundreds to thousands of copies per cell, rapidly evolving, reduced recombination rate, and higher mutation rates compared to those of nuclear genes) that make it useful in population genetics and phylogeny studies. mtDNA analysis in sturgeons is a suitable method for the characterization of species and populations, providing useful information for the management of conservation activities. For intraspecific studies, the most commonly used marker is the mitochondrial control region (D-loop) due to the relatively high degree of nucleotide variation. Studies to discriminate species/populations/aquaculture strains based on the analysis of D-loop region polymorphisms were performed for several species of sturgeon, such as *A. gueldenstaedtii* [66], *A. sinensis* [67], *A. stellatus*and *A. baerii* [68].

The gene coding for *cyt b* is a useful marker for identifying different species of sturgeon based on specific polymorphisms found at this level. A 648 bp fragment of the gene coding for *cytochrome oxidase subunit I* (*CO I*) is considered a real "barcode" for vertebrates and is useful for species identification by *DNA barcoding* technique.

Genes with highly conserved sequence and slow evolution like mitochondrial ribosomal genes (16SrRNA and 12SrRNA genes) are preferred when is about inferring the phylogenetic relationships of fishes at different taxonomic levels. 12SrRNA gene is considered a promising tool for tracing the history of more recent evolutionary events and it has been widely used to study the phylogenetic relationships among different levels of taxa such as families, genera and species [26].

In salmonid species mtDNA has proven to be useful for phylogeographical studies [69] and for analysis of spatial and temporal population structure [70]. Based on the analysis of the mitochondrial D-loop marker Bernatchez [22] showed that there are five different main lineages of *Salmo trutta* in Europe: Atlantic, Danubian, Marmoratus, Mediterranean and Adriatic. Analytical techniques for mtDNA include indirect methods such as the analysis of RFLP markers or the direct analysis of mtDNA sequences.

3. FACTORS INFLUENCING THE GENETIC DIVERSITY OF POPULATIONS

The genes are transferred from one generation to the next, and every individual has two copies of each gene, one which is inherited from the maternal genitor, the other from the paternal genitor. The DNA sequence of a specific locus, either a gene or a non-coding marker, may present some differences, resulting different variants of the same locus. Such variants of a specific locus are called alleles and their existence implies the genetic variation existence.

i. Mutation is in wide sense the process of random change of the DNA sequence and represents a process by which new alleles are created. As the mutation rates at nuclear and mitochondrial genome are very slow, the evolution through mutation is so slow that it is generally impossible to detect it from one generation to another. However, mutation is important as a source of genetic variation. The process of mutation is the only way in which genetic variability is created, and without mutations there would be no biological diversity.

ii. Gene flow. A population acquires new alleles mostly through the immigration of individuals from surrounding populations (gene flow) and through mutations. Apart from these two processes the number of alleles in a population is determined by the size of the population. Gene flow is a change in allele frequency that occurs due to migration of individuals among populations. It is possible that individuals that are moving into a new population to bring new alleles which are not present in that population or that they are in frequencies that differ from the allele frequencies of that population. So the gene flow increases the genetic variation within a population, but tends to make populations genetically similar to each other. If migration between populations occur in large numbers of individuals and the level of gene flow is significantly high, the populations will have the same alleles in the same frequencies and it will be one single population.

iii. Genetic drift represents a random modification in allele frequency of population that occurs if a population size is not infinite. In populations that are of limited size, allele frequencies will change randomly from one generation to the next. In the short term, over a few generations, a result of genetic drift would be the increasing or decreasing of allele frequencies in a random, unpredictable way. In the longer term, the main result of genetic drift is loss of genetic variation as by chance some of the alleles that exist in the parent generation may not be passed on to their offspring. The effects of genetic drift are strongest in small populations because the more impressive is the fluctuation of allele frequencies, and the sooner the loss of genetic variation. Genetic drift also results in different populations becoming genetically different from each other because different alleles will become more frequent or fixed in different populations.

iv. Natural selection is the gradual process by which biological traits become either more or less common in a population as a function of the effect of inherited traits on the differential reproductive success of organisms interacting with their environment. Natural selection appears because different genotypes have different fitness. Individuals with higher fitness survive and reproduce more than other individuals, so these genotypes become increasingly more and more frequent in populations. In different populations, parents of different genotypes pass their genes unequally to the next generation, leading to the genetic differences among isolated populations. So, genetic drift tends to make different populations genetically distinct from each other by chance, whereas natural selection tends to form genetically different populations due to environmental constraints. Consequently, the traits that have high fitness in one population, and evolve through natural selection, will be different from the traits that have high fitness and evolve through natural selection in another population. In generally, the natural selection, genetic drift and gene flow have an effect on genetic variation within populations and between populations. While the genetic drift and selection tend to reduce the variation within populations and increase the differences between populations, the gene flow increases the variation within populations, but makes populations similar.

v. Inbreeding. The birth of offspring resulted from reproduction between close relatives that occurs mainly in small and isolated populations. The consequence of inbreeding is the reduced viability and reproduction, as well as increased occurrences of diseases and defects, so called inbreeding depression [71].

4. MOLECULAR STUDIES FOR SPECIES IDENTIFICATIONS AND GENETIC DIVERSITY ASSESSING IN STURGEONS FROM ROMANIA

The markers analyzed in sturgeon species and populations from Romania were nuclear markers (microsatellites) and mtDNA. The methodologies applied in the studies included microsatellites genotyping, PCR-RFLP and sequencing of mitochondrial markers. Thus, the DNA was isolated from biological samples consisting in small pieces of fin sampled without harming the animals (aspect very important in vulnerable populations) by a classic protocol with phenol-chloroform-isoamylic alcohol.

For mtDNA analysis the primers were designed based on DNA sequences for the interest markers retrieved from GenBank data base. The primers for microsatellites amplification were described in literature in different species and the cross-amplification for the similar species from Romania was tested.

The molecular markers were amplified by PCR or multiplex PCR in specific conditions established consequently several steps of reaction optimization. The microsatellites were analyzed by capillary electrophoresis in ABI Prism 310 Genetic Analyzer (Applied Biosystems) and the mitochondrial markers were sequenced by Sanger method, *dye terminator* variant, in ABI Prism 3130 Genetic Analyzer (Applied Biosystems). The raw data were processed and edited with dedicated computer programs. The interpretation of genotypic data was done by using specialized computer software for population genetics and phylogeny.

In the context of severe decline of sturgeon population from the Lower Danube several studies based on molecular marker analysis were performed in the recent years in Romania. In consequence, the studies were directed to assessing the genetic diversity in Lower Danube sturgeon populations based on microsatellites and to evaluate genetic variability in aquaculture strains in order to sustain the efforts of conservation.

The analyses regarding the genetic diversity were preceded by the correct species identification based on molecular markers for each individual. As mentioned before, the accurate detection of sturgeon species encountered in the Danube River can be very difficult, due to the plasticity of various external morphological features. The number of hybrid individuals occurring in natural conditions is unknown. However, the number of hybrids in natural waters could increase due to escapes from commercial farms where exotic species or genotypes and inter-specific hybrids are regularly used for production purposes. So, decisions to initiate restocking programs for sturgeon conservation need to take into account, besides the socio-economical aspects and the assessment of genetic diversity, the correct diagnostic of fishes included in this type of programs.

For sturgeon species from the Lower Danube diagnostic a molecular methods was set-up by Dudu et al [15] based on the analysis of nuclear markers. In this study 84 individuals belonging to all four species from the Lower Danube (*A. stellatus, A. gueldenstaedtii, A. ruthenus* and *H. huso*) were analyzed. Initially 25 microsatellites were tested from which eight loci (LS19, LS34, LS39, LS54, Aox27, AoxD234, AnacE4, and AnacC11) that have demonstrated a good amplification, results repeatability and interspecific polymorphism in all of the four investigated species were selected. The genotypic data processing involves three successive statistical analyses including Factorial Correspondence Analysis (FCA), STRUCTURE assignation and NewHybrids status determination.

In a first step, the genotypes data were run in a FCA test, using GENETIX software only for the individuals considered being pure species based on morphology analysis. The FCA highlighted the differences between the four analyzed species. Four main clusters, each corresponding to one of the sturgeon species analyzed, were identified (Figure 1).

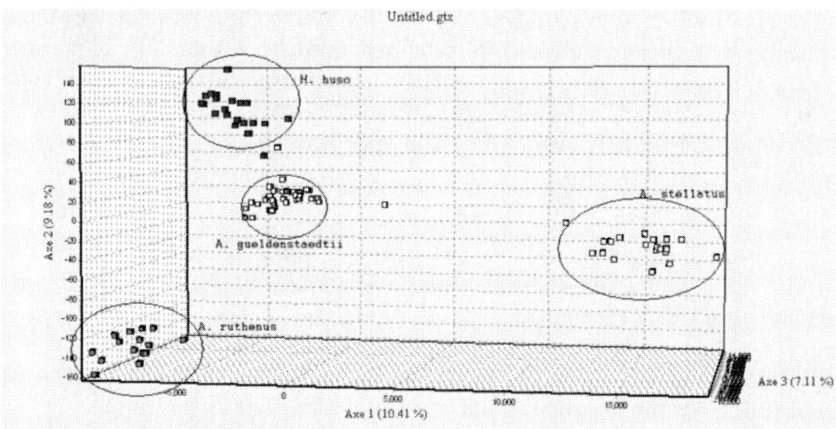

FIGURE 1. Factorial Correspondence Analysis (FCA) based on 8 microsatellite loci in the four pure species of Danube sturgeons [15].

In the second step, the genotypes data of putative hybrids, diagnosed such as based on morphometric indices were included in FCA analysis as supplementary individuals. A fifth category was highlighted including the putative hybrids, but also individuals that were considered as pure species, based on their morphology. The analysis had showed that only three of the four species hybridize each other, the putative hybrids displaying an intermediate position between *A. stellatus, H. huso* and *A. gueldenstaedtii.* Consequently, we deduced that *A. ruthenus* individuals should be eliminated from further tests (Figure 2).

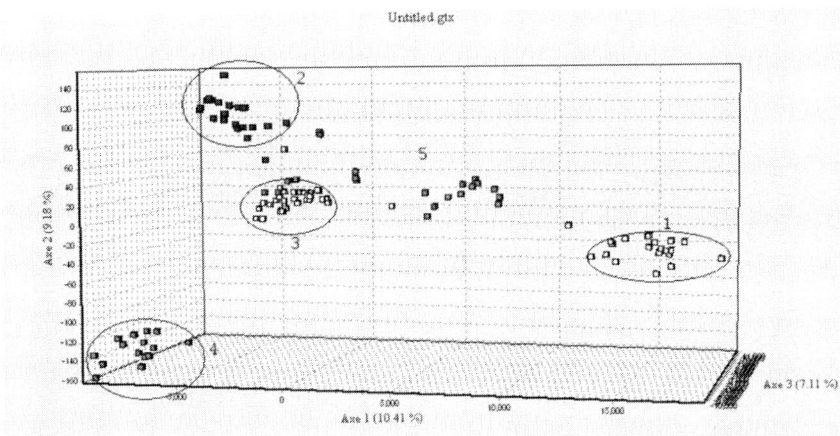

FIGURE 2. Factorial Correspondence Analysis (FCA) based on 8 microsatellite loci in the four pure species and hybrids of Danube sturgeons. (1) A. stellatus; (2) H. huso; (3) A. gueldenstaedtii; (4) A. ruthenus; (5) Hybrids and some pure individuals to be tested with other methods [15].

In consequence in the third step, FCA was performed only on three pure species and hybrid individuals included as supplementary elements in analysis. The analysis grouped the pure species and the hybrids in distinct clusters. The hybrids and some of the individuals labeled as pure were occupying an intermediate position between the pure species groups, some of these being placed approximately in the middle of the triangle delimited by the three pure species, while others appear to be closer to *H. huso*and *A. gueldenstaedtii* (Figure 3).

The second statistical analysis of the genotype data was the assignment test with STRUCTURE, which confirmed the presence of five specific clusters: (1) *A. stellatus* pure species; (2) *H. huso* pure species; (3) *A. gueldenstaedtii* pure species; (4) *A. ruthenus* pure species; (5) hybrids (Figure 4). The individuals classified as pure based on morphology (with three exceptions) were strongly assigned in their corresponding species. The perfect correspondence between morphological and molecular/assignation determinations gives a great reliability to the results.

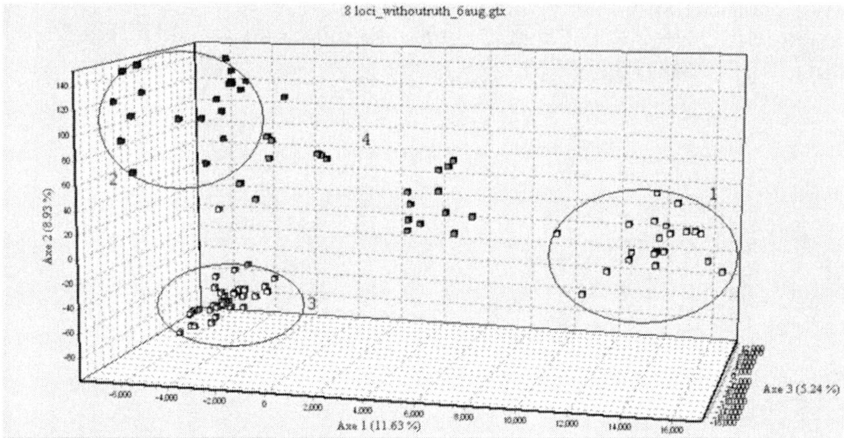

FIGURE 3. Factorial Correspondence Analysis (FCA) based on eight microsatellite loci in three pure species and hybrids of Danube sturgeons: (1) A. stellatus; (2) H. huso; (3) A. gueldenstaedtii; (4) Hybrids and some pure individuals [15].

The hybrids confirmed by the FCA and STRUCTURE assignment test were analyzed together with their two genitor species using the NewHybrids software in order to distinguish between F1 and later hybridization steps.

By the three successive statistical analyses it is possible the diagnostic of surgeon individuals as pure species or hybrids, the method proposed showing a high efficiency in discriminating pure species specimens from F1, F2 and two kinds of backcross.

The molecular analysis for species detection is necessary for all sturgeon individuals captured in the Lower Danube River and implied in restocking programs and from aquaculture. A database with genotypic data can be created as the accuracy of the proposed method increases along with the number of reference individuals analyzed. Also, such database might be of a real use for designing efficient management plans for sturgeon populations.

When a hybrid is detected the origin of maternal species can be identified by analyzing mitochondrial markers. For example the analysis by sequencing or PCR-RFLP of *cytb* gene permits the identification of the maternal genitor involved in hybrid formation as is well known that the mitochondrion genome is exclusively maternally inherited. The species identification based on the analysis of mtDNA by different methods should be handled with precaution when we deal with species that can easily hybridize, like sturgeons. The mitochondrial marker analysis is suitable for pure species, but totally inefficient for hybrids.

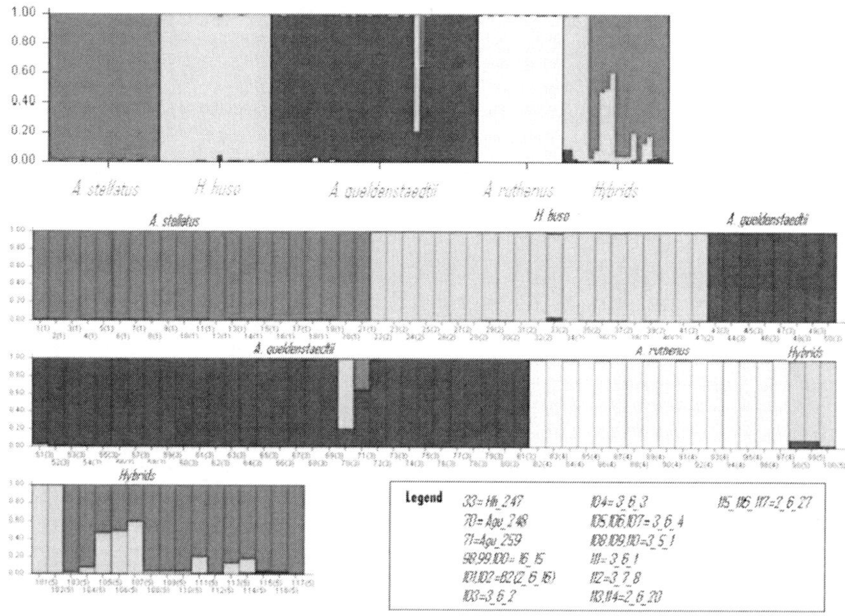

FIGURE 4. Assignation of the 84 sturgeons by STRUCTURE analysis based on eight microsatellite loci in four sturgeon species from Danube. Histograms represent the estimated membership coefficients (Q). Composite bars are expected hybrids [15].

The analysis of genetic diversity in the populations from the Lower Danube by microsatellites was hindered by the complexity of the genome in this group of fish.

All Acipenseriformes are divided into three separate groups depending on the number of chromosomes: (1) species with karyotypes comprising about 120 chromosomes; (2) species with 240 to 270 chromosomes; they are conventionally referred to as 250-chromosomes species; (3) species with around 370 chromosomes [72]. Two scales of Acipenseriformes ploidy have been proposed: (1) the "evolutionary scale": diploid species (extinct), tetraploid species (120-chromosomes), octoploid (250 chromosomes), and 12-ploid (370-chromosomes) species [73]; and (2) the "contemporary scale": diploid (120-chromosomes), tetraploid (250-chromosomes), and hexaploid (370-chromosomes) species [74].

Three of the four sturgeon species from the Lower Danube – *A. stellatus, A. ruthenus* and *H. huso* are considered to be functional diploid as the process of functional reduction of the genome is consider being almost completed in these species, while *A. gueldenstaedtii* is functional tetraploid, with an octaploid ancestor from which some loci are maintained. The analysis of a set of 12 microsatellites loci (AciG93, AciG198, AnacC11, AnacE4, Aox27, AoxD234, As002, LS19, LS34, LS39, LS54, Spl106) in 51 individuals of *A. gueldenstaedtii* from Lower Danube revealed that only two loci were disomic, while the others were polysomic with a number of 3-8 alleles per locus in an individual (Table 1).

TABLE 1.Characteristics of 12 microsatellite loci in A. gueldenstaedtii from Lower Danube.

Locus	Allele size (bp)	Maximum number of alleles per individual
AciG93	383-399	4
AciG198	184 - 188	2
AnacC11	144-204	4
AnacE4	320-356	3
Aox27	110-142	4
AoxD234	188-268	4
As002	92-140	8
LS19	115-154	4
LS34	121-151	4
LS39	85-157	4
LS54	117-237	4
Spl106	234-246	2

The analysis of genetic diversity by using microsatellite with polysomic pattern is facing the problem of correct determination of the genotype. For example, for a locus with tetrasomic pattern for which only three peaks are detected by capillary electrophoresis is hard to determine which allele has two copies. The method of *gene dosage* proposed by Jenneckens et al [75] by which the genotype can be determined by calculating the report of peak areas has proven to be inefficient in our case.

For the *A. stellatus, A. ruthenus* and *H. huso* population from the Lower Danube genetic diversity studies were performed based on microsatellite analysis. From an extended set of microsatellites isolated originally in the North–American sturgeon species *A. fulvescens* and *A. oxyrhinchus* and in Adriatic sturgeon – *A. naccarii*, seven loci (LS19, LS34, LS39, LS54, AnacE4, AnacC11, and AoxD234) that showed good results of amplification and a disomic pattern in all the three species of sturgeon from Lower Danube, were selected for the assessment of genetic diversity.

The analysis was performed in a total number of 158 sturgeon individuals (62 of *A. stellatus*, 54 of *H. huso* and 42 of *A. ruthenus*) that were captured in the river between 2001-2008 as part of a national scientific research study for restocking and monitoring.

The estimation of the genetic diversity was realized by inferring several statistic indices with Genetix software. Thus the average values of expected heterozygosity (H_E), observed heterozygosity (H_O) and the mean number of alleles/locus (MNA) were calculated.

The heterozygosity is the percentage of heterozygous loci in a population. The average values of H_O are similar for the three sturgeon population from the Lower Danube (0.431 – *A. stellatus*; 0.476 – *H. huso*; 0.4017 – *A. ruthenus*), while the average values of H_E are situated between 0.6409 (*H. huso*) și 0.5634 (*A. ruthenus*). The highest values of H_O (0.4760) and MNA per locus (6.2875) were obtained for *H. huso*, which appears to have the highest variability among the analyzed populations.

TABLE 2. Expected heterozygosity (H_E), observed heterozygosity (H_O) and the mean number of alleles per locus.

Locus	Population								
	A. stellatus			*H. huso*			*A. ruthenus*		
	H_E	H_O	MNA	H_E	H_O	MNA	H_E	H_O	MNA
LS19	0.7053	0.5484		0.8033	0.6667		0.4963	0.4524	
LS34	0.1152	0.0862		0.0000	0.0000		0.3501	0.0976	
LS39	0.0000	0.0000		0.5492	0.3889		0.4963	0.4048	
LS54	0.8321	0.7119		0.7758	0.4800		0.4887	0.0476	
AnacE4	0.8100	0.5484		0.6540	0.3889		0.5635	0.1429	
AnacC11	0.7243	0.7833		0.6372	0.4259		0.7735	0.6667	
AoxD234	0.8061	0.3387		0.8146	0.9815		0.7755	1.0000	
Average +/- S.D.	0.5704± 0.3549	0.4310± 0.3010	5.5714	0.6049± 0.2843	0.4760± 0.2989	6.2857	0.5634± 0.1577	0.4017± 0.3454	3.5714

The heterozygosity represents an important index that can give information about the diversity and even about the history of a population. The values of the heterozygosity can range from 0 (absence of heterozygosity) to 1 (significant number of alleles with the same frequency). Thus for the locus LS34 in *H. huso* and for LS39 in *A. ruthenus* we obtained a heterozygosity that is equal to 0, since a fixed allele was highlighted for these loci. The values of the average heterozygosity are correlated in direct proportion with genetic diversity.

The Hardy-Weinberg equilibrium was tested with Genepop v1.2 software. Except the p-value for LS19 and LS39 in *A. ruthenus*, the other loci showed significant departures from equilibrium. In generally, the deviation from the expected values might have several causes like the reduced size of the population, inbreeding or the presence of the null alleles which might lead to a false excess of homozygotes.

Our studies come to complete the data of evaluation and monitoring for the population of sturgeon from the Lower Danube. Thus, the data resulting from the monitoring of the YOYs (Young of the Year) born annually in the Romanian part of the Danube and evaluating the success of the natural recruitment have led to the hypothesis that in case of *H. huso* there is a significant generation born before 1990, which represents the basis of gene pool and which cyclic give birth to new generations with a high number of individuals [31].

5. MOLECULAR STUDIES FOR GENETIC DIVERSITY EVALUATION AND PHYLOGENY INFERRING IN SALMONIDS FROM ROMANIA

The native salmonid species from Romania have been characterized only from a morphological point of view [28], but the studies based on molecular aspects are still at the beginning. Among the few studies that included the molecular analysis of salmonid fishes from Romania, the evaluation of the genetic differentiation of salmonids by PCR-RFLP technique [76] and the phylogenetic classification of Romanian salmonid species by using the 16SrRNA and 12SrRNA gene sequences [26] were performed.

Even if the salmonids are a well-studied group of fish, there are still a number of questions pending with regard to their phylogeny and evolution. So, despite the fact that a large number of studies based on both morphological [77] and molecular data [78-80] were performed, there are still different opinions concerning genus-level relationships [26]. Four salmonid species from Romania (*Salmo trutta fario, Salmo labrax, Salvelinus fontinalis* and *Thymallus thymallus*) were analyzed from molecular point of view using 16SrRNA and 12SrRNA markers with the purpose to position them within the *Salmonidae* family. The biological samples were collected from different rivers from Romania (Dambovita, Bratia, Gilau, Latorita, Cerna, and Nera) and from the Danube Delta. Fragments from 16SrRNA and 12SrRNA mitochondrial genes were amplified with specific primers and sequenced by using Sanger method, *dye terminator* variant. For a more complex phylogenetic evaluation beside the sequences determined from salmonid specimens from Romania, 14 salmonid and 1 osmerid sequences from GenBank were also included in the analysis.

Estimation of phylogenetic relationships was achieved using 16S and 12SrRNAs gene sequences and the concateneted data set, while three methodologies – maximum parsimony (MP), maximum likelihood (ML) and Neighbour-joining (NJ) implemented in PHYLIP v3.68 software were used for phylogenetic reconstructions, in order to compare the consistency of the results produced by different methods.

The phylogenetic analysis revealed that primitive salmonid species such as *Coregonus lavaraetus* and the representatives of genus *Thymallus, T. thymallus* and *T. articus* occupy basal divergence in the tree topology confirming that the *Coregoninae* and *Thymallinae* subfamilies arise from a common ancestry before *Salmoninae* (with the genera *Salmo, Oncorhynchus, Hucho, Brachymystax* and *Salvelinus*). Based on morphological and molecular data, *Coregoninae* and *Thymallinae* were thought to be the earliest branches within the *Salmonidae* family [80].

The species of the *Salmo* genus form a distinct clade, in which the Atlantic salmon, *Salmo salar*occupies a basal divergence. The data reveal a close relationship between *Salmo trutta fario* and the clade formed by sea trout *S. trutta trutta* and the Black Sea trout *S. labrax*. The resulting clade (*S. trutta trutta, S. labrax*) is not surprising, taking into consideration some characteristics of the life history and reproductive behavior of these species [8].

The monophyly of *Salvelinus* was supported by 16SrRNA, 12SrRNA and combined data, but the position of the clade formed by (*S. alpinus, S. fontinallis*) in relationship with *Salmo* and *Onchorhynchus* is dependent on the molecular marker selected for the phylogenetic analysis. The trees resulted for 16SrRNA and 12SrRNA concatenated data has demonstrated a closer relationship between *Salvelinus* and *Oncorhynchus* than between *Salmo* and *Oncorhynchus*. The relationship between *Salvelinus* and *Oncorhynchus* was also supported by previous molecular phylogeny studies based on three other genes (GH1C, VIT and ND3) [80, 81].

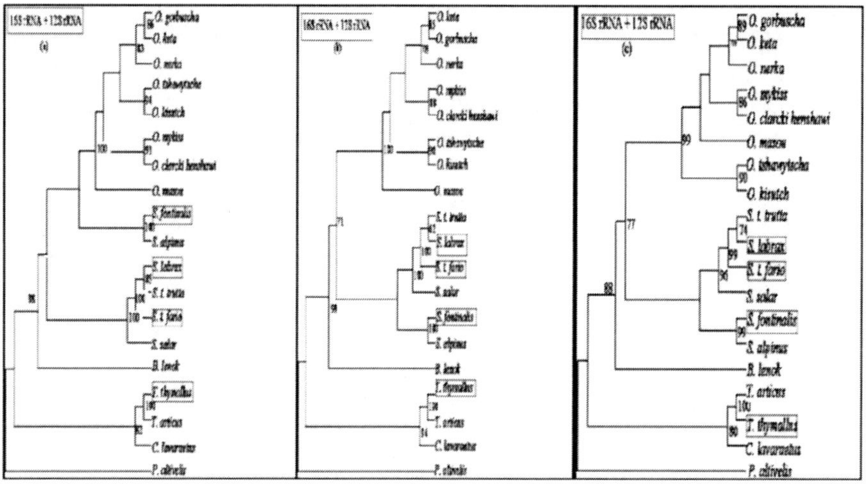

FIGURE 5. Majority with bootstrap support consensus trees for combined data (16SrRNA and 12SrRNA). (a) Combined data Neighbor Joining tree, distance model Kimura 2 Parameters, transition/transversion ratio 2.3; (b) combined data Maximum Parsimony tree; (c) combined data Maximum Likelihood tree [26].

The phylogenetic analysis using mitochondrial ribosomal genes as markers has allowed for the classification of salmonid species from Romania within the *Salmonidae* family. Thus, Romanian *S. trutta fario* and *S. labrax* are placed together within the *Salmo* genus. The *S. labrax*, endemic in the Black Sea, appears to be the sister taxa of the sea trout *S. trutta trutta* from the northwest of Europe (Atlantic coast) and Baltic Sea. The basal divergence in phylogenetic trees occupied by *T. thymallus*, a primitive species, is in agreement with the taxonomic and evolutionary data. Unfortunately, the position of the *Salvelinus* genus relative to the *Salmo* and *Oncorhynchus* genera remains controversial. The data reveal a possible sister taxon relationship between *Oncorhynchus* and *Salvelinus* despite the fact that morphological data support a closer relationship between *Salmo* and *Oncorhynchus*, thus confirming earlier findings of Crespi & Fulton [80] and Oakley & Phillips [81].

Unfortunately, salmonids' natural habitat is disrupted by a series of human activities such as poaching, dams, ballast exploitation and the construction of the micro hydro plants on the water streams from mountain areas. In this context, a preliminary study was directed to towards the assessment of the anthropic impact on the brown trout (*S. trutta fario*) populations from different rivers in Fagaras Mountain, including beside ecological aspects, molecular ones [82]. The area of Fagaras Mountain is the most representative in Romania for the brown trout populations, the tourism and recreational fishing being well known in this area, while the "brown trout of Fagaras" is preferred for alimentary consumption.

In the proposed study 102 individuals of brown trout from four populations from the Meridional Carpathians (Arpas, Ucea and Sambata rivers in the Northern versant of the Fagaras Mountain) were analyzed using mtDNA marker (D-loop) and nuclear markers (nine microsatellite loci - Str73, Str15, Str60, OmyFgt1, Ssa197, Ssa85, Strutta12, Str543, BS131).

The data obtained by sequencing of the complete mitochondrial control region (D-loop) were compared with similar sequences from GenBank. The phylogenetic tree resulted from analysis with MEGA v.5 contained, besides the sequences of D-loop in *S. trutta fario* individuals sampled in the Romanian rivers, complete D-loop sequences from other European lineages in order to observe the affiliation of analyzed Romanian individuals. The dendogram topology showed a classification of analyzed sequences in distinct monophiletic groups corresponding to each evolutive lineage and that the representatives of brown trout from Romania analyzed in the study were placed in the Danubian clade similar to other sequences selected from GenBank and belonging to Danubian lineage (Figure 6).

In Romania, Danubian lineage of brown trout is native and by founding the specific haplotypes for this lineage in our individuals, means that either restocking programs have not been done yet in the area or, if they have, the individuals involved were selected properly [82]. This type of approach can be really useful when the data about restocking and management of the rivers are poor or completely missing.

The genetic differentiation between the four population of brown trout analyzed in this study was evaluated by F-statistics, by using the indices F_{st} and $Gamma_{st}$ calculated with DNAsp software. F_{st}(fixation index) is a Wright statistic index that indicates the genetic variation between the populations and that can take values from 0 to 1. In generally, a higher value than 0.25 is correlated with a high genetic differentiation between populations. $Gamma_{st}$ is similar to F_{st}, making a correction for insufficient sampling. The values obtained for the four populations of brown trout indicate that there is a high differentiation between them from genetic point of view (Figure 7).

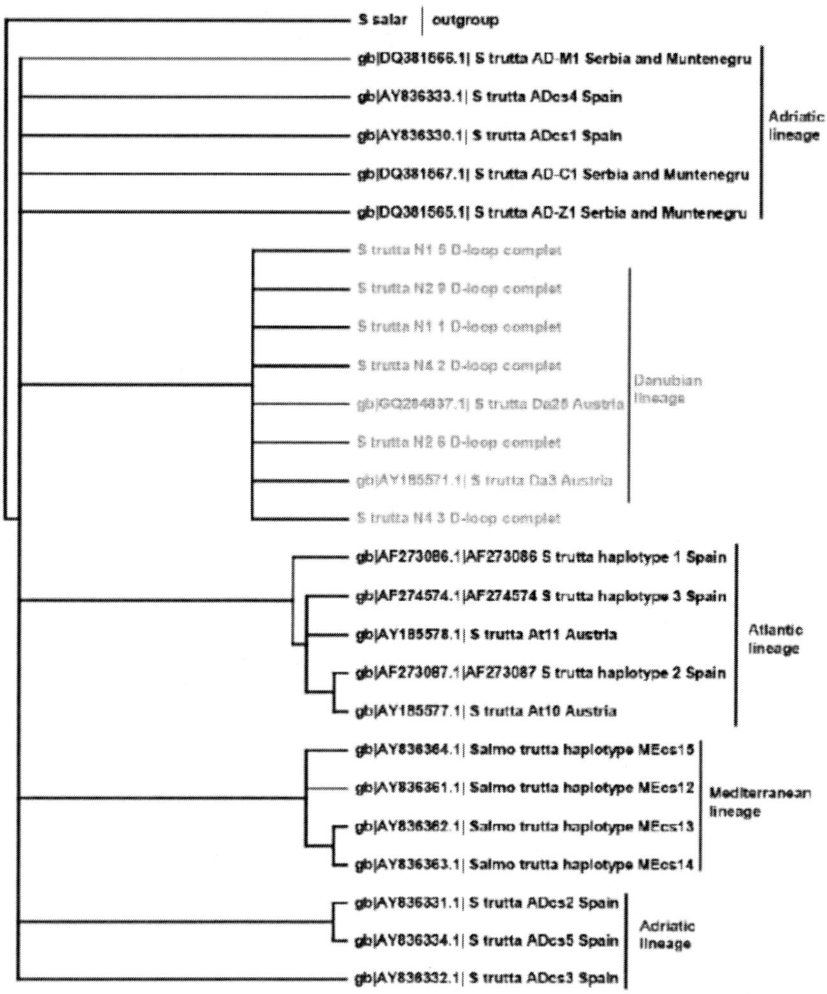

FIGURE 6. Phylogenetic tree obtained with MEGA v5 (NJ method, bootstrap value 1000 replications) illustrating the relationships between the evolutive lineages of brown trout. The haplotypes from Romania are placed in the Danubian lineage along with two sequences from GenBank belonging to the same lineage [82].

Population 1	Population 2	GammaSt	Fst
N1	N2	0.59009	0.72259
N1	N4	0.39013	0.55081
N1	N3	0.44519	0.58468
N2	N4	0.19504	0.30877
N2	N3	0.12098	0.17850
N4	N3	0.18104	0.28120

FIGURE 7. F_{st} and $Gamma_{st}$ values for four populations of brown trout from Romanian rivers [82].

The microsatellite data confirmed the previous results obtained regarding the genetic differentiation. Based on genotypic data resulted consequently microsatellites analysis a Factorial Correspondence Analysis (FCA) was performed by using GENETIX software. This type of analysis assumes the applying of a multidimensional method that permit the conversion of genotypic data obtained by the analysis of the nine microsatellite loci characteristic to each analyzed individual in points distributed in an X, Y, Z axis system. Thus, consequently to this analysis four different clusters were obtained corresponding to the four populations of brown trout, indicating a genetic differentiation between the analyzed populations (Figure 8).

The value of F_{is} index in the analyzed population indicates is positive for each population and indicates a light level of inbreeding within the population (Table 3).

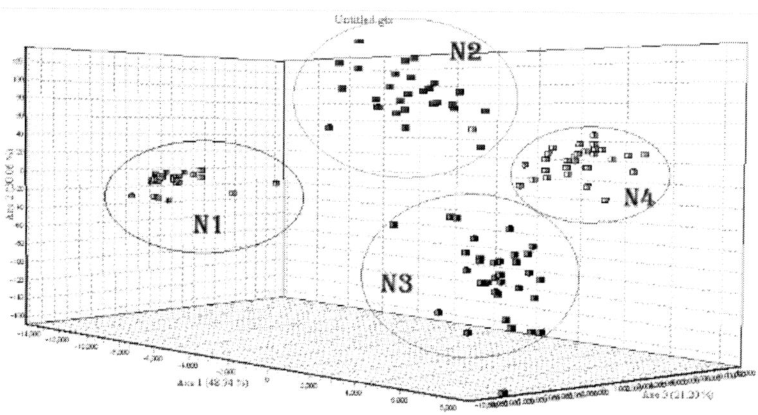

FIGURE 8. Factorial Correspondence Analysis for four populations of brown trout from Romanian rivers.

TABLE 3.F_{is} values calculated with FSTAT software. NA (not assigned).

Locus	Population N1	Population N2	Population N3	Population N4
Str60	0.000	-0.250	0.477	-0.747
Str15	0.618	-0.190	NA	0.444
Str73	NA	NA	NA	NA
OmyFGT	0,143	-0.052	0.224	0.179
Ssa85	-0.500	NA	NA	0.505
Ssa197	0.191	0.254	0.017	0.094
Str543	0.020	0.911	-0.053	0.209
Strutta12	0.094	0.358	0.153	0.399
BS131	1.000	0.259	0.253	-0.127
Average	0.094	0.270	0.139	0.134

Concluding, the analyzed populations are natives (Danubian lineage), differentiated by genetic point of view, but with a light level of inbreeding within population which might conduct in time to a reduced genetic diversity.

The studies are intent to be extended first to the Southern side of Fagaras Mountain, which is considered to be the most affected of the anthropic intervention, then to the main basins in the country populated with brown trout. The information resulted from these studies can be further used in programs of management and conservation, with accent on keeping unmodified the autochthonous species.

6. CONCLUSIONS

The studies developed in our country for analyzing in terms of genetic diversity species and populations of great scientific, ecological, economic and social importance are still at the beginning.

The molecular analyses of Lower Danube sturgeons were directed to species identification and assessing of genetic diversity. Regarding the first topic, a method based on microsatellite markers was set up as it was proven that the identification of species based only on morphology and mtDNA analysis can be misleading and do not serve to hybrid detection.

The populations of sturgeon from Lower Danube appear to be fragile from genetic diversity point of view, but the studies should be extended in a higher number of individuals and using a more significant and informative set of molecular markers

The analysis of microsatellites in these species is complicated by the complexity of the genome, so a better selection of a much significant number of loci with disomic inheritance pattern is recommendable. The decision to initiate restocking programs in order to recover the natural sturgeon populations must take into account not only the social and economic aspects, but also the assessment of genetic diversity.

The studies on salmonid were focused on mainly on phylogenetic classification of species from Romanian fauna. The research aiming on genetic diversity analysis is still in an incipient phase and was directed towards the populations of brown trout from Fagaras Mountain, an important area for salmonid distribution in our country. The results of these studies based on mtDNA and microsatellites analysis showed that the studied populations are pure Danubian brown trout lineage and genetically distinct. In the future, for a better image of salmonids status in Romania the research would be extended in analysis populations from other salmonid species and the area of sampling would be enlarged to the entire Romanian Carpathian Arch.

7. PERSPECTIVES

As our studies are in the preliminary phase, in future we intend to extend them in a higher number of individuals/populations from both fish groups consider of being of highly importance for our country.

Also, we intend to characterize from genetic point of view populations/ stocks/ strains from aquaculture, as we consider that exploitation of genetic data is imperious necessary beside restocking and conservation, for genetic improvement in aquaculture.

Currently the microsatellites and mtDNA are considered to be classical markers. There is an increasing tendency of analyzing the nucleotide variation at the whole genome level by *Next Generation Sequencing (NGS)* techniques. This type of approach started to be applied for different fish species analysis, including sturgeons and salmonids. NGS variants as *RAD (Restriction Sites Associated DNA) sequencing* are used for identification and characterization of a complete panel of SNP markers consider to be extremely useful for genomic analyses at individual, population and species level. This type of markers appears to be highly informative and of real support for conservation programs as is the trend of passing from the conservation genetics to conservation genomics era. The identification by NGS of an extended panel of specific markers should be extremely useful for sturgeons from the Lower Danube in population management and conservation purposes. The analysis should be orientated toward *A. gueldenstaedtii* the most affected sturgeon species from the Danube. This is a polyploid species, presenting a complex genome structure, fact that makes very difficult the analysis using classical markers. Also, the setup of such complex panel of SNP markers would make possible the traceability of caviar in the sturgeon species from Romania.

8. ACKNOWLEDGEMENTS

This work was supported by the PN-II-PT-PCCA Project 116/2012 "Genetic evaluation and monitoring of molecular and biotechnological factors that influence productive performance of Danube sturgeon species bred in intensive recirculating systems". Dudu Andreea was supported by the European Social Funding through the Sectorial Operational Programme for Human Resources Development POSDRU/159/1.5/S/133391.

REFERENCES

1. Global Issues. http://www.globalissues.org/article/170/why-is-biodiversity-important-who-cares, accessed at 23 July 2014.
2. Millennium Ecosystem Assessment, 2005. Ecosystems and Human Well-being: Biodiversity Synthesis. World Resources Institute, Washington DC, http://www.millenniumassessment.org/documents/document.356.aspx.pdf, accesed at 01 September 2014.
3. Frankham R, Ballou JD, Briscoe DA. Introduction to Conservation Genetics. New York: Cambridge University Press; 2002.
4. Wan QH, Wu H, Fujihara T, Fang SG. Which genetic marker for which conservation genetics issue? Electrophoresis 2004; 25:2165–2176.
5. Soule ME, Wilcox BA. Conservation Biology. An Evolutionary–Ecological Perspective. Massachusetts, USA: Sinauer Associates; 1980.
6. FishBase http://www.fishbase.org/search.php, accessed at 23 July 2014.
7. The IUCN Red List of Threatened Species http://www.iucnredlist. org/about /summary-statistics#Tables_1_2 accessed at 23 July 2014.
8. Otel V. Atlasul pestilor din Rezervatia Biosferei Delta Dunarii. Tulcea, Romania: Editura Centrul de Informare Tehnologica Delta Dunarii; 2007.
9. Garrido-Ramos MA, Robles F, de la Herrán R, Martínez-Espín E, Lorente JA, Ruiz-Rejón C, Ruiz-Rejón M. Analysis of Mitochondrial and Nuclear DNA Markers in Old Museum Sturgeons Yield Insights About the Species Existing in Western Europe: *A. sturio*, *A. naccarii* and *A. oxyrinchus*. In Carmona R, Domezain A, Gallego MG, Hernando JA, Rodríguez F, Ruiz-Rejón M. (Ed) Biology, Conservation and Sustainable Development of Sturgeons. Fish and Fisheries Series, 1st Edition, Amsterdam, Netherlands: Springer 2009. p25–50.

10. Bacalbasa-Dobrovici N. Endangered migratory sturgeons of the Lower Danube River and its Delta. In: Balon EK (Ed) Envinronmental Biology of Fishes 1997, 48(1-4):201-207.
11. Kynard B, Zhuang P, Zhang T, Zhang L. Ontogenetic behavior and migration of Volga River Russian sturgeon, *Acipenser gueldenstaedtii*, with a note on adaptive significance of body color. Envinronmental Biology of Fishes 2002; 65:411–421.
12. Navodaru I, Staras M, Banks R. Management of sturgeon stocks of the Lower Danube River system. In: The Delta`s: State-of art protection and management. Conference Proceedings, 26-31 July 1999; Tulcea, Romania.
13. Suciu R. Sturgeons of the NW Black Sea and Lower Danube River countries. At: International Expert Workshop on CITES Non-Detriment Findings; 17-22 November 2008, Cancun, Mexico.
14. Ludwig A, Lippold S, Debus L, Reinartz R. First evidence of hybridization between endangered sterlets (*Acipenser ruthenus*) and exotic Siberian sturgeons (*Acipenser baerii*) in the Danube River. Biological Invasions 2009; 11:753–760.
15. Dudu A, Suciu R, Paraschiv M., Georgescu SE, Costache M, Berrebi P. Nuclear Markers of Danube Sturgeons Hybridization, International Journal of Molecular Sciences 2011; 12(10):6796-6809.
16. Ludwig A. Identification of Acipenseriformes species in trade. Journal of Applied Ichthyology 2008; 24(S1):2–19.
17. Fain S, Straughan D, Hamlin B, Hoesch R, LeMay J. Forensic genetic identification of sturgeon caviars traveling in world trade. Conservation Genetics 2013; 14(4):855-874.
18. Boscari E, Barmintseva A, Pujolar JM, Doukakis P, Mugue N, Congiu L. Species and hybrid identification of sturgeon caviar: a new molecular approach to detect illegal trade. Molecular Ecology Resources 2014; 14:489–498.
19. Nelson JS. Fishes of the World, 4th Edition. New York: John Wiley & Sons; 2006.
20. Romero A. Salmoniformes (Salmons), In: Hutchins M, Thoney DA, Loiselle PV, Schlager N. (ed) Grzimek's Animal Life Encyclopedia, 2nd Edition, Vol. 4, Fishes. Farmington Hilla: Gale Group; 2003. p. 405-420.
21. Kottelat M, Freyhof J. Handbook of European freshwater fishes. Berlin, Germany: Freyhof, 2007; p. 429-430.
22. Bernatchez L. The evolutionary history of brown trout (*Salmo trutta* L.) inferred from phylogeographic, nested clade, and mismatch analysis of mitocondrial DNA variation. Evolution 2001; 55:351-379.
23. Meraner A, Baric S, Pelster B, Dalla Via J. Trout (*Salmo trutta*) mitochondrial DNA polymorphism in the centre of the marble trout distribution area. Hydrobiologia 2007, 579:337–349.
24. Giuffra E, Bernatchez L, Guyomard R. Mitochondrial control region and protein coding genes sequence variation among phenotypic forms of brown trout *Salmo trutta* from northern Italy. Molecular Ecology 1994; 3:161–171.
25. Osinov AG, Bernatchez L. Atlantic and Danubian phylogenetic groupings of brown trout *Salmo trutta* complex: genetic divergence, evolution, and conservation. Journal of Ichthyology 1996; 36:723–746.
26. Dudu A, Georgescu SE, Popa O, Dinischiotu A, Costache M. Mitochondrial 16s and 12s rRNA sequence analysis in four salmonid species from Romania. Acta Zoologica Academiae Scientiarum Hungaricae 2011, 57(3):233-246.
27. Hindar K, Jonsson B, Ryman N, Staahl G. Genetic relationships among landlocked, resident, and anadromous brown trout, Salmo trutta L. Heredity 1991; 66:83-91.
28. Bănărescu P. Fauna Republicii Populare Romîne. Piscies-Osteichthyes (Peşti ganoizi şi osoşi) Vol. XIII. Bucureşti: Editura Academiei R.P.R.; 1964.
29. Bănăduc D. The Hucho hucho (Linnaeus, 1758), (Salmoniformes, Salmonidae), species monitoring in the Viseu River (Maramures, Romania), Transylvanian Review of Systematical and Ecological Research 2008, 5:183–188.

30. Ciolac A, Patriche N. Biological aspects of main marine migratory sturgeons in Romanian Danube River. Migration of fishes in Romanian Danube River. Applied Ecology and Environmental Research 2005; 3(2):101-106.
31. Paraschiv M, Suciu R, Suciu M. Present state of sturgeon stocks in the Lower Danube River, Romania. Proceedings 36th International Conference of IAD, Austrian Committee Danube Research / IAD, Vienna; 2006.
32. Lenhardt M, Cakic P, Kolarevic J, Gacic Z. Morphometric recognition of two morphs in sterlet (*Acipenser ruthenus*) population induced by different reproductive behaviour. Fish as models of behavior. The Fisheries Society of the British Isles. Annual International Symposium, University of East Anglia, Norwich, England; 30 June – 4 July 2003.
33. http://www.merriam-webster.com, accesed at 01 September 2014.
34. Liu ZJ, Cordes JF. DNA marker technologies and their applications in aquaculture genetics. Aquaculture 2004, 238:1-37.
35. O'Brien SJ. Molecular genome mapping: lessons and prospects. Current Opinion in Genetics and Development 1991, 1:105–111.
36. Houston RD, Haley CS, Hamilton A, Guy DR, Tinch AE, Taggart JB, McAndrew BJ, Bishop SC. Major quantitative trait loci affect resistance to infectious pancreatic necrosis in Atlantic salmon (*Salmo salar*). Genetics 2008, 178:1109-1115.
37. Baranski M, Moen T, Vage DI. Mapping of quantitative trait loci for flesh colour and growth traits in Atlantic salmon (*Salmo salar*). GenetIcs Selection Evolution 2010, 42:17.
38. Norman JD, Robinson M, Glebe B, Ferguson MM, Danzmann RG. Genomic arrangement of salinity tolerance QTLs in salmonids: a comparative analysis of Atlantic salmon (*Salmo salar*) with Arctic charr (*Salvelinus alpinus*) and rainbow trout. BMC Genomics 2012, 13:420.
39. Gutierrez AP, Lubieniecki KP, Fukui S, Withler RE, Swift B, Davidson WS. Detection of Quantitative Trait Loci (QTL) related to grilsing and late sexual maturation in Atlantic Salmon (*Salmo salar*). Marine Biotechnology 2014; 16(1):103-109.
40. Szalanski AL, Bischof R, Holland R. Mitochondrial DNA variation in pallid and shovelnose sturgeon. Transactions Nebraskan Academy of Sciences (Affil. Soc.) 2001; 26: 19-21.
41. Waldman JR, Grunwald C, Stabile J. Impacts of life history and biogeography on the genetic stock structure of Atlantic sturgeon Acipenser oxyrinchus oxyrinchus, Gulf sturgeon A-oxyrinchus desotoi, and shortnose sturgeon A. brevirostrum. Journal of Applied Ichthyology 2002; 18:509-518.
42. Wolf C, Hübner P, Lüthy J. Differentiation of sturgeon species by PCR-RFLP. Food Research International 1999; 32:699-705.
43. Panagiotopoulou H, Baca M, Popovic D, Weglenski P, Stankovic A. A PCR-RFLP based test for distinguishing European and Atlantic sturgeons. Journal of Applied Ichthyology 2014, 30:14–17.
44. Ludwig A, Debus L, Jenneckens I. A molecular approach for trading control of black caviar. International Review of Hydrobiology 2002; 87:661-674.
45. Ludwig A, Congiu L, Pitra C, Fickel J. Nonconcordant evolutionary history of maternal and paternal lineages in Adriatic sturgeon. Molecular Ecology 2003; 12: 3253- 3264.
46. Machordom A, Suarez J, Almodovar A, Bautista JM. Mitochondrial haplotype variation and phylogeography of Iberian brown trout populations. Molecular Ecology 2000; 9:1325–1338.
47. Lucentini L, Palomba A, Gigliarelli L, Lancioni H, Viali P, Panara F. Genetic characterization of a putative indigenous brown trout (*Samo trutta fario*) population in the secondary stream of the Nera River Basin (Central Italy) assessed by means of three molecular markers. Italian Journal of Zoology 2006; 73:263–273.
48. Oleinik AG, Skurikhina LA, Brykov VA. Genetic Differentiation of Three Sympatric Charr Species from the Genus Salvelinus Inferred from PCR-RFLP Analysis of Mitochondrial DNA. Russian Journal of Genetics 2003; 39(8):924.

49. Koskinen MT, Ranta E, Piironen J, Veselov A, Titov S, Haugen TO, Nilsson J, Carlstein M, Primmer CR. Genetic lineages and postglacial colonization of grayling (*Thymallus thymallus*, Salmonidae) in Europe, as revealed by mitochondrial DNA analyses. Molecular Ecology 2000; 9(10):1609-24.
50. Tautz D. Hipervariability of simple sequences as a general source for polymorphic DNA markers. Nucleic Acids Research 1989; 17: 6463-6471.
51. Wright JM. DNA fingerprinting in fishes. In Hochachka PW, Mommsen T. (ed.) Biochemistry and Biology of Fishes, Vol. 2, Amsterdam:Elsevier, 1993; p.57-91.
52. Tóth G, Gáspári Z, Jurka J. Microsatellites in different eukaryotic genomes: Survey and analysis. Genome Research 2000; 10(7):967-981.
53. Botstein D, White RL, Skolnick M, Davis RW. Construction of a genetic linkage map in man using restriction fragment length polymorphisms. American Journal of Human Genetics 1980; 32:314–331.
54. Estoup A, Angers B. Microsatellites and minisatellites for molecular ecology: theoretical and empirical considerations. In: Carvalho G. (ed.) Advances in molecular ecology. Amsterdam: IOS Press, 1998; p. 55–86.
55. King TL, Lubinski BA, Spidle AP. Microsatellite DNA variation in Atlantic sturgeon (Acipenser oxyrinchus oxyrinchus) and cross-species amplification in the Acipenseridae. Conservation Genetics 2001; 2: 103-119.
56. Welsh A., May B. Development and standardization of disomic microsatellite loci for lake sturgeon genetic studies. Journal of Applied Ichthyology 2006; 22:337–44.
57. Welsh AB, Blumberg M, May B. Identification of microsatellite loci in lake sturgeon, Acipenser fulvescens, and their variability in green sturgeon, A. medirostris. Molecular Ecology Notes 2003; 3: 47–55.
58. Forlani A, Fontana F, Congiu L. Isolation of microsatellite loci from the endemic and endangered Adriatic sturgeon (Acipenser naccarii). Conservation Genetics 2008; 9: 461–463.
59. Boscari E, Barbisan F, Congiu L. Inheritance pattern of microsatellite loci in the polyploid Adriatic sturgeon (*Acipenser naccarii*). Aquaculture 2011; 321(3–4):223–229.
60. Moghim M, Pourkazemi M, Tan S, Siraj S, Panandam J, Kor D. Development of disomic single-locus DNA microsatellite markers for Persian sturgeon (*Acipenser persicus*) from the Caspian Sea. Iranian Journal of Fisheries Sciences 2013; 12 (2):389-397.
61. Norouzi M, Pourkazemi M. Genetic structure of Caspian populations of stellate sturgeon, *Acipenser stellatus* (Pallas, 1771), using microsatellite markers. International Aquatic Research 2009; 1:61-65.
62. Timoshkina N, Barmintseva AE, Usatov AV, Mugue NS. Intraspecific genetic polymorphism of Russian sturgeon *Acipencer gueldenstaedtii*. Russian Journal of Genetics 2009, 45:1098–1107.
63. Koljonen ML, Tähtinen J, Säisä M, Koskiniemi J. Maintenance of genetic diversity of Atlantic salmon (*Salmo salar*) by captive breeding programmes and the geographic distribution of microsatellite variation. Aquaculture 2003; 212, 69–92.
64. Apostolidis AP, Madeira MJ, Hansen M, Machordom A. Genetic structure and demographic history of brown trout (*Salmo trutta*) populations from the southern Balkans. Freshwater Biology 2008; 53:1555–1566.
65. Nilsson J, Ostergren J, Lundqvist H, Carlsson U. Genetic assessment of Atlantic salmon Salmo salar and sea trout Salmo trutta stocking in a Baltic Sea river. Journal of Fish Biology 2008; 73:1201–1215.
66. Pourkazemi M, Skibinski DO, Beardmore JA. Application of mtDNAd-loop region for the study of Russian sturgeon population structure from Iranian coastline of the Caspian sea. Journal of Applied Ichthyology 1999; 15:23-28.
67. Zhang SM, Deng H, Wang DQ, Zhang YP, Wu QJ. Mitochondrial DNA Length variation and heteroplasmy in Chinese sturgeon (*Acipenser sinensis*). Acta Genetica Sinica 1999; 26, 489–496.

68. Doukakis P, Birstein VJ, DeSalle R. Molecular genetic analysis among subspecies of two Eurasian sturgeon species, Acipenser baerii and A. stellatus. Molecular Ecology 1999; 12:S117–S129.
69. Asplund T, Veselov A, Rimmer C, Bakhmet I, Potutkin A, Titov S, Zubchenko A, Studenov I, Kaluzchin S, Lumme J. Geographical structure and postglacial history of mtDNA haplotypes variation in Atlantic salmon (*Salmo salar* L.) among rivers of the White and Barents Sea basins. Annales Zoologici Fennici 2004, 41:465–475.
70. Laikre L, Järvi T, Johansson L, Palm S, Rubin JF, Glimsater CE, Landergren P, Ryman N. Spatial and temporal population structure of sea trout at the Island of Gotland, Sweden, delineated from mitochondrial DNA. Journal of Fish Biology 2002; 60:49–57.
71. Frankel OH, Soulé ME. Conservation and evolution. Cambridge University Press,1981; Cambridge, UK.
72. Vasil'ev VP. Mechanisms of Polyploid Evolution in Fish: Polyploidy in Sturgeons. In Carmona R, Domezain A, Gallego MG, Hernando JA, Rodríguez F, Ruiz-Rejón M. (Ed) Biology, Conservation and Sustainable Development of Sturgeons. Fish and Fisheries Series, 1st Edition, Amsterdam, Netherlands: Springer 2009. p97–117.
73. Birstein VJ, Bemis WE. How Many Species are there within the Genus Acipenser? In Sturgeon Biodiversity and Conservation, 1st edition; Dordrecht, Netherlands: Kluiwer Academic Publishers, 1997. p157–163.
74. Ludwig A, Belfiore NM, Pitra C, Svirsky V, Jenneckens I. Genome duplication events and functional reduction of ploidy levels in sturgeon (*Acipenser, Huso* and *Scaphirhynchus*). Genetics 2001, 158:1203–1215.
75. Jenneckens I, Meyer JN, Hörstgen-Schwark G, May B. A fixed allele at microsatellite LS-39 is characteristic for the black caviar producer Acipenser stellatus. Journal of Applied Ichthyology 2001; 17:39-42.
76. Dudu A, Georgescu SE, Dinischiotu A, Costache M. PCR-RFLP method to identify fish species of economic importance, Archiva Zootechnica 2010; 13(1):53-59.
77. Stearley RF, Smith GR. Phylogeny of the Pacific trouts and salmons (*Oncorhynchus)* and genera of the family *Salmonidae*. Transaction of American Fisheries Society 1993; 122:1–33.
78. Kitano T, Matsuoka N, Saitou N. Phylogenetic Relationship of the genus *Oncorhynchus* species inferred from nuclear and mitochondrial markers, Genes Genetics and Systematics 1997; 72:25–34.
79. Oohara I, Sawano K, Okazaki T. Mitochondrial DNA Sequence Analysis of the Masu Salmon—Phylogeny in the Genus Oncorhynchus. Molecular Phylogenetics and Evolution 1997; 7:71–78.
80. Crespi BJ, Fulton MJ. Molecular systematics of *Salmonidae*: combined nuclear data yields a robust phylogeny. Molecular Phylogenetics and Evolution 2004; 31:658–679.
81. Oakley TH, Phillips RB. Phylogeny of Salmonine Fishes Based on Growth Hormone Introns: Atlantic (*Salmo*) and Pacific (*Oncorhynchus*) Salmon Are Not Sister Taxa. Molecular Phylogenetics and Evolution 1999; 11(3):381–393.
82. Popa GO, Khalaf M, Dudu A, Curtean-Bănăduc A, Bănăduc D, Georgescu SE, Costache M. Brown trout's populations genetic diversity using mitochondrial markers in relatively similar geographical and ecological conditions – A Carpathian case study, Transylvanian Review of Systematical and Ecological Research 2013, 15(2):125-132.

CHAPTER 5

Sexual Dimorphism of Drumming Muscles in European Hake (*Merluccius Merluccius*)

Anne-Laure Groison & Olav S. Kjesbu Marc Suquet

Department of Biology, University of Bergen, P.O. Box 7803, Thormøhlensgate 55, 5020 Bergen, Norway

Institute of Marine Research, Nordnesgaten 50, P.O. Box 1870, Nordnes 5817 Bergen, Norway

IFREMER, PFOM / ARN, 29840 Argenton, France

ABSTRACT

Dissections of mature and non-mature European hake males and females ($N = 142$) collected in waters off the western coast of Norway and in the Bay of Biscay (France) in 2004–2006 demonstrate for the first time that this gadoid species contains drumming muscles. There were differences in drumming muscles weight with body length, sex and maturity stage. This study shows that the difference between females and males is primarily manifested during the spawning season, seen both in the French and Norwegian samples. For the mature females the drumming muscles dry weight increases only slightly, if at all, with increase in total length. For mature males there is a corresponding rapid increase. There does not seem to be any consistent difference between the average dry weight of the drumming muscles in immature male and immature and mature female hake of the same length, tested on the Norwegian samples. Our results suggest that male hake, like the males of other gadoids studied, may produce sounds in the context of spawning.

Keywords: Drumming muscles *Merluccius merluccius* Sexual dimorphism Sound production Spawning

INTRODUCTION

More than 800 fishes from 109 families are known to produce sounds, though this is likely to be an underestimate (Rountree et al. 2003). It is evident that most of these sounds are deliberate rather than incidental. These sounds have a role in communication, i.e., are used as exchange of information between individual fish as part of their social behaviour (Hawkins and Myrberg 1983). Thus, fish produce sounds in a variety of contexts. Sounds are produced by some species when disturbed or when approached by a predator. Likewise, sounds are also produced by fish which are competing with one another for food or space (Ladich and Myrberg 2006). In many sound-producing fish males produce sounds during courtship of the female to advertise their nest sites, to attract the female, and promote courtship and spawning (Myrberg and Lugli 2006). The gas-filled swimbladder is a characteristic feature of the viscera of teleost fish. It contributes to the ability of a fish to control its buoyancy, and thus to stay at the current water depth without having to waste energy in depth-compensating swimming activity. Another function of the gas bladder is the use as a resonating chamber to produce or receive sound and in some species is equipped with drumming muscles (DM) for sound production. Sounds are produced by contracting DM associated with the swimbladder and thereby vibrating the swimbladder wall (Jones and Marshall 1953; Brawn 1961).

It is known that one particular family, the Gadidae, includes a number of vocal species (Hawkins and Rasmussen 1978), including haddock (*Melanogrammus aeglefinus*) (Hawkins and Chapman 1966), lythe (*Pollachius virens*) (Hawkins and Rasmussen 1978), tadpole fish (*Raniceps raninus*) (Hawkins and Rasmussen 1978) and Atlantic cod (*Gadus morhua*) (Brawn 1961).

In their work on cod Nordeide et al. (2008) found that the DM mass was similar in both sexes a couple of months prior to spawning but became sexually dimorphic at the onset of spawning and continued being sexually dimorphic (bigger in males) for several months after the termination of spawning.

To date, no studies have investigated whether European hake (*Merluccius merluccius*) possess drumming muscles. This is somewhat surprising in view of the importance of this species, and the fact that the presence of drumming muscles have been reported in other gadoid species.

European hake is a semi-demersal, multiple batch spawner found in waters from Mauritania to Norway. It is believed that hake spawning and reproduction occur at depth ranging 100–200 m (Alvarez et al. 2001; Olivar et al. 2003). The peak spawning time of hake is in March in waters south of the Bay of Biscay (France), and occurs progressively later at higher latitudes (Casey and Pereiro 1995).

In the present study we sampled wild hake males and females from French (Bay of Biscay) and Norwegian waters at different times of the year to search for the presence of drumming muscles and, if so, to quantify variation among individuals in drumming muscles size. Specifically, our objectives were to (*a*) record differences in drumming muscle appearance and mass in relation to sex, spawning status, and body size; and (*b*) compare hake drumming muscles with what has been observed in other gadoids.

MATERIAL AND METHODS

Fish collection

A total of 142 wild European hake were sampled offshore Western Norway (Nw) (61°34′N, 5° 56′W) and in the Bay of Biscay, France (Fr) (47°44′N, 4°2′W) (Table 1). Each fishing trip typically lasted 1–3 days. Non-mature fish were captured by trawl in Nw waters while mature fish were captured at both locations by gillnets set overnight at depths of between 30–180 m over sandy sea bottom. Recently dead fish (few hours) were retrieved from the gillnets.

Table 1 Summary of spawning and non-spawning European hake captured in the Bay of Biscay (France, Fr) and in waters western Norway (Nw) at different dates in 2004–2005–2006

		Spawning fish, *N*=		**Non spawning fish, *N*=**	
Date	**Origin**	**Females**	**Males**	**Females**	**Males**
20 March 2006	Fr	9	10	0	0
04 April 2006	Fr	6	14	0	0
17 August 2005	Nw	7	6	0	1
22 August 2006	Nw	1	1	0	0
23 August 2004	Nw	0	0	9	6
23 August 2005	Nw	0	1	0	1
01 September	Nw	6	5	1	0

2005					
12 September 2006	Nw	1	2	0	0
27 September 2004	Nw	0	0	10	13
12 October 2004	Nw	0	0	14	13
13 October 2006	Nw	0	0	2	2
30 November 2004	Nw	0	0	0	1
Total		15Nw + 15Fr	15Nw + 24Fr	36Nw	37Nw

Fish dissection

The fish were transported to laboratory to be dissected within 12–32 h of sampling, and all showed muscles attached to the swimbladder on both sides. For each individual examined we recorded total body length ($N = 140$; TL to the nearest 0.1 cm was measured for all fish except for one individual with damaged tail), total (i.e., ungutted) body mass ($N = 138$; TW) and gonad mass ($N = 61$) (to the nearest 0.1 g consulting only gonads which were not smashed or deteriorated by stripping). Sex and maturity stage (immature, ripening, ripe/spawning, and spent) were recorded. Only two groups of individuals were considered: "spawning" (sp.) for ripening or ripe/spawning individuals ($N = 69$) and "non-spawning" (n. sp.) for immature or spent individuals ($N = 73$). The pair of DM was easily separated from the surrounding tissue using forceps. After excision, DM were dried at 65°C for 3 days to obtain dry weight to the nearest 0.001 mg ($N = 141$). The following fish characteristics were calculated: condition factor (K=[Totalweight/(TL)3]×100K=[Totalweight/(TL)3]×100, $N = 138$); gonadosomatic index (GSI=[Gonadweight/TW]×100GSI= [Gonadweight /TW]×100 in %, $N = 61$), and hepatosomatic index (HSI=[Liverweight /TW]×100HSI=[Liverweight/TW]×100 in %, $N = 71$).

Statistical analysis

Data were presented as means ± SD. Measured and calculated characteristics of the dissected fish were combined (Table 2). Statistical analyses were performed using the software SigmaStat 3.1. Statistical significant difference between two groups were tested at the probability level 0.05 using Student *t*-test (when data were distributed normally and variances were not significantly different) or Mann-Whitney Rank Sum test (if one of these two previous conditions, or both, were invalidated). As the fish dissection resulted in an uneven number of left ($N = 141$) and right DM ($N = 120$) a pilot analysis was run to test for any differences in dry weight between them, which turned out not to be the case (Mann-Whitney rank sum test, $P > 0.05$). Therefore, in the following analysis dry weight of the left DM were used and named DM. ANCOVA with total length as covariate was used to test for differences in drumming muscle mass in relation to sex and spawning status (i.e. "spawning" (sp.) for ripening or ripe/spawning individuals and "non-spawning" (n. sp.) for immature or spent individuals). Relationships between DM dry weight and characteristics of sp. and n. sp. individuals (TL, TW, K, GSI and HSI) were investigated with Pearson correlations. Correlations were investigated separately for Norwegian and French fish and for males and females.

Table 2 Mean ± SD values of fish characteristics measured on spawning and non-spawning hake for male and female (Fem.) captured in the Bay of Biscay (France, Fr) and western Norway waters (Nw): total length (TL), total weight (TW), dry weight of the left drumming muscles (DM), gonadosomatic index (GSI), condition factor (K) and hepatosomatic index (HSI)

Maturation state	**Origin**	**Sex**	***N***	**TL (cm)**	**TW (g)**	**DM (mg)**	**GSI (%)**	**K**	**HSI (%)**
Non-spawning	Nw	Fem.	37	28.7 ± 15.2	327.2 ± 815.1	3.12 ± 5.41	1.56 ± 2.89	0.57 ± 0.06	2.60 ± 1.58
		Male	37	27.1 ± 11.6	143.7 ± 166.1	10.15 ± 23.74	0.17 ± 0.05	0.56 ± 0.05	1.58 ± 0.03
Spawning	Nw	Fem.	15	75.5 ± 7.5	3085.3 ± 1273.1	23.27 ± 8.96	9.38 ± 4.25	0.69 ± 0.08	4.98 ± 5.20
		Male	15	69.1 ± 8.7	2477.1 ± 852.2	225.7 ± 123.6	4.12 ± 2.78	0.69 ± 0.06	2.71 ± 1.06
	Fr	Fem.	15	64.7 ± 13.0	2120.1 ± 1423.1	18.38 ± 9.41	5.10 ± 2.03	0.71 ± 0.06	3.33 ± 0.83
		Male	24	45.7 ± 13.4	784.3 ± 690.3	98.75 ± 78.29	1.44 ± 0.88	0.63 ± 0.07	2.20 ± 1.07

RESULTS

Fish and observations of drumming muscles

Nw-spawning individuals for both sexes showed significant higher GSIs compared to Fr-spawning individuals (Table 2).

The present study showed the presence of DM in hake: a pair of muscular structures is located at the anterior end of the swimbladder, close to its ventral wall (Nw female: Fig. 1a; Nw male: Fig. 1b). Note that the DM are rounded at the posterior end but slightly pointed anteriorly and considerably larger in the male than in the female of similar length (≈80 cm), i.e., in the present example the male DM dry weight was 12 times larger than in the female. The DM of the female sits as a flat structure on the swimbladder wall whereas for the mature male it appears much thicker. For mature specimens there was noticed a difference in the colour of the male and female DM as viewed *in situ*: the DM of the mature male for all size classes tends to be reddish flesh-coloured, while light pinkish for females.

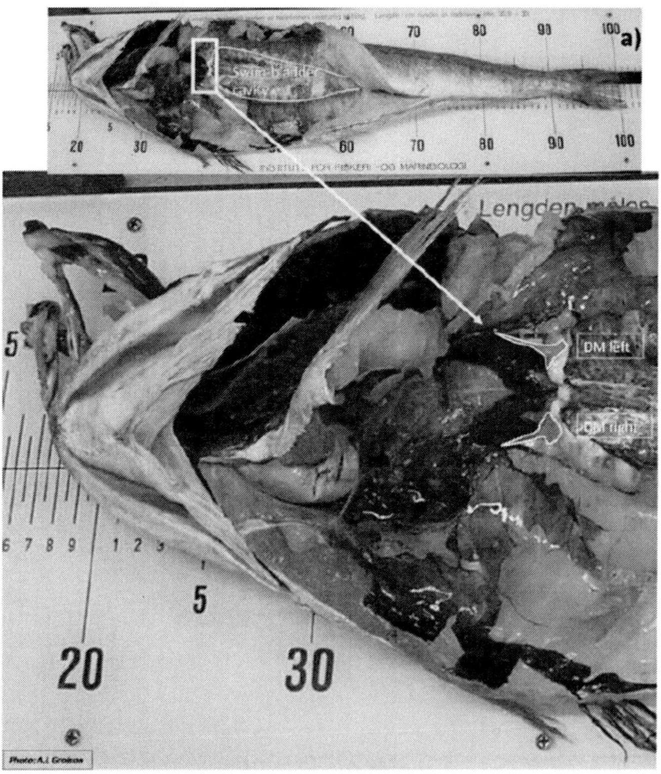

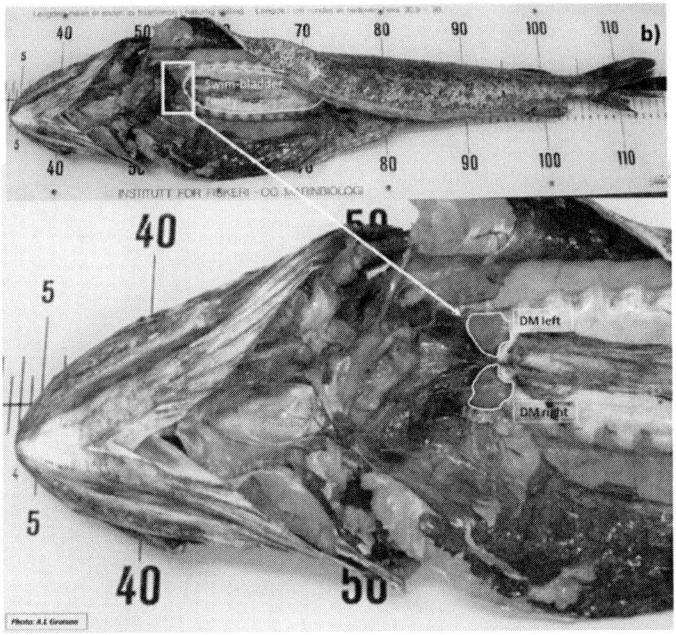

Figure. 1 Dissection of spawning hake caught on 18 August 2005 in waters off Western Norway. Total length (TL), total weight (TW) and the dry weight of the left drumming muscles (DM) are indicated. (**a**) Dissection of a spawning female. TL = 83 cm, TW = 4 270 g and dry weight of left drumming muscles = 22 mg. (**b**) Dissection of a spawning male. TL = 80 cm, TW = 3 440 g and dry weight of left drumming muscles = 266 mg

Relation of drumming muscles dry weight to length, sex and maturity stage

There were examples of significant correlations between DM dry weight and fish total length (Fig. 2; Table 3). However, for spawning females the DM dry weight increased only slightly with length while for spawning males there was a rapid increase at larger lengths. For the same length, Nw spawning males had a significantly heavier DM compared to Nw non-spawning males (Fig. 2b) ($P < 0.001$) (ANCOVA). The same result appeared when compared to Nw non-spawning females (Fig. 2a + b) ($P < 0.001$) and also very much in relation to Nw spawning females (Fig. 2a+b) ($P < 0.01$). The ANCOVA also showed that Fr spawning males had significantly heavier DM compared to Fr spawning females (Fig. 2c+d) ($P < 0.001$).

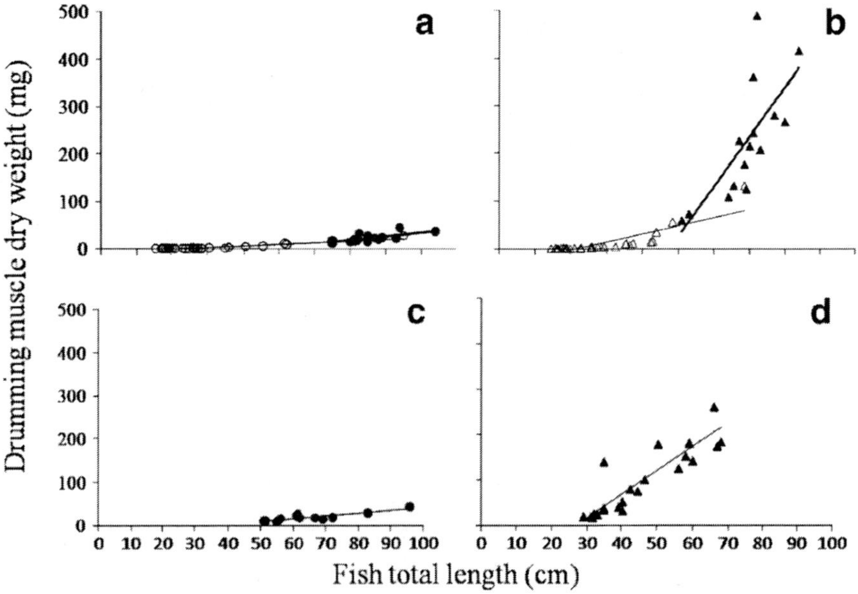

Figure. 2 Variation in left drumming muscle dry weight (DM in mg) in relation to total length (TL in cm) of European hake for females (*circles*) and males (*triangles*), split into non-spawning (*open symbol*) and spawning individuals (*filled symbol*) for Norwegian (**a**) and (**b**), and for French samples (**c**) and (**d**)

Table 3 Regression equations describing left drumming muscle dry weight (*y*) as a function of total length (*x*) for female and male hake from Western Norway (Nw) and France (Fr) (Bay of Biscay) in non-spawning and spawning conditions

Maturation state	**Origin**	**Sex**	**Regression equations**	**Sample size**	r^2	***P*-values**
Non-spawning	Nw	Fem.	y=0.333x−6.45y=0.333x−6.45	37	0.937	$P<0.001$
		Male	y=1.721x−36.96y=1.721x−36.96	36	0.835	$P<0.001$
Spawning	Nw	Fem.	y=0.833x−39.65y=0.833x−39.65	15	0.701	$P<0.01$
		Male	y=10.402x−492.76y=10.402x−492.76	15	0.734	$P<0.01$
	Fr	Fem.	y=0.623x−21.87y=0.623x−21.87	13	0.864	$P<0.001$
		Male	y=5.264x−141.55y=5.264x−141.55	24	0.904	$P<0.001$

DISCUSSION

Observations of the drumming muscles of male and female hake

The present study showed the presence of a sexually dimorphic muscle in hake males and females. The main evidence that this fish produce sounds is provided by the presence of a pair of muscles, one on either side of the swimbladder, similar to those found in known vocalists (as e.g., haddock, cod, lythe (*Pollachius pollachius*), and tadpole-fish (*Raniceps raninus*)) (Sørensen 1884; Hagman 1921; Jones and Marshall 1953; Templeman and Hodder 1958). In cod, Delaroche (1809) and Sørensen (1884) describe the weak, flattened muscles which pass from the sides of the swimbladder at its anterior end to the ribs of the anterior vertebrae. Jones and Marshall (1953) label these external swimbladder muscles of the cod as drumming muscles (DM). In the European haddock, special groups of muscles attached to the ventral wall of the swimbladder defined as well as drumming muscles were described by Templeman and Hodder (1958). These anatomical similarities found between the DM described in cod or in haddock in comparison with the muscles presently described in hake provide more circumstantial evidence for the idea that hake produce sounds. Sound production is caused by rapid contraction and relaxation of the DM attached to the swimbladder, as mentioned earlier (Templeman and Hodder 1958; Brawn 1961; Hawkins and Rasmussen 1978; Hawkins and Amorim 2000).

Relation of drumming muscles dry weight to length, sex and maturity stage

Presently collected Norwegian (Nw) spawning individuals showed significant higher GSI compare to the French (Fr) spawning individuals, respectively for females and for males. The low GSI values for the Fr samples indicate that the sampling in the Bay of Biscay was done at the end of the reproductive season. However, sexual differences in DM were observed also for spawning Fr individuals. Male hake, as described for cod, may present sexually dimorphic DM during and sometime after the termination of spawning (Nordeide et al. 2008). Non-spawning individuals could not be captured in French waters as the two cruises (sampling programmes) were restrained to the supposed reproductive period. Fortunately, in Nw waters both fish from mid-September until late November including spawning and non spawning individuals could be collected.

The drumming muscles of the male hake swimbladder clearly enlarge as the male becomes sexually mature. In female hake, on the other hand, there is no increase in the size of the DM with sexual maturity and no significant increase with increase in the size of the fish. Similarly, controlling for the influence of body size, Rowe and Hutchings (2004) found that male Atlantic cod have larger DM than females and that among males, DM increase in mass before spawning and decline thereafter. In haddock, which also belongs to the cod family (Gadidae), the sexually dimorphic drumming muscles of mature males increase to nearly twice their normal size during the spawning period (Templeman and Hodder 1958; Hawkins et al. 1967). Templeman and Hodder (1958) found no significant difference between the volumes of the DM of female haddock at different seasons. This sex-specific developmental pattern may reflect the important function of the drumming muscles, and thus of sound production, in the reproductive behaviour of haddock (Templeman and Hodder 1958). Both sexes of cod and haddock call during most of the year, whereas only males seem to call during the spawning period (Brawn 1961; Hawkins and Rasmussen 1978; Hawkins and Amorim 2000). Both male and female haddock were observed in tank to produce short sequences of repeated 'knocks' during agonistic encounters. In cod, grunts are produced during defensive and aggressive behaviour by both sexes when examined in aquarium tanks in the laboratory and in netting enclosures in the sea (Hawkins and Rasmussen 1978). During the spawning season, however, male fish produce sounds which vary in their characteristics as courtship proceeds (Hawkins and Amorim 2000). Sexual dimorphism and seasonal variation in sound-producing musculature have been documented for several other fishes, including weakfish (Connaughton and Taylor 1994), plainfin midshipman (*Porichthys notatus*) (Brantley and Bass 1994; Bass 1997) and *Opsanus tau* (Gray and Winn 1961). In some members of the family Sciaenidae only the males possess specially developed DM, e.g. the gray squeteague *Cynoscion regalis* (Fish 1954). For others, both sexes produce sound, as for the sea horse, *Hippocampus brevirostris*, in which both males and females make sounds most intense and most frequent during the breeding period (Dufossé 1874).

We suggest that mature male hake use their drumming muscles more often than either mature females or immatures of either sex, because these muscles in mature males are more highly vascularised (darker red) than those of mature females (light pink). As well, drumming muscle mass in males and females is similar prior to spawning, but during the spawning season it increases significantly in males. According to Lucio et al. (2000), first maturity is reached in *Merluccius merluccius* around 42 cm (both sexes combined). We could therefore expect that hake males and females do not have significant different DM until their first maturity.

CONCLUSION

This study showed for the first time the presence of drumming muscles in hake. During the spawning season only drumming muscles from male individuals are hypertrophied. Based on comparisons established with other gadoids we can thus suppose that sound production by adult males is more frequent during the spawning season than during the rest of the year. It is suggested that differences in size of the drumming muscles of male and female hake reflect changes in sound production with sex, sexual maturity, and season.

ACKNOWLEDGEMENTS

Authors would like to specially thank fishermen for welcoming us on board to collect the biological material and Audrey Geffen for financial support (University of Bergen, Norway). Thanks to Ignacio Serrano for his help in dissecting fish. Finally thanks to Jon Egil Skjæraasen for his advices while initiating this study as well to Anne-Christine Utne Palm (University of Bergen, Norway).

REFERENCES

1. Alvarez P, Motos L, Uriarte A, Egaña J (2001) Spatial and temporal distribution of European hake, *Merluccius merluccius* (L.), eggs and larvae in relation to hydrographical conditions in the Bay of Biscay. Fish Res 50(1–2):111–128
2. Bass AH (1997) Comparative neurobiology of vocal behavior in teleost fishes. Mar Freshw Behav Physiol 29:47–63
3. Brantley RK, Bass AH (1994) Alternative male spawning tactics and acoustic-signals in the plainfin midshipman fish *Porichthys notatus* Girard (Teleostei, Batrachoididae). Ethology 96:213–232
4. Brawn VM (1961) Sound production by the cod (*Gadus callarias* L.). Behaviour 18:239–255
5. Casey J, Pereiro JA (1995) European hake (*M. merluccius*) in the North-east Atlantic. In: Alheit J, Pitcher TJ (eds) Hake: fisheries and markets. Chapman & Hall, London, pp 125–147
6. Connaughton MA, Taylor MH (1994) Seasonal cycles in the sonic muscles of the weakfish, *Cynoscion regalis*. Fish Bull (Wash DC) 92:697–703
7. Delaroche FE (1809) Observations sur la vessie aérienne des poissons. Ann Mus Hist Nat Paris 14:245–298
8. Dufossé M (1874) Recherches sur les bruits et les sons expressifs que font entendre des poisons d'Europe. Ann Sci nat Zool 19(5):1–53, and 20(3):1–134
9. Fish MP (1954) The character and significance of sound production among fishes of the western North Atlantic. Bull Bing Ocean Coll 14(3):1–109
10. Gray GA, Winn HE (1961) Reproductive ecology and sound production of the toadfish, *Opsanus tau*. Ecology 42:274–282

11. Hagman N (1921) Studien über dis Schwimmblase einiger Gadiden und Macruriden 1. Gestalt des Organes und Befestigung desselben binnen Leibeshöhle. Carl Bloms, Lund, p 124
12. Hawkins AD, Amorim MCP (2000) Spawning sounds of the male haddock, *Melanogrammus aeglefinus*. Environ Biol Fish 59:29–41
13. Hawkins AD, Chapman CJ (1966) Underwater sounds of the haddock, *Melanogrammus aeglefinus*. J Mar Biol Assoc U K 46:241–247
14. Hawkins AD, Myrberg AA Jr (1983) Hearing and sound communication underwater. In: Lewis B (ed) Bioacoustics, a comparative approach. Academic, London, pp 347–405
15. Hawkins AD, Rasmussen KJ (1978) The calls of gadoid fish. J Mar Biol Assoc U K 58:891–911
16. Hawkins AD, Chapman CJ, Symonds DJ (1967) Spawning of haddock in captivity. Nature (Lond) 215:923–925
17. Jones FRH, Marshall NB (1953) The structure and functions of the teleostean swimbladder. Biol Rev 28:16–83
18. Ladich F, Myrberg AA Jr (2006) Agonistic behavior and acoustic communication. In: Ladich F, Collin SP, Moller P, Kapoor BG (eds) Communication in fishes, vol. 1. Science Publishers, Enfield, pp 121–148
19. Lucio P, Murua H, Santurtún M (2000) Growth and reproduction of hake (*Merluccius merluccius*) in the Bay of Biscay during the period 1996–1997. Ozeanografika 3:325–354
20. Myrberg AA Jr, Lugli M (2006) Reproductive behavior and acoustical interactions. In: Ladich F, Collin SP, Moller P, Kapoor BG (eds) Communication in fishes, vol. 1. Science Publishers, Enfield, pp 149–176
21. Nordeide JT, Solberg C, Willumsen L, Amble S (2008) Seasonal variation and condition-dependence of the drumming muscle of cod, *Gadus morhua* L.?—An experimental approach. J Exp Mar Biol Ecol 363:66–74
22. Olivar MP, Quílez G, Emelianov M (2003) Spatial and temporal distribution and abundance of European hake, *Merluccius merluccius*, eggs and larvae in the Catalan coast (NW Mediterranean). Fish Res 60(2–3):321–331
23. Rountree RA, Goudey C, Hawkins T (2003) In: Rountree RA, Goudey C, Hawkins T (eds) Listening to fish: proceedings of the international workshop on the applications of passive acoustics to fisheries, April 8–10, 2002, MIT Sea Grant Technical Report, Dedham, MA, pp 4–10
24. Rowe S, Hutchings JA (2004) The function of sound production by Atlantic cod as inferred from patterns of variation in drumming muscle mass. Can J Zool 82:1391–1398
25. Sørensen W (1884) Om Lydorganer hos Fiske. En Physiologisk og Comparativ-anatomisk Undersøgelse. Copenhagen, Thaning pp 245
26. Templeman W, Hodder VM (1958) Variation with fish length, sex, stage of sexual maturity and season, in the appearance and volume of the drumming muscles of the swimbladder in the haddock, *Melanogrammus aeglefinus* L. J Fish Res Board Can 15:355–390

CHAPTER 6

Turnover of Sex Chromosomes and Speciation in Fishes

Jun Kitano & Catherine L. Peichel

Ecological Genetics Laboratory and JST PRESTO, Center for Frontier Research, National Institute of Genetics, Yata 1111, Mishima, Shizuoka 411–8540, Japan

Division of Human Biology, Fred Hutchinson Cancer Research Center, Seattle, WA 98109–1024, USA

ABSTRACT

Closely related species of fishes often have different sex chromosome systems. Such rapid turnover of sex chromosomes can occur by several mechanisms, including fusions between an existing sex chromosome and an autosome. These fusions can result in a multiple sex chromosome system, where a species has both an ancestral and a neo-sex chromosome. Although this type of multiple sex chromosome system has been found in many fishes, little is known about the mechanisms that select for the formation of neo-sex chromosomes, or the role of neo-sex chromosomes in phenotypic evolution and speciation. The identification of closely related, sympatric species pairs in which one species has a multiple sex chromosome system and the other has a simple sex chromosome system provides an opportunity to study sex chromosome turnover. Recently, we found that a population of threespine stickleback (*Gasterosteus aculeatus*) from Japan has an X_1X_2Y multiple sex chromosome system resulting from a fusion between the ancestral Y chromosome and an autosome, while a sympatric threespine stickleback population has a simple XY sex chromosome system. Furthermore, we demonstrated that the neo-X chromosome (X_2) plays an important role in phenotypic divergence and reproductive isolation between these sympatric stickleback species pairs. Here, we review multiple sex chromosome systems in fishes, as well as recent advances in our understanding of the evolutionary role of sex chromosome turnover in stickleback speciation.

Keywords: Multiple, sex, chromosomes, Neo-sex, chromosome, X_1X_2Y, Stickleback, Sexual conflict, Speciation

TURNOVER OF SEX CHROMOSOME IN FISHES

Sex chromosomes have repeatedly and independently evolved in plant and animal species with genetic sex-determination mechanisms (Bull 1983). Although some taxonomic groups appear to have stable sex chromosomes that have been conserved across hundreds of millions of years of evolution, there are many examples of sex chromosome turnover, where even closely related species have different sex chromosomes (White 1973). Heteromorphic sex chromosomes have been found in about 10% of fish species karyotyped (Devlin and Nagahama 2002), and many independent groups of fishes show evidence for sex chromosome turnover (Devlin and Nagahama 2002; Mank et al. 2006). In fishes, there is evidence that sex chromosome turnover has occurred through multiple mechanisms, including the transposition of an existing male-determination locus to an autosome (Woram et al. 2003), the evolution of a new male-determination locus on an autosome (Kondo et al. 2006; Tanaka et al. 2007), and fusions between an autosome and an existing Y-chromosome (Kitano et al. 2009; Ross et al. 2009). In this review, we focus on the evolution of sex chromosome-autosome fusions, which appear to be quite common in fishes. Then, we discuss the mechanisms that might select for these fusions as well as their potential role in phenotypic evolution and speciation.

Multiple sex chromosome systems in fishes

Most species with sex chromosomes either have a simple male heteromorphic (XX female/XY male) or female heteromorphic (ZW female/ZZ male) system. Fusions between an autosome and a sex chromosome (X, Y, Z, or W) create what is commonly referred to as a multiple sex chromosome system. For example, when an autosome fuses to a Y chromosome, this creates an X_1X_2Y sex chromosome system (White 1973). In this case, males have one neo-Y chromosome, one ancestral X chromosome (X_1), and one neo-X chromosome (X_2), while females have two pairs of X chromosomes ($X_1X_1X_2X_2$) (Fig. 1). Similarly, fusions between an autosome and an X, Z, or W chromosome will create an XY_1Y_2, ZW_1W_2, or Z_1Z_2W system, respectively. All of these types of systems have been found in a variety of taxonomic groups. The first multiple sex chromosome system identified in fishes was found by Uyeno and Miller (1971) in a species of Mexican killifish; males have an X_1X_2Y sex chromosome system, with one large metacentric Y chromosome that is missing in females and two small acrocentric X chromosomes that are also present in females (Uyeno and Miller 1971). Since then, multiple sex chromosome systems have been identified in many fish species across diverse families (Table 1). The presence of multiple sex chromosome systems in diverse groups of fishes suggests that they have evolved independently in multiple lineages.

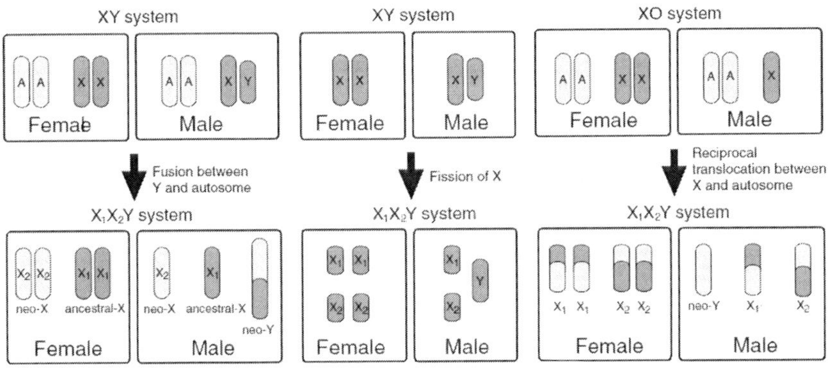

Figure. 1 Three proposed mechanisms for the formation of an X_1X_2Y multiple sex chromosome system, based on White (1983). Autosome, X chromosome, ancestral-X chromosome, neo-X chromosome, and Y chromosome are abbreviated as *A*, *X*, X_1, X_2, and *Y*, respectively

Interestingly, X_1X_2Y systems are more common in fishes than the other types of multiple sex chromosome systems (35/41) (Table 2). X_1X_2Y systems can arise not only through fusions between an autosome and a Y chromosome, but also through centric fission of the X chromosome in species with an XY system or through reciprocal translocations between the X chromosome and an autosome in species with an ancestral XX female/XO male sex chromosome system (Fig. 1) (White 1973). Several lines of evidence support the idea that the multiple sex chromosome systems in fishes arose mainly through fusions between a Y chromosome and an autosome. First, in many species, the neo-Y chromosome is a large metacentric chromosome, while the two X chromosomes are small acrocentric chromosomes, suggesting that the neo-Y chromosome arose through a Robertsonian fusion of two acrocentric chromosomes (Ueno and Takai 2008). Consistent with this hypothesis, the neo-Y is associated with both the X_1 and the X_2 chromosomes during meiosis in all species examined (Ueno and Takai 2008).

Table 1 List of fishes that have multiple sex chromosome systems

Order	Family	Genus	Species	2n (F/M)	Comments[a]	References
$X_1X_1X_2X_2/X_1X_2Y$						
Anguilliformes	Ophichthidae	*Muraenichthys*	*gymnotus*	48/47	Y-A fusion	(Murofushi and Yosida 1984)
Beryciformes	Berycidae	*Beryx*	*splendens*	48/47	Y-A fusion	(Ojima and Kikuno 1986)
Characiformes	Erythrinidae	*Erythrinus*	*erythrinus B*[b]	54/53	Y-A fusion; *E. erythrinus* A population has no obvious sex chromosome (2n=54)	(Bertollo et al. 2004)
Characiformes	Erythrinidae	*Erythrinus*	*erythrinus C*[b]	52/51	Y-A fusion, *E. erythrinus* A population has no obvious sex chromosome (2n=54)	(Bertollo et al. 2004)
Characiformes	Erythrinidae	*Erythrinus*	*erythrinus D*[b]	52/51	Y-A fusion; *E. erythrinus* A population has no obvious sex chromosome (2n=54)	(Bertollo et al. 2004)
Characiformes	Erythrinidae	*Hoplias*	*malabaricus*	40/39	Y-A fusion; other *H. malabaricus* karyomorphs are XX/XY (2n=42) or XX/XY_1Y_2 (2n=41/40)	(Bertollo et al. 2000; Rosa et al. 2009)
Clupeiformes	Clupeidae	*Brevoortia*	*aurea*	46/45	Y-A fusion, *B. pectinata* has no obvious sex chromosome (2n=46)	(Brum and Galetti 1992)
Cypriniformes	Cobitidae	*Cobitis*	*taenia striata*	50/49	Y-A fusion	(Saitoh 1989)
Cyprinodontiformes	Cyprinodontidae	*Garmanella*	*pulchra*	50/49	Y-A fusion	(Levin and Foster 1972)
Cyprinodontiformes	Cyprinodontidae	*Megupsilon*	*aporus*	48/47	first Y-A fusion identified in fishes	(Uyeno and Miller 1971)
Cyprinodontiformes	Goodeidae	*Allodontichthys*	*hubbsi*	42/41	Y-A fusion; no obvious sex chromosomes in family (2n=48)	(Uyeno et al. 1983)
Cyprinodontiformes	Nothobranchiidae	*Nothobranchius*	*guentheri*	36/35	Y-A fusion	(Ewulonu et al. 1985)
Gasterosteiformes	Gasterosteidae	*Gasterosteus*	*aculeatus (JS)*	42/41	Y-A fusion; sympatric *G. aculeatus* population is XX/XY (2n=42)	(Kitano et al. 2009)
Gasterosteiformes	Gasterosteidae	*Gasterosteus*	*wheatlandi*	42/41	Y-A fusion; closely related, sympatric species *G. aculeatus* is XX/XY (2n=42)	(Ross et al. 2009)
Gymnotiformes	Gymnotidae	*Gymnotus*	*pantanal*	40/39	Y-A fusion; two closely related, sympatric species have no obvious sex chromosomes (2n=40 or 54)	(Margarido et al. 2007)
Gymnotiformes	Hypopomidae	*Brachyhypopomus*	*pinnicaudatus*	42/41	Y-A fusion	(Almeida-Toledo et al. 2000a)
Gymnotiformes	Sternopygidae	*Eigenmannia*	sp.2	32/31	Y-A fusion; closely related, sympatric species *E. virescens* is XX/XY (2n=38)	(Almeida-Toledo et al. 2000b; Almeida-Toledo et al. 1984)
Perciformes	Blenniidae	*Parablennius*	*tentacularis*	48/47	Y-A fusion; closely related species have no obvious sex chromosome	(Caputo et al. 2001)
Perciformes	Callionymidae	*Callionymus*	*beniteguri*	38/37	Y-A fusion; closely related species *C. punctatus* and *C. doryssus* are XX/XY	(Murofushi et al. 1983)
Perciformes	Callionymidae	*Callionymus*	*ornatipinnis*	38/37	Y-A fusion; closely related species *C. punctatus* and *C. doryssus* are XX/XY	(Murofushi et al. 1983)
Perciformes	Channichthyidae	*Chaenodraco*	*wilsoni*	48/47	Y-A fusion, closely related species have no obvious sex chromosome	(Morescalchi et al. 1992)
Perciformes	Channichthyidae	*Chionodraco*	*hamatus*	48/47	Y-A fusion; closely related species have no obvious sex chromosome	(Morescalchi et al. 1992)
Perciformes	Channichthyidae	*Chionodraco*	*myersi*	48/47	Y-A fusion; closely related species have no obvious sex chromosome	(Morescalchi et al. 1992)
Perciformes	Channichthyidae	*Chionobathyscus*	*dewitti*	48/47	Y-A fusion; closely related species have no obvious sex chromosome	(Morescalchi et al. 1992)
Perciformes	Channichthyidae	*Pagetopsis*	*macropterus*	48/47	Y-A fusion; closely related species have no obvious sex chromosome	(Morescalchi et al. 1992)
Perciformes	Eleotrididae	*Awaous*	*strigatus*	ND	Species in same family *Dormitator maculatus* is XX/XY	(Oliveira and Almeida-Toledo 2006)
Perciformes	Gobiidae	*Ctenogobius*	*shufeldti*	48/47	Y-A fusion; family has XX/XY, ZW/ZZ and XX/XO sex chromosome systems	(Pezold 1984)
Perciformes	Lutjanidae	*Lutjanus*	*quinquelineatus*	48/47	Y-A fusion; closely related species *L. kasmira* has no obvious sex chromosome (2n=48)	(Ueno and Takai 2008)
Perciformes	Monodactylidae	*Monodactylus*	*sebae*	48/47	Y-A fusion; closely related species *M. argenteus* has no obvious sex chromosome (2n=48)	(Suzuki et al. 1988)
Perciformes	Percidae	*Zingel*	*zingel*	48/47	Y-A fusion; other species in family are XX/XY	(Halacka et al. 2007)
Salmoniformes	Salmonidae	*Oncorhynchus*	*nerka*	58/57	Y-A fusion; closely related species *O. mykiss* is XX/XY	(Thorgaard 1978; Ueda and Ojima 1984)
Scorpaeniformes	Tetrarogidae	*Hypodytes*	*rubripinnis*	48/47	Y-A fusion	(Ueno and Kang 1992)
Siluriformes	Siluridae	*Ompok*	*bimaculatus*	42/41	Y-A fusion	(Rishi 1976)
Tetraodontiformes	Monacanthidae	*Stephanolepis*	*cirrhifer*	34/33	Y-A fusion	(Murofushi et al. 1980)
Tetraodontiformes	Monacanthidae	*Stephanolepis*	*hispidus*	34/33	Y-A fusion	(Brum 1996)
XX/XY_1Y_2						
Characiformes	Erythrinidae	*Hoplias*	*malabaricus*	40/41	Other *H. malabaricus* karyomorphs are XX/XY or XX/X_1X_2Y	(Bertollo et al. 2000; Bertollo et al. 1983)
Siluriformes	Loricariidae	*Ancistrus*	*sp.1*	38/39	Family also has XX/XY, ZW/ZZ, and XX/XO sex chromosome systems	(Alves et al. 2006; Oliveira et al. 2008)
Siluriformes	Loricariidae	*Harttia*	*carvalhoi*	52/53	Closely related species do not have obvious sex chromosomes	(Centofante et al. 2006)
$Z_1Z_2W_1W_2/Z_1Z_1Z_2Z_2$						
Siluriformes	Loricariidae	*Ancistrus*	*sp.2*	52/52	Family also has XX/XY, ZW/ZZ, and XX/XO sex chromosome systems	(Alves et al. 2006; Oliveira et al. 2008)
ZW_1W_2/ZZ						
Aulopiformes	Synodontidae	*Trachinocephalus*	*myops*	27/26	Z-A fusion, other species in family are ZW/ZZ (2n=48)	(Ueno et al. 2001)
Characiformes	Parodontidae	*Apareiodon*	*affinis*	55/54	Fission of W; sympatric population has no obvious sex chromosome (2n=54)	(Jesus et al. 1999; Moreira-Filho et al. 1980)

[a] When mechanisms creating the neo sex chromosomes are suggested in the original paper, they are noted

[b] These three karyomorphs are presented as different species in this table, but might constitute a single species

ND=no data available

Table 2 Number of cases of different multiple sex chromosome systems found in each order

	$X_1X_1X_2X_2/X_1X_2Y$	XX/XY_1Y_2	$Z_1Z_2W_1W_2/Z_1Z_2Z_1Z_2$	ZW_1W_2/ZZ
Anguilliformes	1			
Aulopiformes				1
Beryciformes	1			
Characiformes	4[a]	1		1
Clupeiformes	1			
Cypriniformes	1			
Cyprinidontiformes	4			
Gasterosteiformes	2			
Gymnotiformes	3			
Perciformes	13			
Salmoniformes	1			
Scorpaeniformes	1			
Siluriformes	1	2	1	
Tetraodontiformes	2			

[a]Three of these examples are karyomorphs, which might constitute a single species

Second, XX/XO systems are not as common as XY system in fish, with only twelve examples of XX/XO systems documented to date (Devlin and Nagahama 2002; Alves et al. 2006). Furthermore, only one of these species (*Gobiodon citrinus*) is found in a family (Gobiidae) that also contains a species (*Ctenogobius shufeldti*) with an X_1X_2Y system (Pezold 1984). In seven of the known X_1X_2Y systems, a closely related species in the same genus or other populations of the same species have an XX/XY sex chromosome system, suggesting that this may be the ancestral state (Table 1). In three of these cases, cytogenetic evidence demonstrates that the ancestral Y has indeed fused to an autosome (Kitano et al. 2009; Ross et al. 2009; Cioffi and Bertollo 2010). Although no evidence for heteromorphic sex chromosomes has been found in eleven cases where closely related species have been analyzed cytogenetically (Table 1), it is possible that these species do have an XX/XY sex chromosome system that cannot be distinguished using traditional cytogenetic tools. For example, it was originally reported that the threespine stickleback (*Gasterosteus aculeatus*) had no heteromorphic sex chromosomes, but later studies using fluorescence in situ hybridization revealed that the X and the Y can be distinguished (Ross and Peichel 2008).

Finally, although centric fission of the X chromosome can result in the formation of an X_1X_2Y system (Fig. 1), this is unlikely to be the cause in most fish species with this system. When fissions create an X_1X_2Y system, closely related species with an XY system are expected to have fewer chromosome pairs (2n) than the species with an X_1X_2Y system. However, there are no such cases (Table 1).

Thus, it is likely that most of the X_1X_2Y sex chromosome systems in fishes are derived from fusions between an existing Y chromosome (whether cytogenetically visible or not) and an autosome. However, it is still unknown why these Y-autosome fusions are relatively common in fishes, and whether they have any role in phenotypic evolution or speciation. In order to address these questions, it is particularly useful to identify extant species pairs in which one species has the ancestral simple sex chromosome and the other has the derived multiple sex chromosome system. Indeed, for a few of the species with multiple sex chromosome systems listed in Table 1, such species pairs can be found in sympatry or parapatry. For example, in a single watershed, there are sympatric populations of *Hoplias* aff. *malabaricus*; some populations have a simple XY system, other populations have an X_1X_2Y system, and still others have a XY_1Y_2 system (Bertollo et al. 1983; Bertollo et al. 2000; Rosa et al. 2009). Similarly, two overlapping sub-species of *Eigenmannia* have non-homologous sex chromosomes; *E. virescens*has an XY system and *E. sp.2* has an X_1X_2Y system (Almeida-Toledo et al. 1984; Almeida-Toledo et al. 2000b; Henning et al. 2008). In eastern Hokkaido, Japan, an incipient species of threespine stickleback (*G. aculeatus*) has an X_1X_2Y system; this Japan Sea form is sympatric with the ancestral Pacific Ocean form, which has a simple XY system (Kitano et al. 2009). In the Japan Sea form, cytogenetic studies and linkage analysis revealed that the X_1X_2Y system results from a fusion between an autosome (chromosome 9) and the ancestral Y chromosome (chromosome 19) (Kitano et al. 2009). A similar fusion has occurred in the blackspotted stickleback (*G. wheatlandi*), which is sympatric with the Atlantic Ocean population of threespine stickleback on the eastern coast of North America (FitzGerald and Wootton 1993). Like the Japan Sea form, *G. wheatlandi* also has an X_1X_2Y system, but chromosome 12 has fused to the ancestral Y chromosome (Ross et al. 2009). These data suggest that the fusions have occurred independently in the two stickleback lineages.

What drives sex chromosome turnover?

Theoretical work has suggested that the presence of genes with sexually antagonistic effects (i.e. genes with alleles that have differential fitness effects in males and females) on an ancestral autosome might drive the turnover of sex chromosomes (Charlesworth and Charlesworth 1980; van Doorn and Kirkpatrick 2007). Alleles that enhance fitness in one sex will not necessarily increase, and might even reduce, the fitness of the other sex. If such alleles with sexually antagonistic effects are present on autosomes, they will not easily spread within a population, because selection for an increase in allele frequency in one sex will be counteracted by selection against the allele in the other sex. However, if a sexually antagonistic gene is localized on a sex chromosome and tightly linked to the sex-determination locus, both males and females can increase their fitness by having different allele frequencies at the sexually antagonistic locus. Thus, the presence of genes with sexually antagonistic effects on an autosome might select for the evolution of a new sex-determination locus on that autosome, the transposition of an existing sex-determination locus to that autosome, or the fusion of that autosome to an existing sex chromosome.

Although empirically testing this hypothesis presents a major challenge, the identification of closely related species with different sex chromosome systems provides a unique opportunity to test whether sexual antagonism plays a role in sex chromosome turnover. For example, a new female-determination locus has evolved in species of Lake Malawi cichlids, and a gene involved in sexually dimorphic pigmentation is tightly linked to the female sex-determination locus on the invading W chromosome (Roberts et al. 2009). Because the sexually dimorphic pigmentation pattern may be beneficial in females but detrimental in males of these species, sexually antagonistic selection might have played an important role in the evolution of this new ZW sex chromosome system.

Similarly, we have some preliminary evidence that sexually antagonistic traits are found on the neo-X chromosome created by the Y-autosome fusion in the Japan Sea threespine stickleback. Dorsal spine length is sexually dimorphic in the Japan Sea population (Kitano et al. 2007b) and has been suggested as a sexually antagonistic trait in sticklebacks (Reimchen and Nosil 2004). Because variation in dorsal spine length maps to the Japan Sea neo-X chromosome (Kitano et al. 2009), it is possible that the presence of sexually antagonistic variation for spine length on an ancestral autosome might have selected for the fixation of the sex chromosome fusion. However, we have not yet confirmed that spine length is a sexually antagonistic trait or identified any genes with sexually antagonistic alleles on the ancestral autosome (chromosome 9) of the Pacific Ocean population. Further comparative genomic and functional studies of genes on chromosome 9 in the ancestral Pacific Ocean and the derived Japan Sea sticklebacks will provide insights into whether sexually antagonistic selection plays an important role in sex chromosome turnover.

Role of sex chromosome turnover in speciation

It has long been appreciated that sex chromosomes might play a special role in speciation. There is abundant empirical evidence demonstrating that hybrid sterility genes are preferentially localized on sex chromosomes (Coyne and Orr 2004; Presgraves 2008; Qvarnstrom and Bailey 2009). Such an overrepresentation of hybrid sterility genes on the sex chromosomes may be due to multiple factors, such as faster X evolution, accumulation of sexually antagonistic genes, sex ratio meiotic drive, transcriptional suppression of sex chromosomes during early spermatogenesis, and gene traffic between the X and autosomes (Coyne and Orr 2004; Presgraves 2008; Moyle et al. 2010). The relative role of the sex chromosomes in other isolating barriers is still debated (Qvarnstrom and Bailey 2009), but there are empirical data indicating that the sex chromosomes harbor genes that are important for behavioral isolation between closely related species (Prowell 1998; Reinhold 1998; Lindholm and Breden 2002).

Although there is good evidence that the sex chromosomes do play an important role in speciation, the role of sex chromosome turnover in speciation has been less appreciated. However, genes that play a role in reproductive isolation between populations might accumulate on neo-sex chromosomes for many of the same reasons that they are found on ancestral sex chromosomes. In this case, sex chromosome turnover might actually promote speciation. We have tested this idea by performing genetic linkage analysis of the isolating barriers present between the Japan Sea form (X_1X_2Y) and the Pacific Ocean form (XY) of threespine sticklebacks. These two forms diverged during a period of geographical isolation between the Sea of Japan and the Pacific Ocean about 1.5–2 million years ago. Currently, they are sympatric in eastern Hokkaido, but are reproductively isolated with a low level of hybridization (Higuchi and Goto 1996; Yamada et al. 2001; Kitano et al. 2007a; Yamada et al. 2007). Reproductive isolation between this stickleback species pair results from multiple isolating barriers, including ecogeographical isolation (Kume et al. 2005; Kume et al. 2010), seasonal isolation (Kume et al. 2005), behavioral isolation (Kitano et al. 2007a), reduced fitness of hybrids (Kitano et al. 2009), and hybrid male sterility (Kitano et al. 2007a).

Although the male hybrids resulting from a cross between a Japan Sea female and a Pacific Ocean male are sterile, hybrid males resulting from crosses in the opposite direction and all hybrid females are fertile (Kitano et al. 2007a). Therefore, we were able to make artificial crosses between these two sympatric species and utilize the numerous genomic and genetic tools available for sticklebacks (Kingsley and Peichel 2007) to conduct quantitative trait locus (QTL) mapping of isolating barriers between the species pair (Kitano et al. 2009). Our QTL mapping revealed that hybrid sterility and male courtship display traits mapped to the ancestral X chromosome (chromosome 19) and the neo-X chromosome (chromosome 9) respectively (Kitano et al. 2009). Thus, a neo-X chromosome resulting from a fusion between the ancestral Y chromosome and an autosome contains genes important for the evolution of male courtship displays and can contribute to reproductive isolation between sympatric stickleback species. This is consistent with the idea that the turnover of sex chromosome might promote speciation.

CONCLUSIONS

Species with multiple sex chromosome systems are prevalent in fishes. In some cases, they can be found in sympatry or parapatry with closely related species that have the ancestral, simple sex chromosome system. These species pairs provide a unique opportunity to gain insight into the evolutionary forces that drive the evolution of new sex chromosomes and to the consequences of their evolution on phenotypic divergence and speciation. First, by identifying the genomic loci important for reproductive isolation or phenotypic divergence between closely related species, we can investigate whether the new sex chromosomes created by fusions can contribute to speciation, as we found in the Japanese species pair. Because the gaps between model and non-model organisms are decreasing with the advent of high-throughput sequencing technologies (Rokas and Abbot 2009), it should be possible to conduct similar genetic and genomic analyses on other fish species pairs to investigate the generality of our findings. Second, these systems could be used to test the hypothesis that sexual conflict drives the turnover of sex chromosomes. Although challenging, one approach would be to compare the autosome that has fused to the sex chromosome in both the ancestral population and the derived population to search for sexually antagonistic genes that might have driven the fusion. Finally, neo-sex chromosomes created by fusions provide a great system to study the early stages of sex chromosome evolution. Extensive genomic analyses have been conducted on the *Drosophila miranda* neo-sex chromosome created by a Y-autosome fusion; these studies have provided insights into the evolutionary forces that drive the degeneration of the Y chromosome (Bachtrog and Charlesworth 2002; Bachtrog 2003; Bachtrog 2004; Bachtrog 2006; Bachtrog et al. 2008; Bachtrog et al. 2009). Similar studies will be possible in fishes that have neo-sex chromosomes of a variety of ages, which will deepen our understanding of the process of sex chromosome evolution.

ACKNOWLEDGMENTS

This work was supported by the Uehara Memorial Foundation, JST PRESTO Program, a Grant-in-Aid for Scientific Research (B) from the Ministry of Education, Culture, Sports, Science, and Technology (no. 22770075) to JK, and a Burroughs Wellcome Fund Career Award in the Biomedical Sciences and National Institutes of Health grants R01 GM071854 and P50 HG002568 to CLP.

REFERENCES

1. Almeida-Toledo LF, Daniel-Silva MFZ, Lopes CE, Toledo-Filho SA (2000a) Sex chromosome evolution in fish. II. Second occurrence of an X_1X_2Y sex chromosome system in Gymnotiformes. Chrom Res 8:335–340
2. Almeida-Toledo LF, Foresti F, Daniel MFZ, Toledo SA (2000b) Sex chromosome evolution in fish: the formation of the neo-Y chromosome in *Eigenmannia* (Gymnotiformes). Chromosoma 109(3):197–200
3. Almeida-Toledo LF, Foresti H, Toledo-Filho SA (1984) Complex sex chromosome system in *Eigenmannia* sp. (Pisces, Gymnotiformes). Genetica 64:165–169
4. Alves AL, Oliveira C, Nirchio M, Granado A, Foresti F (2006) Karyotypic relationships among the tribes of Hypostominae (Siluriformes: Loricariidae) with description of XO sex chromosome system in a Neotropical fish species. Genetica 128:1–9
5. Bachtrog D (2003) Accumulation of *spock* and *worf*, two novel non-LTR retrotransposons, on the neo-Y chromosome of *Drosophila miranda*. Mol Biol Evol 20:173–181
6. Bachtrog D (2004) Evidence that positive selection drives Y-chromosome degeneration in *Drosophila miranda*. Nat Genet 36:518–522
7. Bachtrog D (2006) Expression profile of a degenerating neo-Y chromosome in *Drosophila*. Curr Biol 16:1694–1699
8. Bachtrog D, Charlesworth B (2002) Reduced adaptation of a non-recombining neo-Y chromosome. Nature 416:323–326
9. Bachtrog D, Hom E, Wong KM, Maside X, de Jong P (2008) Genomic degradation of a young Y chromosome in *Drosophila miranda*. Genome Biol 9:R30
10. Bachtrog D, Jensen JD, Zhang Z (2009) Accelerated adaptive evolution on a newly formed X chromosome. PLoS Biol 7:e1000082

11. Bertollo LAC, Born GG, Dergam JA, Fenocchio AS, Moreira-Filho O (2000) A biodiversity approach in the neotropical Erythrinidae fish, *Hoplias malabaricus*. Karyotypic survey, geographic distribution of cytotypes and cytotaxonomic considerations. Chrom Res 8:603–613
12. Bertollo LAC et al (2004) Chromosome evolution in the erythrinid fish, *Erythrinus erythrinus*(Teleostei: Characiformes). Heredity 93:228–233
13. Bertollo LAC, Takahashi CS, Moreira-Filho O (1983) Multiple sex chromosomes in the genus *Hoplias* (Pisces: Erythrinidae). Cytologia 48(1):1–12
14. Brum MJI (1996) Cytogenetic studies of Brazilian marine fish. Braz J Genet 19(3):421–427
15. Brum MJI, Galetti PM (1992) Multiple sex chromosomes in south Atlantic fish, *Brevoortia aurea*, Clupeidae. Rev Brazil Genet 15:547–553
16. Bull JJ (1983) Evolution of sex determining mechanisms. Benjamin Cummings, Menlo Park
17. Caputo V, Machella N, Nisi-Cerioni P, Olmo E (2001) Cytogenetics of nine species of Mediterranean blennies and additional evidence for an unusual multiple sex-chromosome system in *Parablennius tentacularis* (Perciformes, Blenniidae). Chrom Res 9:3–12
18. Centofante L, Bertollo LA, Moreira-Filho O (2006) Cytogenetic characterization and description of an XX/XY_1Y_2 sex chromosome system in catfish *Harttia carvalhoi* (Siluriformes, Loricariidae). Cytogenet Genome Res 112(3–4):320–324
19. Charlesworth D, Charlesworth B (1980) Sex differences in fitness and selection for centric fusions between sex-chromosomes and autosomes. Genet Res 35:205–214
20. Cioffi MB, Bertollo LA (2010) Initial steps in XY chromosome differentiation in *Hoplias malabaricus* and the origin of an X_1X_2Y sex chromosome system in this fish group. Heredity 105:554–561
21. Coyne JA, Orr HA (2004) Speciation. Sinauer, Sunderland
22. Devlin RH, Nagahama Y (2002) Sex determination and sex differentiation in fish: overview of genetic, physiological, and environmental influences. Aquaculture 208:191–364
23. Ewulonu UK, Haas R, Turner BJ (1985) A multiple sex chromosome system in the annual killifish, *Nothobranchius guentheri*. Copeia 1985:503–508
24. FitzGerald GJ, Wootton RJ (1993) The behavioural ecology of sticklebacks. In: Pitcher TJ (ed) Behaviour of teleost fishes. Chapman and Hall, London, pp 539–572
25. Halacka K, Vetesnik L, Lusk S, Mendel J, Papousek I (2007) The $X_1X_1X_2X_2/X_1X_2Y$ multiple sex chromosome system in the *Zingel zingel* (Pisces: Perciformes) from the Morava river (Czech Republic). Caryologia 60(3):222–225
26. Henning F, Trifonov V, Ferguson-Smith MA, de Almeida-Toledo LF (2008) Non-homologous sex chromosomes in two species of the genus *Eigenmannia* (Teleostei: Gymnotiformes). Cytogenetic Genome Res 121(1):55–58
27. Higuchi M, Goto A (1996) Genetic evidence supporting the existence of two distinct species in the genus *Gasterosteus* around Japan. Environ Biol Fish 47:1–16
28. Jesus CM, Bertollo LAC, Moreira-Filho O (1999) Comparative cytogenetics in *Apareiodon affinis* (Pisces, Characiformes) and considerations regarding diversification of the group. Genetica 105:63–67
29. Kingsley DM, Peichel CL (2007) The molecular genetics of evolutionary change in sticklebacks. In: Ostlund-Nilsson S, Mayer I, Huntingford FA (eds) Biology of the Three-spined Stickleback. CRC Press, Boca Raton, pp 41–81
30. Kitano J, Mori S, Peichel CL (2007a) Phenotypic divergence and reproductive isolation between sympatric forms of Japanese threespine sticklebacks. Biol J Linn Soc 91:671–685
31. Kitano J, Mori S, Peichel CL (2007b) Sexual dimorphism in the external morphology of the threespine stickleback (*Gasterosteus aculeatus*). Copeia 2007:336–349
32. Kitano J et al (2009) A role for a neo-sex chromosome in stickleback speciation. Nature 461(7267):1079–1083

33. Kondo M et al (2006) Genomic organization of the sex-determining and adjacent regions of the sex chromosomes of medaka. Genome Res 16:815–826
34. Kume M, Kitamura T, Takahashi H, Goto A (2005) Distinct spawning migration patterns in sympatric Japan Sea and Pacific Ocean forms of threespine stickleback *Gasterosteus aculeatus*. Ichthyol Res 52:189–193
35. Kume M, Kitano J, Mori S, Shibuya T (2010) Ecological divergence and habitat isolation between two migratory forms of Japanese threespine stickleback (*Gasterosteus aculeatus*). J Evol Biol 23:1436–1446
36. Levin B, Foster NR (1972) Cytotaxonomic studies in Cyprinodontidae: multiple sex chromosomes in *Garmanella pulchra*. Not Nat Acad Nat Sci Philadelphia 446:1–5
37. Lindholm A, Breden F (2002) Sex chromosomes and sexual selection in Poeciliid fishes. Am Nat 160:S214–S224
38. Mank JE, Promislow DEL, Avise JC (2006) Evolution of alternative sex-determining mechanisms in teleost fishes. Biol J Linn Soc 87:83–93
39. Margarido VP, Bellafronte E, Moreira-Filho O (2007) Cytogenetic analysis of three sympatric *Gymnotus* (Gymnotiformes, Gymnotidae) species verifies invasive species in the Upper Parana River basin, Brazil. J Fish Biol 70:155–164
40. Moreira-Filho O, Bertollo LAC, Galetti PM Jr (1980) Evidence for a multiple sex chromosome system with female heterogamety in *Apareiodon affinis* (Pisces, Parodontidae). Caryologia 33:83–91
41. Morescalchi A, Hureau JC, Olmo E, Ozouf-Costaz C, Pisano E, Stanyon R (1992) A multiple sex-chromosome system in Antarctic ice-fishes. Polar Biol 11:655–661
42. Moyle LC, Muir CD, Han MV, Hahn MW (2010) The contribution of gene movement to the "two rules of speciation". Evolution 64:1541–1557
43. Murofushi M, Nishikawa S, Yosida TH (1983) Cytogenetical studies on fishes. V. Multiple sex chromosome mechanism (XX-Y) found in two dragonet fishes. Proc Jpn Acad B Phys Biol Sci 59:58–61
44. Murofushi M, Oikawa S, Nishikawa S, Yosida TH (1980) Cytogenetical studies on fishes. III. Multiple sex chromosome mechanism in the filefish, *Stephanolepis cirrhifer*. Jpn J Genet 55(2):127–132
45. Murofushi M, Yosida TH (1984) Cytogenetical studies on fishes. VIII. XX-Y sex chromosome mechanism newly found in the snake eel, *Muraenichthys gymnotus* (Anguilloform, Pisces). Proc Jpn Acad Ser B Phys Biol Sci 60(2):21–23
46. Ojima Y, Kikuno T (1986) A heteromorphic chromosome of *Beryx splendens*, Berycidae (Pisces). Proc Jpn Acad B Phys Biol Sci 62:317–320
47. Oliveira C, Almeida-Toledo LF (2006) Evidence of an XX/XY sex chromosome system in the fish *Dormitator maculatus* (Teleostei, Eleotrididae). Genet Mol Biol 29:653–655
48. Oliveira RR, Feldberg E, dos Anjos MB, Zuanon J (2008) Occurrence of multiple sexual chromosomes (XX/XY_1Y_2 and $Z_1Z_1Z_2Z_2/Z_1Z_2W_1W_2$) in catfishes of the genus *Ancistrus*(Siluriformes, Loricariidae) from the Amazon Basin. Genetica 134(2):243–249
49. Pezold F (1984) Evidence for multiple sex chromosomes in the freshwater goby, *Gobionellus shufeldti* (Pisces: Gobiidae). Copeia 1984:235–238
50. Presgraves DC (2008) Sex chromosomes and speciation in *Drosophila*. Trends Genet 24:336–343
51. Prowell DP (1998) Sex linkage and speciation in Lepidoptera. In: Howard DJ, Berlocher SH (eds) Endless forms: Species and speciation. Oxford Univ Press, New York, pp 309–319
52. Qvarnstrom A, Bailey RI (2009) Speciation through evolution of sex-linked genes. Heredity 102:4–15
53. Reimchen TE, Nosil P (2004) Variable predation regimes predict the evolution of sexual dimorphism in a population of threespine stickleback. Evolution 58:1274–1281
54. Reinhold K (1998) Sex linkage among genes controlling sexually selected traits. Behav Ecol Sociobiol 44:1–7

55. Rishi KK (1976) Mitotic and meiotic chromosomes of a teleost, *Callichrous bimaculatus* (Bloch) with indications of male heterogamety. Cienc Cult 28:1171–1173
56. Roberts RB, Ser JR, Kocher TD (2009) Sexual conflict resolved by invasion of a novel sex determiner in Lake Malawi cichlid fishes. Science 326:998–1001
57. Rokas A, Abbot P (2009) Harnessing genomics for evolutionary insights. Trends Ecol Evol 24:192–200
58. Rosa R, Caetano-Filho M, Shibatta OA, Giuliano-Caetano L (2009) Cytotaxonomy in distinct populations in *Hoplias* aff. *malabaricus* (Characiformes, Erythrinidae) from lower Paranapanema River basin. J Fish Biol 75:2682–2694
59. Ross JA, Peichel CL (2008) Molecular cytogenetic evidence of rearrangements on the Y chromosome of the threespine stickleback fish. Genetics 179:2173–2182
60. Ross JA, Urton JR, Boland J, Shapiro MD, Peichel CL (2009) Turnover of sex chromosomes in the stickleback fishes (Gasterosteidae). PLoS Genet 5:e1000391
61. Saitoh K (1989) Multiple sex-chromosome system in a loach fish. Cytogenet Cell Genet 52:62–64
62. Suzuki A, Taki Y, Takeda M, Akatsu S (1988) Multiple sex chromosomes in a monodactylid fish. Jpn J Ichthyol 35(1):98–101
63. Tanaka K, Takehana Y, Naruse K, Hamaguchi S, Sakaizumi M (2007) Evidence for different origins of sex chromosomes in closely related *Oryzias* fishes: substitution of the master sex-determining gene. Genetics 177:2075–2081
64. Thorgaard GH (1978) Sex chromosomes in the sockeye salmon: a Y-autosome fusion. Can J Genet Cytology 20:349–354
65. Ueda T, Ojima Y (1984) Sex chromosomes in the Kokanee salmon *Oncorhynchus nerka*. Bull Jpn Soc Sci Fish 50:1495–1498
66. Ueno K, Kang J-H (1992) Multiple sex chromosomes in the redfin velvetfish, *Hypodytes rubripinnis*. Jpn J Ichthyol 39(2):170–173
67. Ueno K, Ota K, Kobayashi T (2001) Heteromorphic sex chromosomes of lizardfish (Synodontidae): focus on the ZZ-ZW_1W_2 system in *Trachinocephalus myops*. Genetica 111:133–142
68. Ueno K, Takai A (2008) Multiple sex chromosome system of $X_1X_1X_2X_2/X_1X_2Y$ type in lutjanid fish, *Lutjanus quinquelineatus* (Perciformes). Genetica 132(1):35–41
69. Uyeno T, Miller RR (1971) Multiple sex chromosomes in a Mexican cyprinodontid fish. Nature 231:452–453
70. Uyeno T, Miller RR, Fitzsimons JM (1983) Karyology of the Cyprinodontoid fishes of the Mexican family Goodeidae. Copeia 1983:497–510
71. van Doorn GS, Kirkpatrick M (2007) Turnover of sex chromosomes induced by sexual conflict. Nature 449:909–912
72. White MJD (1973) Animal Cytology and Evolution. Cambridge Univ Press, Cambridge
73. Woram RA et al (2003) Comparative genome analysis of the primary sex-determining locus in salmonid fishes. Genome Res 13:272–280
74. Yamada M, Higuchi M, Goto A (2001) Extensive introgression of mitochondrial DNA found between two genetically divergent forms of threespine stickleback, *Gasterosteus aculeatus*, around Japan. Environ Biol Fish 61:269–284
75. Yamada M, Higuchi M, Goto A (2007) Long-term occurrence of hybrids between Japan Sea and Pacific Ocean forms of threespine stickleback, *Gasterosteus aculeatus*, in Hokkaido Island, Japan. Environ Biol Fish 80:435–443

CHAPTER 7

Appearance Traits in Fish Farming: Progress from Classical Genetics to Genomics, Providing Insight into Current and Potential Genetic Improvement

Nelson Colihueque[1*], andCristianAraneda[2]

[1]*Laboratorio de Biología Moleculary Citogenética, Departamentode Ciencias Biológicasy Biodiversidad, Universidadde Los Lagos, Osorno, Chile*

[2]*Laboratorio de Biotecnologíay Genética Aplicadaala Acuicultura, Departamentode Producción Animal, Facultadde Ciencias Agronómicas, Universidadde Chile, Santiago,Chile*

ABSTRACT

Appearance traits in fish, those external body characteristics that influence consumer acceptance at point of sale, have come to the forefront of commercial fish farming, as culture profitability is closely linked to management of these traits. Appearance traits comprise mainly body shape and skin pigmentation. Analysis of the genetic basis of these traits in different fish reveals significant genetic variation within populations, indicating potential for their genetic improvement. Work into ascertaining the minor or major genes underlying appearance traits for commercial fish is emerging, with substantial progress in model fish in terms of identifying genes that control body shape and skin colors. In this review, we describe research progress to date, especially with regard to commercial fish, and discuss genomic findings in model fish in order to better address the genetic basis of the traits. Given that appearance traits are important in commercial fish, the genomic information related to this issue promises to accelerate the selection process in coming years.

Keywords: skin pigmentation, body shape, appearance traits, fish farming, quantitative trait loci

INTRODUCTION

Over the past few decades, body shape and skin pigmentation have become valuable appearance traits in commercial fish (Table 1). Due to increasing market sophistication, fish size, meat quality, and other traditional traits are not the only attributes that influence consumer choice at point of sale, especially when fish are sold whole.

TABLE 1. Examples of appearance traits for body shape and skin pigmentation used in fish farming.

Category of trait	Species	Strains	Trait characteristic	Practical application	Reference
Body shape	Common carp (*Cyprinus carpio*)	Aischgrund and Galician	High-backed	To increase esthetics of body shape	Ankorion et al. (1992)
		var. *wuyuanensis* (also called Purse red carp)	Broadly elliptical body (red skin); standard length to body height ratio: 2.3	To increase esthetics, creating a desirable "wallet" shape and all-red body color	Zhang et al. (2013)
		var. *haematopterus* (also called Amur wild carp)	Spindle-shaped body (steel-gray skin); standard length to body height ratio: 3.5	To increase desirability for sport fishing (its body is elongated, making it an excellent sport fish)	Zhang et al. (2013)
	Rainbow trout (*Oncorhynchus mykiss*)	Finnish national breeding program	Slender body (i.e., low body height to length ratio; silvery skin with fewer spots)	To produce fish more visually appealing desirable for the whole carcass market	Kause et al. (2003)
Skin color	Tilapia (*Oreochromis niloticus*)	Red strains (red Stirling, red Yumbo, etc)	Red body color without signs of normal black pigmentation of wild-type fish	To add value to the final product	McAndrew et al. (1988), Moreira et al. (2005)
	Rainbow trout (*Oncorhynchus mykiss*)	Blue Back	Intense bluish back; whitish belly; reduced number of dark spots, both on the back and below the lateral line	To satisfy market demands	Colihueque et al. (2011)

On the subject of skin pigmentation, previous work conducted in other livestock species has demonstrated that proper handling of pigmentation traits allows for response to consumer demands for various food products, such as skin color in pigs and egg shell, yolk, and skin color in chickens (see review by Hudon, 1994). This topic is relevant for producers because the color of a food product is a quality attribute for the consumer. For example, consumers perceive redder salmon filets as being fresher, better-tasting, and higher quality as compared with paler salmon, and, therefore, they are willing to pay significantly more for the product (Anderson, 2000; Alfnes et al., 2006). Therefore, to satisfy modern market demands and increase profitability, producers are forced to manage external traits more intensively on an industrial scale, in particular body shape and skin color.

However, this is not an easy task, because body shape and skin color in fish are complex traits, involving numerous genetic and environmental factors. Thus, progress in this field will depend in part on dissecting the underlying genetics of these traits for future implementation of modern selection strategies, such as marker-assisted selection based on molecular data.

In commercial fish, such as the common carp, tilapia, sea bream, and salmonids (Pillay and Kutty, 2005), this strategy will complement progress made to date based solely on breeding values estimated with phenotypic and genealogical information or classical genetics which, for example, has enabled the development of new strains (Figure 1).

Further understanding of this issue may be gained from progress achieved in model and ornamental fish, where characterization of the inheritance mode of mutation, genes, and quantitative trait loci (QTL) for external traits is more advanced.

In this review, we describe efforts made to improve the external traits in commercial fish based on classical genetic approaches, as well as recent progress in genomics, the latter initially aimed at identifying the specific region that harbors genes controlling quantitative traits. This information, together with data available on this issue in model fish, will enhance progress in this field, an objective of tremendous importance for producers who need to increase the competitiveness of their cultures by managing external characteristics that give added value to cultured fish.

Fish Body Shape

In common carp, one of the first domesticated fish in the world (Balon, 2004), the long process of domestication has produced a domesticated phenotype very different from the wild-type phenotype (Ankorion et al., 1992; Zhang et al., 2013). Many of these changes have arisen due to intentional selection of traits, but it is equally true that many traits are the result of an unintentional selection process. This phenomenon emerged as a result of the adaptation of fish to captive conditions, quite different than the natural environment inhabited by wild-type fish.

A
B
C
D

FIGURE 1. Examples of commercial fish strains with improved skin pigmentation and body shape. (A) A wild-type tilapia (*Oreochromis niloticus*) with normal black pigmentation; (B) a red strain tilapia (red Yumbo) with improved red skin pigmentation; (C) a wild-type rainbow trout (*Oncorhynchus mykiss*) with normal pigmentation and (D) a Blue Back rainbow trout with improved intense bluish back, whitish belly, and a reduced number of dark spots; (E) a common carp (*Cyprinus carpio*) of var. *haematopterus* (or Amur wild carp) with a spindle-shaped body and steel-gray skin color; and (F) a common carp of var. *wuyuanensis* with improved broadly elliptical body (red skin color).

The domesticated phenotype phenomenon arises because the body shape of an organism results from the integration of morphological, behavioral, and physiological traits (Reid and Peichel, 2010), where different genetic and environmental pressures can lead to functional trade-offs (Reznick and Ghalambor, 2001; Ghalambor et al., 2003; Walker, 2010). This creates functional constraints, where those changes with the greatest positive and fewest negative effects on fitness will be selected (Reid and Peichel, 2010). For example, in natural populations, there is a relationship between body shape and swimming performance, but body shape is also influenced by foraging behavior, the risk of predation, and stream velocity (Webb, 1984; Walker, 1997). The trade-off for body shape also operates in captive populations. For instance, cultured populations of rainbow trout selected for rapid growth result in more rotund fish, given the existence of a positive genetic correlation of body mass with body shape and condition factor (Gjerde and Schaeffer, 1989; Kause et al., 2003); that is, mass gain in fish achieved by increasing body width and height rather than by increasing body length.

It has been shown that other factors, such as water velocity (Pakkasmaa and Piironen, 2001), rearing environment (Swain et al., 1991), fish density, and diet (Higgs et al., 1992; Einen et al., 1998; Jenkins et al., 1999) may also modify body shape in fish. This phenomenon occurs given that many morphological growth-related traits show phenotypic and genetic correlations in fish (Kause et al., 2003; Martyniuk et al., 2003), whose origins are related to the genetic architecture of traits such as covariation in QTL location or conservation of chromosomal regions homologous across species (Reid et al., 2005).

In common carp, this process has produced various phenotypes of commercial value that are currently used in fish farming, such as the high-backed and elliptical body shape morphs, typical of the Galician and Wuyuanensis strains, respectively (Ankorion et al., 1992; Zhang et al., 2013). Even this process of domestication can reach a high level of body shape modification, such as it has been observed in the ornamental goldfish (*Carassius auratus*), where various morphological traits have been modified (e.g., body shape, fins, and eyes). Several of these modified traits can be found in the same individual, giving rise to popular strains called "monstrosities" (Balon, 1990).

Alterations of morphology characteristics, mainly body length and fins (Haffter et al., 1996) can also be seen in mutants of zebrafish (*Danio rerio*). In the guppy (*Poecilia reticulate*), variation in male body shape occurs in association with mating success (Tripathi et al., 2009). Therefore, available evidence in fish indicates that these organisms are highly amenable to morphological modification, already widely explored in ornamental as well as in model fish.

The underlying genetics of phenotypic variation is beginning to be understood in various commercial fish (Massault et al., 2009; Loukovitis et al., 2013; Zhang et al., 2013). These studies are focusing on finding significant QTL for morphometric traits based mostly on geometric morphometry; in these analyses, different types of molecular markers have been used. These investigations are contributing to our understanding of the genetic architecture of divergence in body shape, by means determining the number of genes or QTLs that contribute to a particular trait, or the number of traits that a particular gene or QTL affects, i.e., the pleiotropic effect, and the location of genes or QTLs within the genome that affect body shape, along with their interaction.

Progress in model fish should be mentioned here, particularly in zebrafish (Haffter et al., 1996) where a set of dominant Mendelian loci affecting body shape and fins in induced mutants have been identified. For example, loci that affect body shape may cause a reduction of overall body length in the adult fish, due to a reduction either in the length of vertebrae (*stöpsel*mutant) or number of vertebrae (*däumling* mutant). Interestingly, mechanisms of body shape variation involving axial length modification also occur naturally across several fish species (Ward and Mehta, 2010), which indicates that this mechanism has been of evolutionary significance for body form differentiation in fish.

Recent work on QTL searching in commercial fish clearly supports the existence of major genes underlying the quantitative genetic variation of morphological and body shape-related traits (Table 2). In Gilthead seabream (*Sparus aurata*), Loukovitis et al. (2013), using half-sib regression analysis, found significant morphology QTLs, e.g., distances from pectoral fin to dorsal fin or from pectoral fin to anal fin (see Table 2), in three linkage groups (9, 21, and 25) identified at genome-wide level that explain 18.5 to 27.1% of trait variation. This result suggests the existence of one locus in each linkage group affecting several traits in this fish. Moreover, given that QTLs affecting body weight were located at the same positions for the linkage groups 9 and 21 (Loukovitis et al., 2011), the authors conclude that there might be only one pleiotropic QTL in each LG affecting overall body size. This is in accordance with the high genetic correlations ($r_G > 90\%$) observed between all traits analyzed (see Table 2), including body weight. These results, combined with those obtained from previous studies (Boulton et al., 2011), underline highly significant loci affecting overall morphology in *S. aurata*.

On the other hand, using half-sib regression analysis and variance component analysis at the genome-wide level in sea bass (*Dicentrarchus labrax*), six significant QTLs for a combination of morphometric traits (standard length, head length, body length, pre-anal length, abdominal length, post-anal length, head depth, body depth; see Table 2) on linkage groups 1B, 4, 6, 7, 15, and 24 were reported by Massault et al. (2009). These QTLs explain between 9.4 and 16% of phenotypic variance. In this study, a body weight QTL was discovered at the same linkage groups (linkage groups 4 and 6) and at similar positions as morphology QTLs, which might explain the high correlation observed between body weight and all morphometric traits studied in this fish.

TABLE 2. Examples of QTLs for body shape-related traits mapped on different commercial fish genomes.

Species	Trait	Linkage group (LG) where QTLs were detected. Name of QTL (bold), linked markers, and respective phenotypic variation noted in parentheses	Reference
Gilthead seabream (*Sparus aurata*)	Standard length, body length, pectoral dorsal 1, pectoral dorsal 2, body length 2, pectoral anal 2, pectoral anal 1, dorsal fin length, belly length, body depth 2, body depth 3, head length	LG-9 (**SL**, **BL**, **PecDor1**, **Pecdors2**, **BL2**, **PectAnal1**, **PecAnal2**, Bd20, 18.5–23%) LG-21 (**DFL**, **BellyL**, **BD2**, **BD3**, ELD36 and SAGT1, 18.6–27.1%) LG-25 (**HL**, 22.6%)	Loukovitis et al. (2013)
Sea bass (*Dicentrarchus labrax*)	Combination of morphometric traits (standard length, head length, body length, pre-anal length, abdominal length, post-anal length, head depth, body depth)	LG-1B (**MORPH**, 14%) LG-4 (**MORPH**, 13%) LG-6 (**MORPH**, 9.4%) LG-7 (**MORPH**, 16%) LG-15 (**MORPH**, 12%) LG-24 (**MORPH**, 13%)	Massault et al. (2009)
Common carp (*Cyprinus carpio* var. *haematopterus*)	Body height Body width Standard length	LG-1 (**qbh1**, snp0163, 20.4%; **qbw1**, snp0163, 20.7%) LG-12 (**qsl12**, snp0315, 21.1%; **qbh12**, snp1133, 18.9%) LG-20 (**qbh20**, hlj1717, 19.5%)	Zhang et al. (2013)

QTLs were detected at genome-wide level using permutation tests at a significance threshold value of $P < 0.05$ for S. aurata and D. labrax, and of $P < 0.01$ for C. carpio.

Moreover, in common carp (*Cyprinus carpio*), in a primary genome-wide scan using single nucleotide polymorphisms (SNPs) and microsatellite markers, Zhang et al. (2013) found five significant QTLs for body-shape related traits (body height, body width and standard length) located at linkage groups 1, 12, and 20, which explain 20.4 to 20.7%, 18.9 to 21.1%, and 19.5% of phenotypic variance, respectively. Given that QTLs of linkage group 1 were located in the same interval, it was concluded that only one QTL produced pleiotropic effects on these traits, which was not the case for QTLs found in linkage group 12, indicating that different factors control the traits. Importantly, this study provides strong evidence that the marked body shape differences of *Cyprinus carpio* populations, in particular between *Cyprinus carpio* var. *wuyuanensis* and *Cyprinus carpio* var. *haematopterus*, depend on quantitative genetic variations that control different body shape-related traits that may have originated through the process of selective breeding that has occurred for decades in this species.

In salmonids, another important group of commercial fish (Pillay and Kutty, 2005), progress has been made in this field through QTLs searching, mainly for growth-related traits, including fork length, body weight, and Fulton's condition factor, and also for meristic traits (for review see Araneda et al., 2008). For example, in Atlantic salmon (*Salmo salar*), four QTLs for condition factor and two for body weight were detected in comparative studies with rainbow trout (*Oncorhynchus mykiss*) and Arctic charr (*Salvelinus alpinus*). One strong QTL explaining 20.1% of variation in body weight was found on linkage group AS-8, while another QTL with a strong effect on condition factor accounting for 24.9% of trait variation was found on linkage group AS-14. This result suggests that a significant portion of quantitative variation in body weight and condition factor in this species is under the control of a few QTLs with relatively large effects.

However, to date no study has been specifically undertaken to search QTLs in salmonids based on the geometric morphology method. It is noteworthy that in some species of this group, such as rainbow trout, a marked intra- and inter-population differentiation in body shape has been observed (Kause et al., 2003; Hecht et al., 2012; Pulcini et al., 2013). For example, Pulcini et al. (2013), in a common-garden experiment, found a marked morphological variation in body shape traits such as body profile, head length, dorsal and anal fin length, and caudal peduncle size, using geometric morphometry among wild, semi-wild and domestic lines of this species. Domestic lines have a deeper body profile, with longer dorsal and anal fins and shorter and deeper caudal peduncles than wild lines. This differentiation, attributed to exposure of domestic lines to captive conditions, suggests that the variations may result from fixed genetic differences among lines due to the existence of QTL. Therefore, further QTL analysis in rainbow trout would be useful in clarifying the underlying genetics of this striking differentiation in body shape. To achieve this goal the use of SNPs it is possible given that these markers are considered to be the most desirable molecular markers for developing high-density genome scan to discover and locate target genes underlying the quantitative traits (Wang et al., 1998). This approach has been demonstrated to be efficient to discover several QTLs in guppy that control the complex patterns of skin pigmentation of males (Tripathi et al., 2009). However, it needs to use next-generation sequencing analysis to discover thousands of such SNPs (Miller et al., 2007; Davey et al., 2011) and in several cases developing SNP chips to perform genome-wide scans.

Skin Pigmentation

The cellular basis of skin pigmentation in fish is well known. Skin color depends on five types of pigment cells (or chromatophores) known as melanophores, xanthophores, erythrophores, iridophores, and leucophores, each producing a different color (black or brown, yellow or orange, red, iridescent, blue, silver or gold, and white, respectively; Fujii, 1969, 1993). The underlying genetics of skin pigmentation phenotype, however, has been explored mostly for qualitative traits by means of large-scale analyses of natural or induced color mutants, mainly in model fishes (see review by Colihueque, 2010). Evidence from studies in these and in other fish species indicates that the inheritance mode of qualitative traits for skin pigmentation has a simple genetic basis (Tave, 1986), i.e., a monogenetic control, which may be recessive, completely/incompletely dominant, and co-dominant or sex-linked. Moreover, these studies indicate that several genes participate in producing a specific skin color or color pattern that may be involved in chromatophore development, pigment synthesis, and pigment expression. For example, about 90 and 40 genes of this type have been identified in zebrafish and medaka (*Oryzias latipes*), respectively, that control specification, proliferation, survival, differentiation, and distribution of chromatophores, among other processes.

However, recent studies emphasize that skin pigmentation in fish can also possess a more complex genetic architecture, characterized by specific genome regions that harbor genes controlling quantitative traits (Table 3). For example, in the threespine sticklebacks (*Gasterosteus aculeatus*), two significant QTLs on linkage groups 1 and 6 that control the degree of barring and explain 26.6% of the variance of the trait were found (Greenwood et al., 2011). Given that these QTLs were associated with spatial variation in melanophore number (linkage group 6) and degree of melanization of melanophores (linkage group 1), this finding reveals the existence of different loci underlying variation in pigment patterns of this fish, which shows striking diversity of among freshwater (barred) and marine (unbarred) populations. Moreover, the number of dorsal and ventral melanophores is also controlled by different loci in this fish, since they were mapped to linkage group 7 and linkage group 1, respectively.

Through synteny analysis, Greenwood et al. (2011) identified the *Gja5* gene contained in the barring QTL on linkage group 6, which encodes a gap junction protein whose mutation disrupts the normal pigmentation pattern in zebrafish, in which spots form in place of the typical horizontal stripes, caused by an altered melanophore distribution (Watanabe et al., 2006). Moreover, the region on linkage group 1 that mapped QTLs associated with both barring and degree of melanization contains the *tyrosinase* gene, which encodes a key enzyme in melanin synthesis, whose mutation eliminates all pigmentation in zebrafish and medaka (Koga et al., 1995; Iida et al., 2004). Therefore, these results suggest that a few genes with large effects underlie the pigmentation pattern variation in the threespine sticklebacks.

In guppy males, a more complex control of pigmentation pattern has been observed (Tripathi et al., 2009), including a phenotype characterized by multi-colored areas with an ornamental function involved in female choice and in male mating success, and therefore, important for male fitness. In the genome of this fish, using interval mapping and the multiple-QTL model, 49

TABLE 3. Examples of QTLs for skin pigmentation traits mapped on different fish model genomes.

Table 3 | Examples of QTLs for skin pigmentation traits mapped on different fish model genomes.

Species	**Trait**	**Linkage group (LG) where QTLs were detected. Name of QTL (bold), linked markers, and respective phenotypic variation noted in parentheses**	**Reference**
Threespine sticklebacks (*Gasterosteus aculeatus*)	Degree of barring Degree of melanization Number of dorsal melanophores Number of ventral melanophores	LG-1 (**barring**, chrI:3310077, 6.6%) LG-1 (**melanization**, chrI:4816374, 11.7%) LG-6 (**barring**, chrVI:15780594, 20%) LG-1 (**ventral melanophores**, chrI:21909727, 8.9%) LG-7 (**dorsal melanophores**, chrVII:1728753, 11.6%)	Greenwood et al. (2011)
Guppy (*Poecilia reticulate*)	Dorsal fin black area Dorsal fin orange area Central blue white spot Anterior main black stripe Anterior orange spot Black spot by gonopodium Central orange spot Posterior main black stripe Posterior ventral black stripe Posterior orange spot Hind fin lower orange area	LG-1, LG-2, LG-4, LG-9, LG-12 (**Dorsal fin black area**, 22.9%) LG-1, LG-2, LG-4, LG-6, LG-7, LG-16 (**Dorsal fin orange area**, 26.8%) LG-2, LG-12, LG-17, LG-21, LG-23 (**Central blue white spot**, 26.5%) LG-8, LG-19 (**Anterior main black stripe**, 9.4%) LG-6, LG-7, LG-9, LG-12, LG-20 (**Anterior orange spot**, 23%) LG-01, LG-04, LG-07, LG-12, LG-16 (**Black spot by gonopodium**, 19.1%) LG-05, LG-08, LG-20, LG-22 (**Central orange spot**, 15.3%) LG-04, LG-10, LG-12, LG-13, LG-22, LG-23 (**Posterior main black stripe**, 21.6%) LG-01, LG-02, LG-13, LG-16, LG-18 (**Posterior ventral black stripe**, 17.7%) LG-08, LG-09, LG-15, LG-16, LG-18 (**Posterior orange spot**, 26.2%) LG-03, LG-04, LG-06, LG-12, LG-20 (**Hind fin lower orange area**, 23.3%)	Tripathi et al. (2009)

QTLs were detected at genome-wide level using permutation tests at a significance threshold value of $P < 0.05$ for G. aculeatus and P. reticulate.

QTLs for 11 areas of pigmentation traits were found (see Table 3), which explain 9.4 to 26.8% of phenotypic variation in these traits. In addition, these QTLs were mapped in 19 out of 24 linkage groups of this species, although mainly on linkage group 12 and 4. QTLs located on linkage group 12, which corresponds to the sex chromosome of the guppy, indicate that loci responsible for polymorphisms in guppy color patterns are clustered on this chromosome. This result coincides with previous knowledge regarding physical linkage of major color pattern loci to sex chromosomes in this species (Winge and Ditlevsen, 1947; Khoo et al., 1999). In summary, the results obtained in this fish strongly suggest that multiple QTLs with minor effects contribute to each color trait in guppy males.

In commercial fish, most color phenotypes of commercial value are qualitative traits known to be under Mendelian control, such as the Red Stirling strain of tilapia (*Oreochromis niloticus*; dominant inheritance, McAndrew et al., 1988), or the iridescent metallic blue variant of rainbow trout (recessive inheritance, Kincaid, 1975); therefore, given its simple inheritance mode, they could be more easily subjected to selective breeding for new stocks with particular colors.

However, there are some rainbow trout skin pigmentation phenotypes of commercial value, such as the Blue Back (Colihueque et al., 2011) and Finnish national breeding program (Kause et al., 2003) traits with complex pigmentation patterns comprising several attributes (skin color, number, size, and position of dark spots) that vary continuously at the intrapopulation level. A substantial quantitative genetic component for the different attributes that compose these traits has been reported (Kause et al., 2003; Díaz et al., 2011). As it has been seen in model fish, it is possible that these skin pigmentation traits may possess a complex genetic architecture, with the existence of a variable number of quantitative loci with a minor or major effect for the different trait attributes. Further analysis of these traits will clarify their particular genetic architecture.

Concluding Remarks

In farmed fish, several traits are taken into account in order to obtain a quality fish harvest suitable for marketing. These traits include body shape and skin pigmentation, both of which affect consumer acceptance of marketed fish at the point of sale. A fish with an improved appearance has greater consumer acceptance and, therefore, has a higher sale value than a fish with a normal appearance.

There has been some progress in this area with commercial fish, including traditional and new cultures, mainly through selective breeding or classical genetic analysis. This selection strategy has resulted in new fish stocks whose market participation is constantly increasing, contributing to the improved profitability of fish cultures. This trend is expected to continue over the next few years due to the sophistication of the market in many areas of the world. Therefore, there is interest in fish selection to ensure specimens that are visually appealing, for example, tilapia, rainbow trout, common carp, gilthead sea bream, and sea bass.

However, to meet this challenge, fish farmers must adapt and align their selective breeding goals with market demands. One tool that may be explored to achieve this objective derives from the discovery of QTLs or genes that underlie body shape and skin pigmentation, in which continuous variation of the different attributes that compose these traits is usually observed. This information could be used to implement selective breeding based on molecular markers tightly linked to QTLs that control various appearance traits of commercial interest, that is, marker-assisted selection.

This strategy may offer a more rapid response, yielding fish with a specific external appearance to satisfy market demands.

ACKNOWLEDGMENTS

The suggestions and constructive comments of all those who helped to improve the final version of this manuscript, are gratefully acknowledged. The publication fee of this work was supported by the Departamento de Ciencias Biológicas y Biodiversidad of the Universidad de Los Lagos.

REFERENCES

1. Alfnes, F., Guttormsen, A. G., Steine, G., and Kolstad, K. (2006). Consumers'willingness to pay for the color of salmon: a choice experiment with real economic incentives. *Am. J. Agric. Econ.* 88, 1050–1061. doi: 10.1111/j.1467-8276.2006.00915.x
2. Anderson, S. (2000). Salmon color and the consumer. *Paper Presented at: IIFET 2000 International Institute of Fisheries Economics and Trade. Proceedings of the International Institute of Fisheries Economics and Trade 2000*, Corvallis, Oregon.
3. Ankorion, Y., Moav, R., and Wohlfarth, G. W. (1992). Bidirectional mass selection for body shape in common Carp. *Genet. Sel. Evol.* 24, 43–52. doi: 10.1186/1297-9686-24-1-43
4. Araneda, C., Neira, R., Lam, N., and Iturra, P. (2008). "Salmonids," in *Genome Mapping and Genomics in Animals*, eds T. D. Kocher and C. Kole (Berlin: Springer-Verlag), 1–43. doi: 10.1007/978-3-540-73837-4_1
5. Balon, E. K. (1990). Epigenesis of an epigeneticist: the development of some alternative concepts on the early ontogeny and evolution of fishes. *Guelph Ichthyol. Rev.* 1, 1–42.
6. Balon, E. K. (2004). About the oldest domesticates among fishes. *J. Fish Biol.* 65(Suppl. A), 1–27. doi: 10.1111/j.0022-1112.2004.00563.x
7. Boulton, K., Massault, C., Houston, R. D., de Koning, D. J., Haley, C. S., Bovenhuis, H.,et al. (2011). QTL affecting morphometric traits and stress response in the Gilthead Seabream (*Sparus aurata*). *Aquaculture* 319, 58–66. doi: 10.1016/ j.aquaculture. 2011.06.044
8. Colihueque, N. (2010). Genetics of skin pigmentation: clues and prospects for improving the external appearance of farmed salmonids. *Rev. Fish Biol. Fisher.* 20, 71–86. doi: 10.1007/s11160-009-9121-6
9. Colihueque, N., Parraguez, M., Estay, F. J., and Díaz, N. F. (2011). Skin color characterization in rainbow trout by use of computer-based image analysis. *N. Am. J. Aquac.* 73, 249–258. doi: 10.1080/15222055.2011.581578
10. Davey, J. W., Hohenlohe, P. A., Etter, P. D., Boone, J. Q., Catchen, J. M., and Blaxter, M. L. (2011). Genome-wide genetic marker discovery and genotyping using next-generation sequencing. *Nat. Rev. Genet.* 12, 499–510. doi: 10.1038/nrg3012
11. Díaz, N. F., Estay, F., Araneda, C., and Colihueque, N. (2011). Heritability of Blue Back skin color and correlations with growth in rainbow trout, *Oncorhynchus mykiss*. *Paper Presented at: Australian Society for Fish Biology 2011 Conference. Proceeding of the Australian Society for Fish Biology 2011 Conference*, Townsville, Australia.

12. Einen, O., Waagan, B., and Thomassen, M. S. (1998). Starvation prior to slaughter in Atlanticsalmon (*Salmo salar*). I. Effects on weight loss, body shape, slaughterand fillet-yield, proximate and fatty acid composition. *Aquaculture* 166, 85–104. doi: 10.1016/S0044-8486(98)00279-8
13. Fujii, R. (1969). "Chromatophores and pigments," in *Fish Physiology*, Vol. III, eds W. S. Hoar and D. J. Randall (New York: Academic Press), 307–354.
14. Fujii, R. (1993). Cytophysiology of fish chromatophores. *Int. Rev. Cytol.* 143, 191–255. doi: 10.1016/S0074-7696(08)61876-8
15. Ghalambor, C. K., Walker, J. A., and Reznick, D. N. (2003). Multi-trait selection, adaptation, and constraints on the evolution of burst swimming performance. *Integr. Comp. Biol.* 43, 431–438. doi: 10.1093/icb/43.3.431
16. Gjerde, B., and Schaeffer, L. R. (1989). Body traits in rainbow trout. II. Estimates of heritabilities and of phenotypic and genetic correlations. *Aquaculture* 80, 25–44. doi: 10.1016/0044-8486(89)90271-8
17. Greenwood, A. K., Jones, F. C., Chan, Y. F., Brady, S. D., Absher, D. M., Grimwood, J.,et al. (2011). The genetic basis of divergent pigment patterns in juvenile threespine sticklebacks. *Heredity* 107, 155–166. doi: 10.1038/hdy.2011.1
18. Haffter, J., Mary, O., Lin, S., Farrell, M. J., Vogelsang, E., Haas, F.,et al. (1996). Mutations affecting pigmentation and shape of the adult Zebrafish. *Dev. Genes Evol.* 206, 260–276. doi: 10.1007/s004270050051
19. Hecht, B. C., Thrower, F. P., Hale, M. C., Miller, M. R., and Nichols, K. M. (2012). Genetic architecture of migration-related traits in rainbow and steelhead trout, *Oncorhynchus mykiss. G3 (Bethesda)* 2, 1113–1127. doi: 10.1534/g3.112.003137
20. Higgs, D. A., Dosanjh, B. S., Uin, L. M., Himick, B. A., and Eales, J. G. (1992). Effects of dietary lipid and carbohydrate levels and chronic 3,5,30-triiodo-L-thyronine treatment on growth, appetite, food and protein utilization and body composition of immature rainbow trout, *Oncorhynchus mykiss*, at low temperature. *Aquaculture* 105, 175–190. doi: 10.1016/0044-8486(92)90129-9
21. Hudon, J. (1994). Biotechnological applications of research on animal pigmentation. *Biotech. Adv.* 2, 49–69. doi: 10.1016/0734-9750(94)90290-9
22. Iida, A., Inagaki, H., Suzuki, M., Wakamatsu, Y., Hori, H., and Koga, A. (2004). The *tyrosinase* gene of the i^b albino mutant of the medaka fish carries a transposable element insertion in the promoter region. *Pigment Cell Res.* 17, 158–164. doi: 10.1046/j.1600-0749.2003.00122.x
23. Jenkins, T. M., Diehl, S., Kratzet, K. W., and Cooper, S. D. (1999). Effects of population density on individual growth of brown trout in streams. *Ecology* 80, 941–956. doi: 10.1890/0012-9658 (1999)080[0941:EOPDOI]2.0.CO;2
24. Kause, A., Ritola, O., Paananen, T., Eskelinen, U., and Mantysaari, E. (2003). Big and beautiful? Quantitative genetic parameters for appearance of large rainbow trout. *J. Fish Biol.* 62, 610–622. doi: 10.1046/j.1095-8649.2003.00051.x
25. Khoo, L., Lim, T. M., Chan, W. K., and Phang, V. P. E. (1999). Linkage analysis and mapping of three sex-linked color pattern genes in the guppy, *Poecilia reticulate. Zool. Sci.* 16, 893–903. doi: 10.2108/zsj.16.893
26. Kincaid, H. L. (1975). Iridescent metallic blue color variant in rainbow trout. *J. Hered.* 66, 100–102.
27. Koga, A., Inagaki, H., Bessho, Y., and Hori, H. (1995). Insertion of a novel transposable element in the *tyrosinase* gene is responsible for an albino mutation in the medaka fish, *Oryzias latipes. Mol. Gen. Genet.* 249, 400–405. doi: 10.1007/BF00287101
28. Loukovitis, D., Batargias, C., Sarropoulou, E., Apostolidis, A. P., Kotoulas, G., Magoulas, A.,et al. (2013). Quantitative trait loci affecting morphology traits in gilthead seabream (*Sparus aurata* L.). *Anim. Genet.* 44, 480–483. doi: 10.1111/age.12027
29. Loukovitis, D., Sarropoulou, E., Tsigenopoulos, C. S., Batargias, C., Magoulas, A., Apostolidis, A. P.,et al. (2011). Quantitative trait loci involved in sex determination

and body growth in the gilthead seabream (*Sparus aurata* L.) through targeted genome scan. *PLoS ONE* 6:e16599. doi: 10.1371/journal.pone.0016599

30. Martyniuk, C. J., Perry, G. M. L., Moghadam, H. K., Ferguson, M. M., and Danzmann, R. G. (2003). The genetic architecture of correlations among growth-related traits and male age at maturation in rainbow trout (*Oncorhynchus mykiss*). *J. Fish Biol.* 63, 746–764. doi: 10.1046/j.1095-8649.2003.00188.x
31. Massault, C., Hellemans, B., Louro, B., Batargias, C., Van Houdt, J. K. J., Canario, A.,et al. (2009). QTL for body weight, morphometric traits and stress response in European Sea Bass *Dicentrarchus labrax*. *Anim. Genet.* 41, 337–345. doi: 10.1111/j.1365-2052.2009.02010.x
32. McAndrew, B. J., Roubal, F. R., Roberts, R. J., Bullock, A. M., and McEwen, I. M. (1988). The genetics and history of red, blond and associated color variants in *Oreochromis niloticus*. *Genetica* 76, 127–137. doi: 10.1007/BF00058811
33. Miller, M. R., Dunham, J. P., Amores, A., Cresko, W. A., and Johnson, E. A. (2007). Rapid and cost-effective polymorphism identification and genotyping using restriction site associated DNA (RAD) markers. *Genome Res.* 17, 240–248. doi: 10.1101/gr.5681207
34. Moreira, A. A., Marques Moreira, H. L., and Silva Hilsdorf, A. W. (2005). Comparative growth performance of two Nile tilapia (Chitralada and Red-Stirling), their crosses and the Israeli tetra hybrid ND-56. *Aquac. Res.* 36, 1049–1055. doi: 10.1111/j.1365-2109.2005.01318.x
35. Pakkasmaa, S., and Piironen, J. (2001). Water velocity shapes juvenile salmonids. *Evol. Ecol.* 14, 721–730. doi: 10.1023/A:1011691810801
36. Pillay, T. V. R., and Kutty, M. N. (2005). *Aquaculture Principles and Practices*. Oxford: Blackwell Publishing Ltd.
37. Pulcini, D., Wheeler, P. A., Cataudella, S., Russo, T., and Thorgaard, G. H. (2013). Domestication shapes morphology in rainbow trout *Oncorhynchus mykiss*. *J. Fish Biol.* 82, 390–407. doi: 10.1111/jfb.12002
38. Reid, D. P., Szanto, A., Glebe, B., Danzmann, R. G., and Ferguson, M. M. (2005). QTL for body weight and condition factor in Atlantic salmon (*Salmo salar*): comparative analysis with rainbow trout (*Oncorhynchus mykiss*) and Arctic charr (*Salvelinus alpinus*). *Heredity* 94, 166–172. doi: 10.1038/sj.hdy.6800590
39. Reid, D. T., and Peichel, C. L. (2010). Perspectives on the genetic architecture of divergence in body shape in sticklebacks. *Integr. Comp. Biol.* 50, 1057–1066. doi: 10.1093/icb/icq030
40. Reznick, D. N., and Ghalambor, C. K. (2001). The population ecology of contemporary adaptations: what empirical studies reveal about the conditions that promote adaptive evolution. *Genetica* 112–113, 183–198. doi: 10.1023/A:1013352109042
41. Swain, D. P., Riddell, B. E., and Murray, C. B. (1991). Morphological differences between hatchery and wild populations of coho salmon (*Oncorhynchus kisutch*): environmental versus genetic origin. *Can. J. Fish. Aquat. Sci.* 48, 1783–1791. doi: 10.1139/f91-210
42. Tave, D. (1986). *Genetics for Hatchery Managers*. Westport, CT: AVI Publishing Company Inc.
43. Tripathi, N., Hoffmann, M., Willing, E. M., Lanz, C., Weigel, D., and Dreyer, C. (2009). Genetic linkage map of the guppy, *Poecilia reticulata*, and quantitative trait loci analysis of male size and colour variation. *Proc. Biol. Sci.* 276, 2195–2208. doi: 10.1098/rspb.2008.1930
44. Walker, J. A. (1997). Ecological morphology of lacustrine threespine stickleback *Gasterosteus aculeatus* L. (Gasterosteidae) body shape. *Biol. J. Linn. Soc.* 61, 3–50. doi: 10.1006/bijl.1996.9999
45. Walker, J. A. (2010). An integrative model of evolutionary covariance: a symposium on body shape in fishes. *Integr. Comp. Biol.* 50, 1051–1056. doi: 10.1093/icb/icq014
46. Wang, D. G., Fan, J. B., Siao, C. J., Berno, A., Young, P., Sapolsky, R.,et al. (1998). Large-scale identification, mapping, and genotyping of single-nucleotide polymorphisms in the human genome. *Science* 280, 1077–1082. doi: 10.1126/science.280.5366.1077

47. Ward, A. B., and Mehta, R. S. (2010). Axial elongation in fishes: using morphological approaches to elucidate developmental mechanisms in studying body shape. *Integr. Comp. Biol.* 50, 1106–1119. doi: 10.1093/icb/icq029
48. Watanabe, M., Iwashita, M., Ishii, M., Kurachi, Y., Kawakami, A., Kondo, S.,et al. (2006). Spot pattern of *leopard Danio* is caused by mutation in the zebrafish *connexin41.8 gene. EMBO Rep.* 7, 893–897. doi: 10.1038/sj.embor.7400757
49. Webb, P. W. (1984). Body form, locomotion and foraging in aquatic vertebrates. *Am. Zool.* 24, 107–120.
50. Winge, O., and Ditlevsen, E. (1947). Colour inheritance and sex determination in Lebistes. *Heredity* 1, 65–83. doi: 10.1038/hdy.1947.4
51. Zhang, Y., Wang, S., Li, J., Zhang, X., Jiang, L., Xu, P.,et al. (2013). Primary genome scan for complex body shape-related traits in the common carp *Cyprinus Carpio. J. Fish Biol.* 82, 125–140. doi: 10.1111/j.1095-8649.2012.03469.x

CHAPTER 8

Genetic Variation and Breeding Signature in Mass Selection Lines of the Pacific Oyster (*Crassostrea gigas*) Assessed by SNP Markers

ABSTRACT

In breeding industries, a challenging problem is how to keep genetic diversity over generations. To investigate genetic variation and identify breeding signatures in mass selected lines of Pacific oyster (*Crassostrea gigas*), three sixth-generation selected lines and four wild populations were assessed using 103 single nucleotide polymorphism (SNP) markers. The genetic diversity data indicated that the selected lines exhibited a significant reduction in the observed heterozygosity and observed number of alleles per locus compared with the wild populations ($P \leq 0.05$), indicating the selected lines tended to lose genetic diversity contrasted with the wild populations. The unweighted pair-group method with arithmetic mean (UPGMA) analysis showed that the wild populations and selected lines were not separated into two groups. Using four outlier tests, a total of 17 loci were found under selection at two levels. The global outlier detection suggested that 4 common outlier loci were subject to selection using both the hierarchical island model and Bayesian likelihood approaches. At regional level, 3 SNPs were detected as outlier using at least two outlier tests and one outlier SNP (CgSNP309) was overlapped in the two wild-selected population comparisons. The candidate outlier SNPs provide valuable resources for future association studies in *C. gigas*.

FUNDING: This study was supported by grants from the National High Technology Research and Development Program (2012AA10A405-6), National Natural Science Foundation of China (31372524), and Seed Improvement Project of Shandong Province.

INTRODUCTION

The Pacific oyster (*Crassotrea gigas*), naturally distributed around Japan, China and Korea, is now an important cultivated oyster species worldwide [1]. Many countries has started introducing *C. gigas* since 1940s, mainly because of its rapid growth rate, high disease resistance and strong environmental adaptability [2]. The *C. gigas* is one of the most popular oyster species in China, and its main places of production are Shandong and Liaoning provinces [3, 4]. In 2012, China produced 3.94 million tons of oysters with *C. gigas* as one of the most dominant species [4, 5]. However, the broodstock used today remains largely unselected and thus *C. gigas* has gained little from heredity improvement by selective breeding [6].

Several selective breeding projects have been launched in couple of countries and obtained encouraging results [7–9]. With the aim of improving the productivity traits of *C. gigas*, a breeding program selected for fast growth has also been initiated in China. The first generation of selection was carried out by mass selection on three cultured stocks from China, Japan and Korea in 2007 and the average improvement (increase in shell height) was 10% [4]. An increase in growth rate is detected over the successive six-generation selection, but little is known about the effects of strong directional selection on genetic level in the process of cultivation. The main focus of selected lines is on inbreeding and the decrease of genetic diversity [10, 11]. Yu and Guo [11] detected rare alleles of the 4 selected strains decreased significantly compared with a wild population in Eastern oyster (*Crassostrea virginica*). Moreover, Cruz et al. [12] found selected strains through breeding programs tended to lose genetic diversity compared with wild populations in Pacific white shrimp (*Litopenaeus vannamei*). The reduction of genetic diversity in a population may possibly reduce disease resistance and decrease adaptability to environmental changes [13, 14]. Therefore, it is crucial to survey the genetic diversity within or between selected lines and wild populations for successful hatchery management of *C. gigas*.

In recent years, the single nucleotide polymorphisms (SNPs) and simple sequence repeats (SSRs) developed from EST sequences have been extensively used in population genetic studies. Microsatellites have been chosen in many studies for their codominant inheritance and high allelic variability [15]. However, SNP genotyping results can be compared across different platforms and laboratories more easily than microsatellite data, therefore facilitating the integration and interpretation of genotyping data across different databases. Moreover, mutations resulting in some SNPs can be responsible for an adaptive phenotype or the direct target of selection and have been shown to be correlated with economic traits in many aquatic animals [16–18]. Therefore, SNP markers offer a valuable chance to explore the genetic basis of phenotypic variation during the selective breeding program in China.

In the present study, we used 103 SNP markers to evaluate the genetic diversity level, assess the genetic differentiation, and identify putative loci under selection in three sixth-generation selected lines and four wild populations of *C. gigas*. The main goal is to demonstrate the effects of cultivation on genetic diversity in the selective breeding program of *C. gigas*, and to identify candidate SNPs under selection through the use of outlier analysis methods.

MATERIALS AND METHODS

Ethics Statement

The *C. gigas* does not belong to endangered species and the sample locations are not in protection.

Oyster Collections and DNA Extraction

Three selected lines and four wild populations of *C. gigas* were surveyed in our study. Collection details for the 7 *C. gigas* populations were shown in Fig 1 and Table 1. The cultivated oysters were randomly selected from three sixth-generation selected lines. In 2007, Pacific oysters from three cultivated populations in Pusan, South Korea (Stock K, 35.1°N, 129.1°E), Onagawa Bay, Miyagi Prefecture, Japan (Stock J, 38.3°N, 141.3°E), and Rushan, Shandong, China (Stock C, 36.4°N, 121.3°E) were used to establish selected lines for fast growth [4]. In each line, 30–60 pairs of these individuals from the top end of the shell height distribution were used as broodstock oysters for the next generation [19]. In July 2012, 91 (50 ♀×41 ♂) and 107 (59 ♀×48 ♂) individuals from the fifth-generation selected lines of Japan and Korea (JS5 and KS5) were selected as parental oysters for the sixth-generation selected lines of Japan and Korea (JS6 and KS6), respectively. In July 2013, 85 (50 ♀×35 ♂) individuals in the fifth-generation selected lines of China (CS5) were selected as parental oysters for the sixth-generation selected lines of China (CS6). Fertilization was conducted at a ratio of 50 sperm per egg with 10^7 oocytes for each female [4, 19]. Oysters were cultivated in Weihai Bay, Shandong, China (37.3°N, 122.1°E). One-year-old oysters of the three sixth-generation selected lines (CS6, 48 individuals; JS6, 56 individuals and KS6, 48 individuals) were randomly selected for the analysis. The wild oysters were sampled from four locations: Rushan, Shandong Province, China (RS; 36.4°N, 121.3°E; 48 individuals; shell length, 3.45±0.48 cm, shell height, 5.63±0.71 cm, shell width, 1.87±0.35 cm); Dongying, Shandong Province, China (DY; 37.6°N, 119.0°E; 48 individuals; shell length, 3.93±0.68 cm, shell height, 4.96±0.90cm, shell width, 1.67±0.39 cm); Miyagi Prefecture, Japan (MG; 38.3°N, 141.3°E; 40 individuals; shell length, 4.22±0.78 cm, shell height, 7.32±0.91 cm, shell width, 2.01±0.37 cm) and Inchon, Korea (RC; 37.4°N, 126.6°E; 48 individuals; shell length, 3.62±0.78 cm, shell height, 5.73±0.98 cm, shell width, 1.63±0.39 cm). The adductor muscle was sampled and kept at −80°C. DNA was extracted using the phenol–chloroform method with a modification [20].

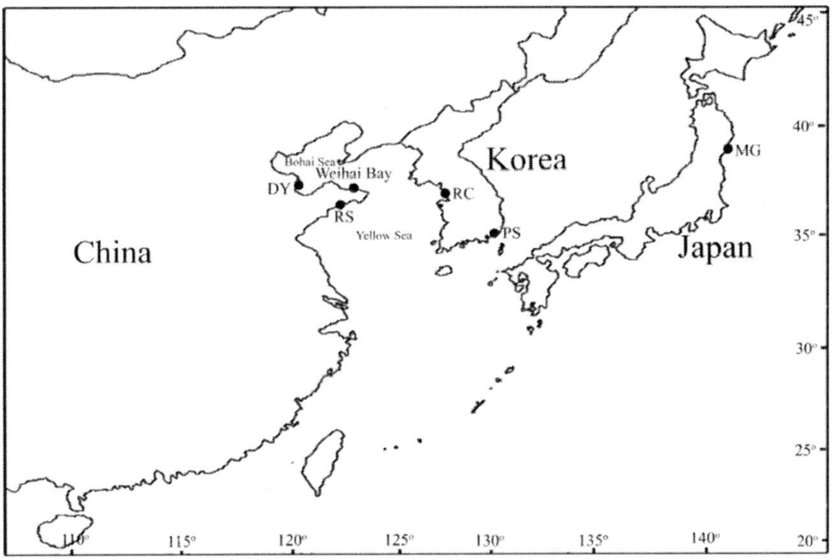

Figure 1. Approximate location of sampling sites shown with shaded circles. Populations are marked by abbreviations that correspond to Table 1.

Table 1. Sample information for the Pacific oyster *Crassostrea gigas*.

Source	Abbreviation	Sample type (size)	Geographical coordinates
Rushan, Shandong, China	Stock C, RS	Base stock, wild population (48)	36.4°N, 121.3°E
Onagawa Bay, Miyagi Prefecture, Japan	Stock J, MG	Base stock, wild population (40)	38.3°N, 141.3°E
Pusan, South Korea	Stock K/PS	Base stock	35.1°N, 129.1°E
Dongying, Shandong, China	DY	Wild population (48)	37.6°N, 119.0°E
Inchon, Korea	RC	Wild population (48)	37.4°N, 126.6°E
Weihai Bay, Shandong, China	CS6, JS6, KS6	Selected lines (48, 56, 48)	37.3°N, 122.1°E

SNP Genotyping

A total of 103 SNP markers (from CgSNP4 to CgSNP457) were used in the study [21, 22]. SNPs were genotyped using the high resolution melting (HRM) method on the LightCycler 480 real-time PCR machine (Roche Diagnostics, Burgess Hill, UK) according to the procedure described by Zhong et al. [21]. The 10-μl mixture included 0.25 U Taq DNA polymerase (Takara, Dalian, China), 10 × PCR buffer, 0.2 mM dNTP mix, 0.2 μM of each primer set, 1.5 mM $MgCl_2$, 5 μM SYTO9 (Invitrogen Foster City, CA, USA) and 10 ng DNA [21]. The PCR cycling conditions were as follows: an activation step at 95°C, 5 min; 45–50 cycles of 95°C, 20 s; a touch down of 68°C to 58°C, 20 s (0.5°C /cycle) and extension for 72°C, 20 s. Before melting, the amplicons were heated at 95°C for 60s, and then cooled at 40°C for 60s. Melting curves were produced by heating amplicons from 60°C to 90°C [21].

Data Analysis

The observed number of alleles (*Na*), effect number of alleles (*Ne*), shannon's information index (*I*), observed heterozygosity (*Ho*), expected heterozygosity (*He*), minor allele frequency (*MAF*), inbreeding coefficients (*Fis*) and chi-square analysis of Hardy–Weinberg equilibrium (*HWE*) were evaluated by POPGENE 1.32 [23]. All comparable statistics (*Na, Ne, I, Ho* and *He*) were calculated between wild populations and selected lines using Mann-Whitney U-Test [24] implemented with SPSS 16.0.

Pairwise *Fst* values were calculated with Arlequin 3.5.1.3 [25]. *Fst* values were calculated for significance using 10000 permutations. Gene flow among wild populations was calculated following the formula [26]:

$$Nm = \frac{(1 - Fst)}{4Fst}.$$

The Nei's genetic distance [27] was also estimated by POPGENE 1.32. The unweighted-pair-group method with arithmetic-mean (UPGMA) tree was constructed by POPTREE2 [28]. Bootstrap analyses were conducted with 1000 replicates.

The population genetic structure analysis was estimated by Structure 2.3.4 with a model-based Bayesian procedure [29, 30]. The analysis was conducted using a burn-in period of 10^5iterations and a run length of 10^6 MCMC replications. The number of clusters (K) was set from 1 to 10 with three replicates. The most probable K value was evaluated by STRUCTURE HARVESTER [31, 32].

To explore candidate SNPs under selection, *Fst*-based outlier tests were conducted using two island models including a finite island model (FDIST approach) and a hierarchical island model implemented in Arlequin 3.5.1.3 [25, 33], and Bayesian likelihood approach implemented in BayeScan [34]. To detect additional evidence of selection, lnRH tests were also conducted in population pairwise comparisons.

The outlier analysis was implemented at two levels. (1) An overall analysis encompassed all populations under the hierarchical island models. Three groups were set (CS6+RS+DY+RC vs. JS6+MG vs. KS6) based on the Structure result. The analysis was conducted using 50000 coalescent simulations with 10 groups of 100 demes. (2) An analysis was conducted between selected lines and wild populations from which the selected lines originated (CS6 and RS; JS6 and MG) under a finite island-model. We conducted 50000 coalescent simulations with 100 demes to generate the joint distribution of *Fst* versus heterozygosity. Loci which were outside the 99% confidence intervals were treated as outliers potentially under selection.

The outlier tests at the two levels were also calculated with the BayeScan software. The analyses were estimated using a pilot run of 20, a burn-in of 50000 with 100000 iterations each, a sample size of 5000, a thinning interval of 10, and an FDR of 0.05. Loci with $\log_{10}$ values of the posterior odds (PO) >0.5 and 2.0 were regarded as candidate SNPs under selection with substantial and decisive evidence [35].

The lnRH test was applied to estimate the ratio of gene diversity (heterozygosity) of all loci in the population pairwise comparisons as

$$\ln RH = \ln\left(\frac{\left(\frac{1}{1-H_{pop1}}\right)^2 - 1}{\left(\frac{1}{1-H_{pop2}}\right)^2 - 1}\right)$$

follows:

where H_{pop1} and H_{pop2} denote expected heterozygosity for population 1 and population 2 [36]. The lnRH values were standardized for each population comparisons and therefore the standardized distributions had a mean of zero and a standard deviation of one [37]. Those SNPs fell out of the 95% interval were taken as candidate loci under selection.

Sequence annotation was conducted by BLASTx software in NCBI database (http://www.ncbi.nlm.nih.gov/) and OysterBase (http://www.oysterdb.com), and the critical *E*value was set as 1.0×10^{-6}. The NCBI ORF finder (http://www.ncbi.nlm.nih.gov/gorf/gorf.html) was used to distinguish synonymous SNPs, non-synonymous SNPs or SNPs from untranslated regions (UTRs). The putative function of genes was identified by using the Gene Ontology (GO) annotation by mining the Swiss-Prot database (http://us.expasy.org/sprot/sprot-top.html) and OysterBase.

RESULTS

Genetic Diversity

Genetic variability indices for the four wild populations and three selected lines of *C. gigas* were shown in Table 2. For all the genetic diversity parameters, the selected lines had lower values (mean *Na*, *Ne*, *I*, *Ho* and *He*) than the wild populations, but no significant loss of *Ne*, *I* or *He*were detected ($P > 0.05$). The observed heterozygosities ranged from 0.2703 to 0.2939 with a mean of 0.2806 in the wild populations, and ranged from 0.2479 to 0.2733 with a mean of 0.2599 in the selected lines. There is a significant reduction of observed heterozygosities in the selected lines in contrast with the wild populations ($P = 0.05$). The observed number of alleles per locus varied from 1.9417 to 1.9709 in the selected lines, and varied from 1.9806 to 2.000 in the wild populations. There was a significant reduction of the observed number of alleles per locus in the selected lines compared with the wild populations ($P < 0.05$).

Table 2. Genetic diversity parameters of selected lines and wild populations.

Population	*Na*	*Ne*	*I*	*Ho*	*He*
CS6	1.9709±0.1690	1.5031±0.3206	0.4599±0.2034	0.2733±0.1766	0.3053±0.1585
JS6	1.9515±0.2160	1.4780±0.3472	0.4332±0.2247	0.2586±0.1846	0.2867±0.1729
KS6	1.9417±0.2354	1.4539±0.3464	0.4146±0.2346	0.2479±0.1969	0.2743±0.1772
RS	1.9903±0.0985	1.5138±0.3033	0.4774±0.1748	0.2939±0.1613	0.3154±0.1416
MG	2.000±0.0000	1.5362±0.3025	0.4892±0.1742	0.2801±0.1576	0.3261±0.1426
RC	1.9806±0.1387	1.5052±0.3206	0.4639±0.1945	0.2703±0.1637	0.3070±0.1548
DY	1.9806±0.1387	1.5012±0.3032	0.4664±0.1879	0.2780±0.1601	0.3083±0.1481
Selected	1.9547±0.2068	1.4783±0.3381	0.4359±0.2209	0.2599±0.1860	0.2888±0.1695
Wild	1.9879±0.0940	1.5141±0.3074	0.4742±0.1829	0.2806±0.1607	0.3142±0.1468
Total	1.9713±0.1504	1.4962±0.3228	0.4551±0.2019	0.2703±0.1734	0.3015±0.1582

Note: *Na*, observed number of alleles; *Ne*, effective number of alleles; *I*, shannon's information index; *Ho*, observed heterozygosity; *He*, expected heterozygosity.

Information of the 103 SNPs evaluated from the three selected lines and four wild populations were summarized in S1 Table. All loci had two alleles in the 7 populations except that 19 loci had only one allele in one or more populations, including 3 loci (CgSNP252, CgSNP265 and CgSNP283) in CS6, 5 loci (CgSNP149, CgSNP176, CgSNP254, CgSNP265 and CgSNP305) in JS6, 6 loci (CgSNP14, CgSNP140, CgSNP176, CgSNP187, CgSNP252 and CgSNP319) in KS6, 1 locus (CgSNP283) in RS, 2 loci (CgSNP254 and CgSNP309) in RC and 2 loci (CgSNP176 and CgSNP283) in DY. A total of 83 of the 721 single-locus exact tests showed significant departures from *HWE* after sequential Bonferroni correction ($P < 0.05/103$), and all loci showed heterozygote deficiencies except for CgSNP261 in CS6 and JS6, CgSNP36 in KS6 and MG, and CgSNP402 in RS.

Genetic Differentiation

Most pairwise *Fst* values were significant ($P<0.001$), except for that estimated between wild populations RS and RC. The lowest value was observed between RS and RC (*Fst* = 0.00689), however the highest value was detected between JS6 and KS6 (*Fst* = 0.11737) (Table 3). Within the selected lines, the values varied from 0.06941 (KS6 and CS6) to 0.11737 (JS6 and KS6). Within the wild populations, the values ranged from 0.00689 (RS and RC) to 0.07046 (RC and MG). Moreover, gene flow among the wild populations ranged from 3.29811 (MG and RC) to 36.03447 (RS and RC).

Table 3. Pairwise *Fst* values (lower diagonal) and Nei's genetic distance (upper diagonal) among seven populations.

	CS6	JS6	KS6	RS	DY	MG	RC
CS6	0	0.0407	0.0349	0.0190	0.0323	0.0451	0.0193
JS6	0.07944	0	0.0571	0.0407	0.0357	0.0226	0.0443
KS6	0.06941	0.11737	0	0.0382	0.0542	0.0552	0.0346
RS	0.03067	0.07742	0.07460	0	0.0160	0.0413	0.0083
DY	0.05764	0.06847	0.10685	0.02380	0	0.0354	0.0191
MG	0.07851	0.04069	0.10513	0.06930	0.06006	0	0.0408
RC	0.03149	0.08571	0.06802	0.00689	0.03035	0.07046	0

The pairwise Nei's genetic distances among populations were also shown in Table 3. The lowest value was observed between RS and RC (0.0083), while the highest value was detected between KS6 and JS6 (0.0571). The UPGMA tree was constructed based on pairwise genetic distance (Fig 2). The UPGMA tree separated the 7 populations into two clusters. The KS6 originated from Korea belonged to one cluster while the other cluster included the other 6 *C. gigas* populations. Moreover, this cluster was further divided into 2 subgroups; the first subgroup contained JS6 and MG originated from Japan, and the second subgroup contained CS6, RS and DY originated from China, and RC from Korea.

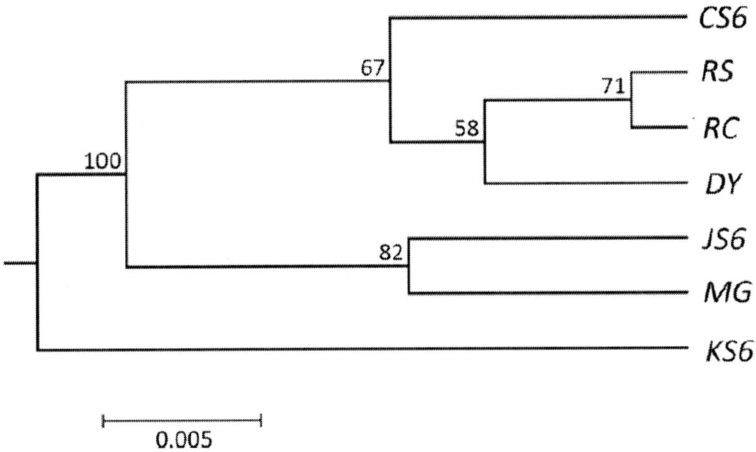

Figure 2. Phylogenetic tree of four wild populations (RS, DY, RC and MG) and three selected lines (CS6, KS6 and JS6) using the unweighted pair-group method with arithmetic mean (UPGMA) based on Nei's genetic distance derived from 103 SNPs.

Numbers above branches indicate bootstrap values.

Genetic Structure

The genetic structure test indicated that the most likely number of genetic groups was K = 3. The first group consisted of CS6, RS and DY originated from China, and RC from Korea. The second group consisted of JS6 and MG originated from Japan. Moreover, the selected line KS6 originated from Korea constituted the third genetic group (S1 Fig).

Outlier SNPs

Both the hierarchical island model and Bayesian likelihood approach revealed 4 loci (CgSNP140, CgSNP158, CgSNP176 and CgSNP209) under selection across all populations (Table 4; Fig 3). To detect specific footprint of artificial selection at regional level, we conducted pairwise tests between selected lines and wild populations from which the selected lines originated. The Fdist method revealed 5 candidate SNPs under selection, including 3 loci (CgSNP40, CgSNP225 and CgSNP236) for populations originated from China (CS6+RS) and 2 loci (CgSNP176 and CgSNP197) for populations originated from Japan (JS6+MG) (Table 4). The same two SNPs (CgSNP 225 and CgSNP 176) were also detected by BayeScan. The lnRH test deteced 12 outlier SNPs, including 5 SNPs (CgSNP23, CgSNP82, CgSNP83, CgSNP225 and CgSNP236) for CS6-RS comparison, 6 SNPs (CgSNP33, CgSNP140, CgSNP194, CgSNP283, CgSNP375 and CgSNP397) for JS6-MG comparison and one common SNP (CgSNP 309) for the two population comparisons. Totally, 17 candidate outlier SNPs were found under selection using Arlequin, BayeScan and lnRH tests. Among the 17 SNPs, 16 were detected in the coding region, and 1 in the UTR. Among the 16 SNPs located in the coding region, 11 SNPs were synonymous and 5 nonsynonymous. Nine SNPs could be annotated by BLASTx software. The CGI_10015432, CGI_10017873, NADH dehydrogenase [ubiquinone] iron-sulfur protein 2 and Collagen alpha-5(VI) chain were involved in G-protein coupled receptor protein signaling pathway, lipid transport, oxidation-reduction process and lipoprotein metabolic process, however, the putative function of other five proteins (CGI_10028477, flotillin 2, CGI_10002462, UPF0451 protein C17 or f61-like protein and CGI_10010736) could not be identified using GO searches.

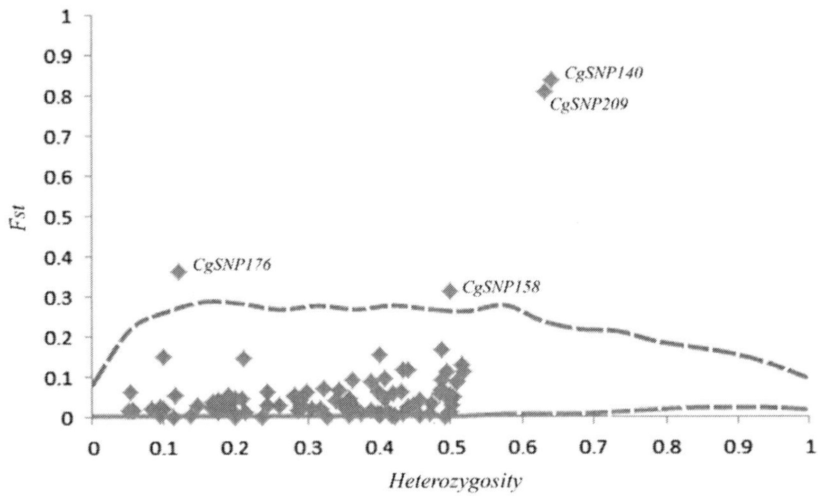

(a)

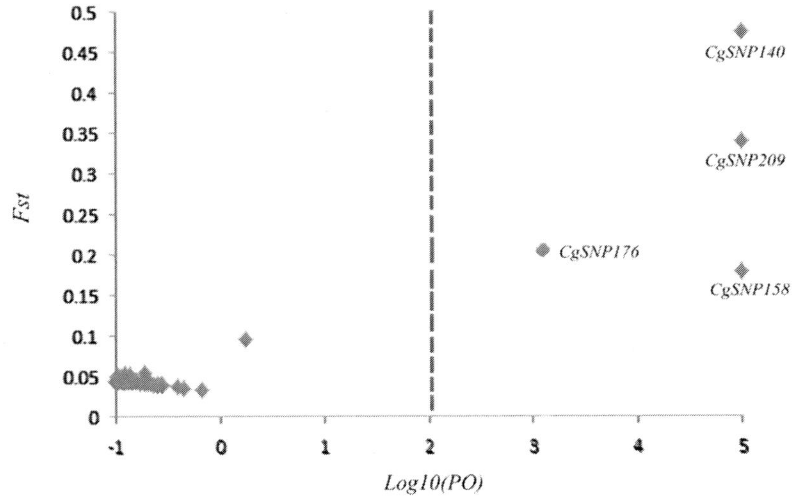

(b)

Figure 3. Results of two outlier tests in all populations.

Table 4. Outlier SNPs detected using Arlequin, BayeScan and lnRH tests.

	Arlequin				BayeScan				LnRH test		
	Locus	*Ho*	*Fst*	*P*	$Log_{10}(PO)$	*Fst*	*Probi*	*Alpha*	LnRH	Type	Annotation
All populations	CgSNP140	0.640	0.837	0.000	1000	0.476	1.000	3.268		S (Ala)	CGI_10015432
	CgSNP158	0.500	0.311	0.002	1000	0.179	1.000	1.566		UTR	Unknown
	CgSNP176	0.121	0.363	0.002	3.097	0.205	0.999	1.732		N (Cys/Tyr)	CGI_10028477
	CgSNP209	0.631	0.809	0.000	1000	0.341	1.000	2.588		S (Glu)	Unknown
CS6+RS	CgSNP23	-	-	-	-	-	-	-	-2.258	N (The/Ala)	Unknown
	CgSNP40	0.570	0.239	0.008	-	-	-	-	-	S (Arg)	Unknown
	CgSNP82	-	-	-	-	-	-	-	2.265	S (Ser)	Unknown
	CgSNP83	-	-	-	-	-	-	-	2.265	S (Met)	Unknown
	CgSNP225	0.328	0.266	0.001	0.628	0.110	0.809	1.124	-3.628	S (Leu)	Hypothetical protein flotillin 2
	CgSNP236	0.379	0.250	0.003	-	-	-	-	2.947	N (Pro-Thr)	Hypothetical protein CGI_10002462
	CgSNP309	-	-	-	-	-	-	-	-2.084	S (Asn)	Collagen alpha-5(VI) chain
JS6+MG	CgSNP33	-	-	-	-	-	-	-	-3.137	N (Gln/Arg)	Unknown
	CgSNP140	-	-	-	-	-	-	-	-2.092	S (Ala)	CGI_10015432
	CgSNP176	0.415	0.497	0.000	3.398	0.241	1.000	1.995	Not fit	N (Cys/Tyr)	CGI_10028477
	CgSNP194	-	-	-	-	-	-	-	2.150	S (Lys)	Unknown
	CgSNP197	0.438	0.230	0.007	-	-	-	-	-	N (Glu/Lys)	CGI_10017873
	CgSNP283	-	-	-	-	-	-	-	-2.249	S (Ser)	NADH dehydrogenase [ubiquinone] iron-sulfur protein 2
	CgSNP309	-	-	-	-	-	-	-	-2.370	S (Asn)	Collagen alpha-5(VI) chain
	CgSNP375	-	-	-	-	-	-	-	-2.928	S (His)	UPF0451 protein C17or f61-like protein
	CgSNP397	-	-	-	-	-	-	-	-2.092	S (Ser)	Hypothetical protein CGI_10010736

Note: *Ho*, observed heterozygosity; *PO*, posterior odds; *Alpha*, positive values indicate positive selection, while negative values indicate putative balancing selection; *Probi*, probability; *Not fit*, CgSNP176 was monomorphic in JS6, thus not fit for the lnRH test; *S*, synonymous; *N*, non-synonymous; *UTR*, untranslated region.

Locus names of putative outliers potentially affected by selection are indicated. (a) Hierarchical island model: empirical distribution of *Fst* against heterozygosity. The upper and lower lines are the 99% confidence intervals. (b) BayeScan: *Fst* estimates plotted against $\log_{10}$ of the posterior odds (PO). The dashed lines correspond to the posterior odds 100 ($\log_{10}(PO) = 2$).

DISCUSSION

Genetic Diversity

Maintenance of genetic variation is known to be important for long-term survival of populations because the level of variation determines their adaptability to environmental changes [38, 39]. Heterozygosity in many instances is considered as the primary parameter to reflect overall genetic variability of populations [40, 41]. In our study, the selected lines showed a significant reduction in the observed heterozygosity contrasted with the wild populations. Similar results have been detected in cultivated Atlantic salmon (*Salmo salar*) and Pacific abalone (*Haliotis discus hannai*) populations [42, 43].

Many researches have indicated loss of alleles is often more easily observed than decrease of heterozygosity in cultivated or selected populations [10, 11, 44]. Yu and Guo [11] detected the decrease in allele number was significant for selected lines NEH (by 36.8%) and CTS (by 35.1%) contrasted with the wild population of *C. virginica*. Wang et al. [45] found the cultured populations lost 9 of the 45 alleles (20%) compared with the wild population of bay scallop (*Argopecten irradians*). Compared with wild populations, 20% to 48% fewer alleles were detected in farmed salmon strains from Ireland and Norway [46]. In our study, the selected lines exhibited a significant reduction in observed number of alleles in contrast with the wild populations ($P<0.05$). Moreover, 0.98% (2/205) and 2.43% (5/206) fewer alleles were found in selected lines CS6 and JS6 than their wild progenitors (RS and MG), respectively. The CS6 stock seemed to be missing two rare alleles at 2 loci (CgSNP252 and CgSNP265) compared with RS. No individuals with the rare allele of the CgSNP265 found rarely in RS, were sampled from CS6. There was a possibility that this rare allele existed in CS6, but was not sampled in the population. As this rare allele frequency was 0.0521 in RS, the probability of completely missing this rare allele in the CS6 (48 samples) should be estimated as $0.9479^{96} = 0.0059$. Therefore, it is most likely that the CS6 has indeed lost this rare allele at the CgSNP265. Similarly, the JS6 may have indeed lost rare alleles at 4 loci (CgSNP149, CgSNP176, CgSNP254 and CgSNP305). Taken together, we can make a conclusion that the selected lines through breeding programs tended to lose genetic diversity compared with the wild populations. The decrease of genetic diversity may result from genetic drift, bottlenecks, and inbreeding caused by a reduced effective population size.

Significant deviations from *HWE* were detected both in selected lines and wild populations. In selected lines, heterozygote deficiency may be mainly because of a limited number of founders and artificial selection [47]. In the wild populations, heterozygote deficiency may be cause by Wahlund effect and natural selection. As heterozygotes with null alleles may be taken as homozygotes, null alleles may also account for the heterozygote deficiency.

Genetic Differentiation

The UPGMA analysis indicated that the selected lines and wild populations were not separated into two groups and even the wild populations and selected lines from the same origin country were culstered into the same subgroup except for KS6 and RC, suggesting there was no clear division between the wild populations and selected lines. The result is also supported by the Structure analysis.

Within the selected lines, moderate genetic differentiation ($0.05<Fst<0.15$) was detected. The UPGMA analysis also demonstrated the three selected lines fell into three groups. The significant genetic differentiation among the three selected lines may result from three different founder populations, thus providing the genetic basis for the establishment of selected lines for faster growth using three *C. gigas* stocks from China (Rushan), Japan (Miyagi Prefecture) and Korea (Pusan).

Within the wild populations, slight genetic differentiations ($Fst<0.05$) were detected among the three populations RS, DY and RC. The *Nm* values among these populations ranged from 7.98723 (RC and DY) to 36.03447 (RS and RC), which appeared to be sufficiently large to swamp potential for large genetic differences. Since oyster adults are sedentary, larval dispersal becomes a decisive factor affecting gene flow among wild populations. Therefore, the long pelagic larval phase (14–21 days) and high fecundity (10 to 50×10^6 oocytes per female) of the Pacific oyster may account for the relatively large gene flow detected among RS, DY and RC [48, 49]. However, moderate genetic differentiations were found between population MG and other three populations RS, DY and RC, suggesting potential barriers to gene exchange may exist. Many studies showed a significant relationship between gene flow and geographic distance, suggesting that a pattern of isolation by distance (IBD) explained a great proportion of gene flow in *Mactra chinensis* and *Crassostrea ariakensis* [50, 51]. The IBD pattern indicates long geographic distance can restrict migration among wild populations, which may explain the restricted gene flow between population MG and other three populations RS, DY and RC. Overall, the genetic relationship among the four wild populations could be visualized using the UPGMA tree, which indicated that RS and RC had the closest relationship with DY being the sister group, while MG was in another clade.

Outlier SNPs

In order to improve the productivity traits of *C. gigas*, successive generations of mass selection for fast growth were conducted in three *C. gigas* stocks from China, Japan and Korea since 2007. As there are similarities in controlled selective regimes, the selected lines may provide a valuable opportunity to study parallel evolution, which is taken as one of the most convincing manifestations of the effect of selection in driving adaptive change [52, 53]. Our study provided support for the hypothesis of parallel evolution at the DNA level as one outlier SNP (CgSNP309) was overlapped in the two wild-selected population comparisons. The parallel evolution is also shown in two ways in transcription level, one by the same DNA sequence polymorphisms with direct changes in gene expression, and the other by the different DNA sequence polymorphisms with the same downstream pathways during the cultivation [54]. Flori et al. [55] detected that different DNA sequence polymorphisms but the same physiological pathways were influenced during the cultivation in dairy cattle. The similar results were also found in rainbow trout (*Oncorhynchus mykiss*) and brook charr (*Salvelinus fontinalis*) [56, 57]. However, we can not conclude that parallel evolution at functional levels may account for the result in our study as only four outliers could be annotated with GO terms.

Four different tests (hierarchical island model, fdist island model, Bayesian likelihood approach and lnRH test) were used to detect outlier SNPs potentially under selection during the selective breeding program. These approaches rely on the rationale that selection will lead to increased genetic differentiation between populations and should exhibit a reduction or elevation in genetic variation contrasted with neutral genes. Actually, selection is not the only cause of variation changes at particular loci, reduced variation or increased differentiation can also arise from genetic drift, bottlenecks or founder events [58]. The effective population size for these selected lines is likely to have been less than 100, allowing for substantial random genetic drift over six generations.

The highly consistent results from the hierarchical island model and Bayesian likelihood approach for global outlier analyses strongly suggested that the identified outlier SNPs were subject to selection. Moreover, the same three SNPs (CgSNP140, CgSNP176 and CgSNP209) were also detected as outliers using selected lines and wild populations from which the selected lines originated (CS6, RS, JS6 and MG) (data not shown). At regional level, 3 SNPs (CgSNP225, CgSNP 236 and CgSNP 176) was detected as outlier using at least two analysis methods. Those outlier SNPs found in only one test should be considered with caution. The inconsistent results obtained by the three methods (fdist island model, Bayesian likelihood approach and lnRH test) were probably due to the different measures of variability and different assumptions [59]. Although the outlier analysis provides an encouraging result, association genetics and functional studies are ultimately required to confirm the outlier loci are involved in the artificial selection during the selective breeding program in China.

In this study, SNPs has been demonstrated to be a good marker of choice for monitoring the genetic variation in wild populations and selected lines of *C. gigas*. The genetic diversity analysis showed that the selected lines tended to lose genetic diversity contrasted with the wild populations during the successive six-generation selection. Moreover, the UPGMA results suggested there was no clear division between the wild populations and selected lines. In addition, a total of 17 loci were found under selection at two levels using four outlier tests. Further functional studies are needed to confirm the role of the candidate outlier SNPs during the domestication process in *C. gigas*.

ACKNOWLEDGMENTS

We wish to thank Qi Li for his help on the project designment and comments on previous drafts of this manuscript. We also thank Hong Yu, Lingfeng Kong and Dandan Feng to help us accomplish this project. This work was supported by the National High Technology Research and Development Program, National Natural Science Foundation of China, and Seed Improvement Project of Shandong Province.

REFERENCES

1. Liu YG, Liu LX. Isolation and characterization of twenty two polymorphic simple sequence repeat markers from AFLP sequences of *Crassostrea gigas*. Conserv Genet Resour. 2015; 7: 659–661
2. Dundon WG, Arzul I, Omnes E, Robertb M, Magnaboscoc C, Zambonc M, et al. Detection of type 1 Ostreid Herpes variant (OsHV-1 μvar) with no associated mortality in French-origin Pacific cupped oyster *Crassostrea gigas* farmed in Italy. Aquaculture 2011; 314: 49–52

3. Wang Y, Ren R, Yu Z. Bioinformatic mining of EST-SSR loci in the Pacific oyster, *Crassostrea gigas*. Anim Genet. 2008; 39: 287–289. pmid:18307582
4. Li Q, Wang QZ, Liu SK, Kong LF. Selection response and realized heritability for growth in three stocks of the Pacific oyster *Crassostrea gigas*. Fisheries Sci. 2011; 77: 643–648
5. DOF. China fisheries statistic yearbook (in Chinese). China Agriculture Press; 2013
6. Sauvage C, Boudry P, Lapegue S. Identification and characterization of 18 novel polymorphic microsatellite makers derived from expressed sequence tags in the Pacific oyster *Crassostrea gigas*. Mol Ecol Resour. 2009; 9: 853–855. pmid:21564767
7. Ward RD, English LJ, McGoldrick DJ, Maguire GB, Nell JA, Thompson PA. Genetic improvement of the Pacific oyster *Crassostrea gigas* (Thunberg) in Australia. Aquac Res. 2000; 31: 35–44
8. Langdon C, Evans F, Jacobson D, Blouin M. Yields of cultured Pacific oysters *Crassostrea gigas* Thunberg improved after one generation of selection. Aquaculture 2003; 220: 227–244
9. Dégremont L, Bédier E, Boudry P. Summer mortality of hatchery-produced Pacific oyster spat (*Crassostrea gigas*). II. response to selection for survival and its influence on growth and yield. Aquaculture 2010; 299: 21–29
10. Hedgecock D, Sly F. Genetic drift and effective population sizes of hatchery-propagated stocks of the Pacific oyster, *Crassostrea gigas*. Aquaculture 1990; 88: 21–38
11. Yu ZN, Guo XM. Genetic analysis of selected strains of Eastern oyster (*Crassostrea virginica* Gmelin) using AFLP and microsatellite markers. Mar Biotechnol. 2005; 6: 575–586
12. Cruz P, Ibarra AM, Mejia-Ruiz H, Gaffney PM, Pérez-Enríquez R. Genetic variability assessed by microsatellites in a breeding program of Pacific white shrimp (*Litopenaeus vannamei*). Mar Biotechnol. 2004; 6: 157–164. pmid:14595549
13. Kong L, Li Q. Genetic comparison of cultured and wild populations of the clam *Coelomactra antiquata* (Spengler) in China using AFLP markers. Aquaculture 2007; 271: 152–161
14. Allendorf FW, Phelps SR. Loss of genetic variation in a hatchery stock of *Cutthroat trout*. Trans Am Fish Soc. 1980; 109: 537–543
15. Chistiakov DA, Hellemans B, Volckaert FAM. Microsatellites and their genomic distribution, evolution, function and applications: a review with special reference to fish genetics. Aquaculture 2006; 255: 1–29
16. Prasertlux S, Khamnamtong B, Chumtong P, Klinbunga S, Menasveta P. Expression levels of RuvBL2 during ovarian development and association between its single nucleotide polymorphism (SNP) and growth of the giant tiger shrimp *Penaeus monodon*. Aquaculture 2010; 308: 83–90
17. Thanh NM, Barnes AC, Mather PB, Li Y, Lyons RE. Single nucleotide polymorphisms in the actin and crustacean hyperglycemic hormone genes and their correlation with individual growth performance in giant freshwater prawn *Macrobrachium rosenbergii*. Aquaculture 2010; 301: 7–15
18. .Cong RH, Kong LF, Yu H, Li Q. Association between polymorphism in the insulin receptor-related receptor gene and growth traits in the Pacific oyster *Crassostrea gigas*. Biochem Syst Ecol. 2014; 54: 144–149
19. Jiang Q, Li Q, Yu H, Kong LF. Genetic and epigenetic variation in mass selection populations of Pacific oyster *Crassostrea gigas*. Genes Genom. 2013; 35: 641–647
20. Li Q, Park C, Kijima A. Isolation and characterization of microsatellite loci in the Pacific abalone, *Haliotis discus hannai*. J Shellfish Res. 2002; 21: 811–815
21. Zhong XX, Li Q, Yu H, Kong LF. Development and validation of single nucleotide polymorphism markers in the Pacific oyster, *Crassostrea gigas*, using high-resolution melting analysis. J World Aquacult Soc. 2013; 44: 455–465
22. Zhong XX, Li Q, Guo X, Yu H, Kong LF. QTL mapping for glycogen content and shell pigmentation in the Pacific oyster *Crassostrea gigas* using microsatellites and SNPs. Aquacult Int. 2014; 22: 1877–1889

23. Yeh FC, Yang RC, Boyle TBJ, Ye ZH, Mao JX. POPGENE, version 1.32: the user friendly software for population genetic analysis. Molecular Biology and Biotechnology Centre, University of Alberta; 1999
24. Sokal RR, Rohlf FJ. Biometry: The principles and practice of statistics in biological research, 3rd edn. W.H. Freeman; 1995
25. Excoffier L, Lischer HEL. Arlequin suite ver 3.5: A new series of programs to perform population genetics analyses under Linux and Windows. Mol Ecol Resour. 2010; 10: 564–567. pmid:21565059
26. Wright S. The interpretation of population structure by F-statistics with special regard to systems of mating. Evolution 1965; 19: 395–420
27. Nei M. Genetic distance between populations. Am Nat. 1972; 106: 283–292
28. Takezaki N, Nei M, Tamura K. POPTREE2: Software for constructing population trees from allele frequency data and computing other population statistics with Windows-interface. Mol Biol Evol. 2010; 27: 747–752. pmid:20022889
29. Pritchard JK, Stephens M, Rosenberg NA, Donnelly P. Association mapping in structured populations. Am J Hum Genet. 2000; 67: 170–181. pmid:10827107
30. Falush D, Stephens M, Pritchard JK. Inference of population structure using multilocus genotype data: linked loci and correlated allele frequencies. Genetics 2003; 164: 1567–1587. pmid:12930761
31. Evanno G, Regnaut S, Goudet J. Detecting the number of clusters of individuals using the software Structure, a simulation study. Mol Ecol. 2005; 14: 2611–2620. pmid:15969739
32. Earl DA, vonHoldt BM. STRUCTURE HARVESTER: a website and program for visualizing STRUCTURE output and implementing the Evanno method. Conserv Genet Resour. 2012; 4: 359–361
33. Excoffier L, Hofer T, Foll M. Detecting loci under selection in a hierarchically structured population. Heredity 2009; 103: 285–298. pmid:19623208
34. Foll M, Gaggiotti O. A genome-scan method to identify selected loci appropriate for both dominant and codominant markers: a Bayesian perspective. Genetics 2008; 180: 977–993. pmid:18780740
35. Jeffreys H. Theory of probability, 3rd edn. Oxford University Press; 1961
36. Schlötterer C, Dieringer D. In Selective sweep: Nurminsky D (ed) A novel test statistic for the identification of local selective sweeps based on microsatellite gene diversity. Plenum Publishers; 2005
37. Kauer MO, Dieringer D, Schlötterer C. A microsatellite variability screen for positive selection associated with the "Out of Africa" habitat expansion of *Drosophila melanogaster*. Genetics 2003; 165: 1137–1148. pmid:14668371
38. Li Q, Xu KF, Yu RH. Genetic variation in Chinese hatchery populations of the Japanese scallop (*Patinopecten yessoensis*) inferred from microsatellite data. Aquaculture 2007; 269: 211–219
39. Fisher RA. The genetical theory of natural selection, 2nd revised edition. Dover; 1958
40. Leberg PL. Effects of population bottlenecks on genetic diversity as measured by allozyme electrophoresis. Evolution 1992; 46: 471–494
41. Liu FL, Yao JT, Wang XL, Repnikova A, Galanin DA, Duan D. Genetic diversity and structure within and between wild and cultivated *Saccharina japonica* (Laminariales, Phaeophyta) revealed by SSR markers. Aquaculture 2012; 358: 139–145
42. Clifford SL, Mcginnity P, Ferguson A. Genetic changes in an Atlantic salmon population resulting from escaped juvenile farm salmon. J Fish Biol. 1998; 52: 118–127.
43. Li Q, Park C, Endo T, Kijima A. Loss of genetic variation at microsatellite loci in hatchery strains of the Pacific abalone (*Haliotis discus hannai*). Aquaculture 2004; 235: 207–222

44. Dillon RT, Manzi JJ. Hard clam, *Mercenaria mercenaria*, broodstocks: genetic drift and loss of rare alleles without reduction in heterozygosity. Aquaculture 1987; 60: 99–105
45. Wang LL, Zhang H, Song LS, Guo X. Loss of allele diversity inintroduced populations of the hermaphroditic bay scallop *Argopecten irradians*. Aquaculture 2007; 271: 252–259
46. Norris AT, Bradley DG, Cunningham EP. Microsatellite genetic variation between and within farmed and wild Atlantic salmon (*Salmo salar*) populations. Aquaculture 1999; 180: 247–264.
47. Kohlmann K, Kerten P, Flajshans M. Microsatellite-based genetic variability and differentiation of domesticated, wild and feral common carp (*Cyprinus carpio* L.) populations. Aquaculture 2005; 247: 253–266
48. Thorson G. Reproductive and larval ecology of marine bottom invertebrates. Biol Rev. 1950; 25: 1–45. pmid:24537188
49. Suquet M, Labbé C, Puyo S, Mingant C, Quittet B, Boulais M, et al. Survival, growth and reproduction of cryopreserved larvae from a marine invertebrate, the Pacific oyster (*Crassostrea gigas*). PLoS One 2014; 9: e93486. pmid:24695576
50. Xiao J, Cordes JF, Wang HY, Guo X, Reece KS. Population genetics of *Crassostrea ariakensis* in Asia inferred from microsatellite markers. Mar Biol. 2010; 157: 1767–1781
51. Ni L, Li Q, Kong LF. Microsatellites reveal fine-scale genetic structure of the Chinese surf clam *Mactra chinensis* (Mollusca, Bivalvia, Mactridae) in Northern China. Mar Ecol. 2011; 32: 488–497
52. Harvey PH, Pagel MD. The comparative method in evolutionary biology. Oxford University Press; 1991
53. Roberge C, Einum S, Guderley H, Bernatchez L. Rapid parallel evolutionary changes of gene transcription profiles in farmed Atlantic salmon. Mol Ecol. 2006; 15: 9–20. pmid:16367826
54. Vasemägi A, Nilsson J, McGinnity P, Cross T, O'Reilly P, Glebe B, et al. Screen for footprints of selection during domestication/captive breeding of Atlantic Salmon. Comp Funct Genomic. 2012; 2012: 628204
55. Flori L, Fritz S, Jaffrézic F, Boussaha M, Gut I, Heath S, et al. The genome response to artificial selection: a case study in dairy cattle. PLoS One 2009; 4: e6595. pmid:19672461
56. Tymchuk W, Sakhrani D, Devlin RH. Domestication causes large-scale effects on gene expression in rainbow trout: analysis of muscle, liver and brain transcriptomes. Gen Comp Endocr. 2009; 164: 175–183. pmid:19481085
57. Sauvage C, Derôme N, Normandeau E, Cyr JS, Audet C, Bernatchez L. Fast transcriptional responses to domestication in the brook charr *Salvelinus fontinalis*. Genetics 2010; 185: 105–112. pmid:20194962
58. Kane NC, Rieseberg LH. Selective sweeps reveal candidate genes for adaptation to drought and salt tolerance in common sunflower, *Helianthus annuus*. Genetics 2007; 175: 1823–1834. pmid:17237516
59. Li MH, Iso-Touru T, Laurén H, Kantanen J. A microsatellite-based analysis for the detection of selection on BTA1 and BTA20 in northern Eurasian cattle (*Bos taurus*) populations. Genet Sel Evol. 2010; 42: 32. pmid:20691068

CHAPTER 9

Sex Control in Fish: Approaches, Challenges and Opportunities for Aquaculture

Alyssa M. Budd*, Quyen Q. Banh, Jose A. Domingos and Dean R. Jerry

Centre for Sustainable Tropical Fisheries and Aquaculture, James Cook University, Townsville, Queensland 4810, Australia

ABSTRACT

At present, aquaculture is the fastest growing sector of animal food production and holds great potential as a sustainable solution for world food security. The ability to control sex is one of the most important factors for the commercialisation and efficient propagation of fish species, due to influences on reproduction, growth and product quality. Accordingly, there is a large body of research that targets sexual development in commercially important species in an attempt to understand and control fish sex and reproductive function. In this review, we provide an introduction to sex determination and differentiation in fish, including the genetic, epigenetic and environmental factors that can influence fish sex ratios. We also summarise the major approaches used to control sex in fish and discuss their application in commercially important species. Specifically, we discuss the use of exogenous steroid hormones, chromosome ploidy, environmental manipulations, sex-linked genetic markers, selection for altered sex ratios, and transgenics and comment on the challenges associated with controlling sex in a commercial environment.

Keywords: sex ratio; reproductive control; triploidy; hormonal manipulation; epigenetics; environment; QTL

1. INTRODUCTION

Sex control is one of the most important and highly targeted areas of aquaculture research due to influences on husbandry management, productivity and economics. Without the ability to regulate sexual differentiation, maturation, and reproduction, farmers have little control over breeding processes, both in the hatchery and throughout grow-out. Arguably, in aquaculture species that have become global commodities, control over sex and reproduction has been the primary facilitator for large-scale industrial production. In species that are yet to reach industrial scale production, elucidation of sex differentiation and improved reliability of reproduction remains a key area of applied research.

The Need for Sex Control

Several broad goals in aquaculture can be reached through a better understanding of sex control. These include: (i) prevention of precocious maturation and uncontrolled reproduction (e.g., in tilapia); (ii) the desire to farm monosex populations due to differences in growth rate and economic value of the sexes (e.g., tilapia, shrimp); (iii) reducing the impact of phenotypic sex on product quality (e.g., Atlantic salmon, oysters); (iv) increasing stability of mating systems (e.g., sex change in groupers) and (v)) environmental and/or intellectual property protection (e.g., non-indigenous species, or genetically improved strains). The relative importance of each of these goals depends upon the reproductive biology and culture system of the species concerned.

Precocious maturation occurs in several farmed species including Nile tilapia (*Oreochromis niloticus*) [1], freshwater crayfish (*Cherax destructor*) [2] and Atlantic salmon (*Salmo salar*), which have a tendency to sexually mature and reproduce before attaining a body size that is suitable for harvest. This precocious maturation leads to slow growth as energy is diverted into reproduction, creates large variance in product size at harvest and results in overpopulation of ponds and, therefore, an inability to control animal densities and feeding rates. Furthermore, deterioration in flesh quality is often observed in female Atlantic salmon as they reach sexual maturity through the diversion of energy (e.g., lipids) towards reproductive processes resulting in differences in economic value between males and females [3]. The desire to farm monosex populations may also be provoked by sex-specific growth rates. Male Nile tilapia, for example, grow faster and have lower feed conversion rates than females [4], while female Kuruma prawns (*Penaeus japonicus*) are generally larger than males at the time of harvest [5]. As a result, farmers have adopted both manual (e.g., hand sexing and selective removal) and/or various technological (e.g., exogenous hormone treatment, chromosome ploidy manipulation, molecular tools, or hybridisation) approaches to produce monosex populations for culture (Table 1).

Table 1. Common approaches used to manipulate sex in aquaculture fishes.

Approach	Technique	Purpose	Representative example species
Hormonal manipulation	Administration of exogenous hormones (e.g., 17β-estradiol, 11-α-methyltestosterone)	Monosex	Atlantic cod, *Gadus morhua* [6] Nile tilapia (*Oreochromis niloticus*) [7]
	Administration of aromatase inhibitor (e.g., Fadrozole)	Monosex	Nile tilapia (*Oreochromis niloticus*) [8] Honeycomb grouper (*Epinephelus merra*) [9]
Hybridisation	Cross breeding	Monosex	Tilapia (*O. aurea x O. niloticus*) [10] Bass (*Morone saxatilis x M. mississippiensis*) [11]
Chromosome Ploidy	Gynogenetics	Monosex	Rainbow trout (*Oncorhynchus mykiss*) [12]
	Triploidy	Sterility	Atlantic salmon (*Salmo salar*) [13,14]
Envionmental manipulation	Manipulation of social factors	Production of male broodstock	Orange-spotted grouper (*Epinephelus coioides*) [15]
	Temperature treatment during gonadal differentiation	Monosex populations	European Seabass (*Dicentrarchus labrax*) [16–18]
Selection	Marker assisted selection (MAS)	Monosex populations	Nile tilapia (*Oreochromis niloticus*) [19,20] Turbot (*Scophthalmus maximus*) [21]

An ability to control sex and breeding is also important for hatcheries in order to produce seedstock, particularly if the purpose is reliable production of specific family combinations for selective breeding. Many fishes such as Atlantic salmon, rainbow trout (*Oncorhynchus mykiss*) and channel catfish (*Ictalurus punctatus*) can be reliably dry stripped to obtain eggs and sperm, which are subsequently mixed in buckets to produce fertilized eggs. This ability to strip spawn fish allows for easy creation of large numbers of either half- or full-sib families, depending on the mating design. However, in many other aquaculture species, strip spawning is inadequate for industrial scale production, and accordingly, natural reproduction is relied upon. Barramundi (*Lates calcarifer*) (also known as Asian seabass), for example, is a mass-spawning, sequential protandrous (male to female sex-changing) hermaphroditic species farmed throughout South-East Asia and Australia. In this species, female broodstock are extremely valuable due both to the maintenance and extensive holding time until a fish changes sex (approximately 4 or more years of age), as well as the high fecundity of female barramundi (2–32 million eggs depending on size; [22]). Although there are published accounts of dry-stripping in this species [23], industry has not adopted this approach due to unpredictable timing of sperm hydration and final maturation of eggs, as well as the risk of egg rupture and or damage to blood vessels in the ovary leading to septicemia in females used for repeated spawning [24]. As a result, hormonally induced natural mating is practiced in this species. In other species, such as sex-changing marine groupers (*Epinephelus* spp.), optimal, stable sex ratios are difficult to maintain in mating tanks due to the dependence of sexual differentiation on an individual's position within social hierarchies. For example, a fish that was introduced to a mating tank as a male may sex revert into a female, depending on its relative dominance in the tanks existing social hierarchy [25,26]. In the case of grouper, stability of males in mating groups is achieved by implants containing methyl-testosterone [27].

Given the importance of sex control to aquaculture production, the intent of this review is to introduce the major approaches currently adopted by farmers in the endeavor to have control over sexual differentiation, maturation and breeding in commercially farmed fish, as well as to highlight some of the challenges in controlling sex within a commercial aquaculture environment. Several excellent reviews on sex determination and differentiation in fish have previously been published (see [28,29,30,31,32,33,34] for further reading), we begin with a brief introduction.

2. SEX DETERMINATION AND DIFFERENTIATION IN FISH

Sex determination and differentiation in fish is an evolutionarily diverse and highly plastic developmental process [33]. Such diversity leads to great challenges when trying to develop a general understanding of sex in fish. At the individual or species level, however, heightened plasticity of sexual phenotypes can lead to increased opportunity for sex control, which is particularly important for farmed fish. In many species, the sex of an individual is subject to modification by external factors, such as environmental temperature, which can interact with hormonal, genetic and/or epigenetic regulatory pathways to alter fish sexual development [35].

The term sex determination can be used to describe the genetic or environmental cue(s) that ultimately govern the sex of an individual [30]. For example, in most mammals, inheritance of the Y chromosome determines that an individual will develop as male [36]. In many reptiles, however, temperatures experienced during embryonic development, rather than genetic factors, provide the sex-determining cue. Sexual differentiation, on the other hand, can refer to the *physical*realisation of sex determination cues, and largely pertains to the development of the testicular or ovarian tissues (e.g., phenotypic sex) that follows on from a sex-determining cue [30]. In some cases, processes of sex determination and gonadal differentiation may partially overlap and, as a result, the terms are often used interchangeably [28,34]. This is especially evident in teleost fish, in which the molecular mechanisms of sexual development are evolutionarily plastic, and require complex regulatory pathways that are governed by genetic factors (genetic sex determination; GSD), environmental cues (environmental sex determination; ESD), or on an interaction between the two (genotype x environment interactions; GxE) [37].

2.1. Genetic Factors Involved in Fish Sex Determination and Differentiation

Genetic sex determination can be grouped into two broad categories: Chromosomal sex determination (CSD), where sex is governed by the inheritance of sex-related genes located on specific chromosomes (*i.e.*, the sex chromosomes), or polygenic sex determination (PSD) in which sex determining genes are distributed throughout the entire chromosome complement, although major genetic effects are found in most cases. In fish, CSD can include both male (XX/XY) and female (ZZ/ZW) heterogamety, similar to that seen in mammals and birds. Some fish species, however, have undergone a loss of the derived chromosome (termed X0 or Z0) as in the Chilean galaxiid (*Galaxias platei*) [38], while others demonstrate translocations or fusions of sex chromosomes with autosomes (X1X1X2X2/X1X2Y) as in the wolf fish (*Hoplias malabaricus*) [39], or even exhibit multiple, fully derived chromosomes (WXZ) as in the swordtail platyfish (*Xiphophotus maculates*) [30]. In the majority of fish species, however, sex chromosomes are weakly differentiated, with only 7% of species showing sex-associated chromosome heteromorphisms [28,40]. Whilst this lack of chromosomal differentiation is suggestive of a high occurrence of PSD in fish, there are very few examples in which this has been experimentally demonstrated. These include the southern platyfish (*Xiphophorus maculatus*) [41], domesticated zebrafish [42] and likely the European seabass [43]. The lack of heteromorphism between sex chromosomes of fish may instead be explained by the inability of traditional cytogenetic techniques to identify small scale differences, such as the inversions and deletions in the Y chromosome of threespine sticklebacks (*Gasterosteus aculeatus*) [44], or the single SNP variation between X and Y chromosomes in the tiger puffer fish (*Takifugu rubripes*) [45]. Furthermore, whilst domesticated zebrafish are thought to exhibit PSD, sex chromosomes have been identified in their wild counterparts, indicating a loss, which has occurred in less than 100 generations [46].

This illustrates the rapid evolutionary turnover of sex determination systems in fish, and is an important consideration for the domestication of farmed populations.

The sex determining genes themselves may be considered as either upstream "master" switches, or downstream differentiators, depending on their relative roles in sex determination and/or differentiation. Whilst it appears that in mammals, the master sex determining gene, sex determining region-Y (*SRY*), appears relatively ubiquitous, master sex-determining genes in fish show huge variety [34]. Six master sex-determining genes have been isolated in fish thus far, including PG17: DM-domain gene on the Y chromosome (*dmY*), the major testis-determining factor in the Japanese medaka (*Oryzias latipes*) (XX-XY) [47,48] anti-Müllerian hormone (*amhY*) and anti-Müllerian hormone receptor, type II (*amhr2*) in the Patagonian pejerrey (*Odentesthes hatchery*) [49,50] and tiger puffer fish [45], respectively, gonadal somatic cell derived factor (*gsdf*) in the Luzon ricefish (*Oryzias luzonensis*) [51], sexually dimorphic on the Y-chromosome gene (*sdY*) in the rainbow trout [52] and finally SRY-related HMG-Box gene 3 (*sox3*) in the medaka (*Oryzias dancena*) [53]. Many of these master sex determining genes are thought to have been "up-recruited" from the network of downstream genes involved more specifically in gonadal differentiation [35]. As a general model for non-mammalian vertebrates, male sex differentiation can be achieved through up-regulation of a highly conserved transcription factor, doublesex and mab-3 related transcription factor 1 (*dmrt1*), which acts in combination with transcription factor *sox-9* (*sox9*) to promote testis formation [54,55]. Alternatively, female sex differentiation is stimulated by cytochrome P450 aromatase (*cyp19a*) through a positive feedback loop involving the female-associated transcription factor known as forkhead box protein L2 (*foxl2*) [31,56]. *Cyp19a* encodes for gonadal aromatase, which catalyses the conversion of androgens into estrogens and is thought to play a pivotal role in sex differentiation and sex change in fish [31]. Other genes, such as r-spondin 1 (*rspo*) and wnt-signalling protein (*wnt*) in the ovary, as well as *amh* and *gsdf* in the testis, are thought to play roles in β-catenin and TGF-β signaling pathways, respectively, to promote sexual differentiation and subsequent gonadal development. Precisely how these genes interact to coordinate gonadal development, particularly in the ovaries, remains a major question of sex differentiation and is likely a reflection of the variability in their relative roles between species [34,57,58].

2.2. Epigenetic Mechanisms Involved in Fish Sex Determination and Differentiation

Recently it has been proposed that epigenetic mechanisms play an important role in sex determination and differentiation in fish and other vertebrates [59]. Epigenetics is a rapidly expanding field that investigates changes in gene function that cannot be explained by changes in the DNA sequence [60]. These changes largely act to enable or inhibit the activity of transcriptional machinery on the DNA to regulate patterns of gene expression and can be influenced by environmental stimuli, such as temperature [61]. It is becoming increasingly obvious that, in many fish, differences between the sexes cannot be fully explained by differences in the genes alone, particularly in cases where sex ratio shifts occur in response to temperature, but also in serial and sequential hermaphrodites. Epigenetic modifications may be a key mechanism allowing for interactions between the environment and sex determination and differentiation processes in fish [16,59,62] (see also section 3.3 Environmental Manipulations and Opportunities for Sex Control).

2.3. Environmental Factors Involved in Fish Sex Determination

In contrast to GSD, true environmental sex determination (ESD) does not rely on sexual dimorphism at the genomic level [42]. This is because the major determinant of sex is not genetic, but environmental. The best-known environmental determinant of sex is temperature and has been well studied in reptiles that exhibit temperature dependant sex determination (TSD). In fish, the first definitive evidence for the effect of environmental temperature on sex differentiation was reported in the Atlantic silverside (*Menidia menidia*), a species which shows TSD characteristics both in the wild and in captivity [63]. TSD is also thought to exist in the high-potential South American aquaculture species, the Patagonian pejerrey, where high temperature treatments can be used to achieve all male populations [64]. In most fish, however, even the application of extreme temperature treatment rarely produces monosex populations. What is more commonly seen is an incomplete shift in the sex ratio towards either the male or female phenotype [65]. For example in the European seabass (*Dicentrarchus labrax)* temperature treatments of 20 °C and 15 °C can be applied to achieve 73% male and 77% female populations, respectively [66]. In such cases, genotype is thought to be inhibiting complete control of the sex differentiation pathway by external factors, indicating that both genomic and environmental factors (GxE) contribute to sex determination, likely through epigenetic modifications to the DNA. The fine line between TSD and GSD + temperature effects (GSD+TE) in fish necessitates careful consideration of the ecological relevancy of temperature manipulations when attempting to classify fish sex determination systems [65]. For farmed fish, however, temperature treatments that are considered to be ecologically irrelevant can still provide important information towards the development of methods for sex control [32].

2.4. Sex Reversal and the Labile Period

Sensitivity to environmental factors in fish allows for sexual reversal, a phenomenon whereby phenotypic sex no longer corresponds to genotypic sex following experimental manipulation. For example, a temperature treated fish with ZW chromosomes (genotypically female) may exhibit testis formation as well as other male-specific secondary sexual characteristics (phenotypically male) [62]. The type and timing of treatments that are able to induce sexual reversal vary greatly between species. For example, treatments influencing sex determination and differentiation in fish include temperature, pH, density, exogenous hormone treatment, social factors, or a combination of these. Furthermore, there is a discrete time period during development, often referred to as the labile period (also sensitive window or period), during which the gonad undergoes differentiation towards the male or female state [32]. Often temperature, hormone or other treatments become ineffective following this period of differentiation, once the gonads have become established and sex is stabilised [65,67]. In the case of naturally sex-changing fish the gonads can remain responsive to either endogenous or exogenous stimuli well into adulthood. For example, social cues transmitted along the brain-pituitary-gonadal axis are thought to stimulate adult sex change in many species of reef fish [68,69]. Information on the specific cues able to influence sexual development, as well as knowledge of the timing of gonadal development and the labile period, allows for the most accurately timed and therefore effective treatments in a given species.

As the list of master sex determining genes isolated from different fish species grows, and we become more aware of the complexity of the genes in the downstream network and their interaction with the environment, it is becoming increasingly more apparent that in order to develop appropriate intervention and control strategies for sex in farmed fish, species-specific studies will be necessary. By targeting the specific developmental time periods, or environmental cues to which fish are most sensitive, such intervention can reduce the production liability of uncontrolled sex processes.

3. APPROACHES USED TO MANIPULATE SEX IN FARMED FISH

Given the complexity and variability in sex-determination and differentiation mechanisms in fish, there is no single approach that has proven effective in the manipulation of sex in all species. Rather sex control has relied on several major approaches including the addition of exogenous hormones, chromosomal ploidy manipulation, hybridisation (not discussed further, see [70] for a recent review), varying environmental and social parameters, and selection for sex ratios using a molecular and/or quantitative genetic approach (summarised in Figure 1).

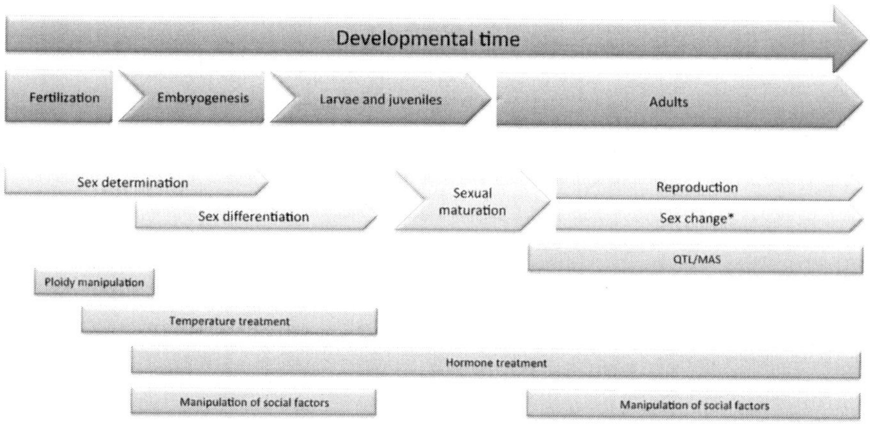

Figure 1. Generalised model of targeted developmental time periods for applying strategies to manipulate sex in farmed fish. Screening of quantitative trait loci (QTL) and marker assisted selection (MAS) is used to identify broodstock that will produce the desired sex ratio in offspring; chromosome (ploidy) manipulation to produce triploids takes place usually within 30 min of fertilisation; temperature treatments are generally best applied prior to the onset or during sex differentiation; hormonal treatments are most often applied during the period of sex differentiation, but, in some species, may remain effective throughout juvenile and adult stages; manipulation of social hierarchies are often undertaken during adult sex change (* only occurs in some fish), but can also be effective during the period of sexual differentiation of sex changing fish, or gonochoristic species where stocking density is important.

3.1. The Use of Exogenous Hormones and Other Chemicals to Control Sex

Of the external factors known to control fish sexual development, administration of exogenous sex hormones is the most frequently utilized approach due to ease of application at a commercial scale and consistency in producing monosex populations. Sex steroids (*i.e.*, those steroid hormones that are known to interact with androgen and estrogen receptors) are of critical importance to the natural process of phenotypic sex differentiation, thus providing the basis for the administration of exogenous sex steroids to alter sex ratios in farmed fish [71]. The first successful effort to artificially induce sex reversal was achieved in the medaka, through the administration of estrogens and androgens to sexually undifferentiated fish, and resulted in both functional females and males, respectively [72,73]. Following these early experiments, similar treatments have been applied to a variety of fish species demonstrating that, using sex steroid therapy, it is possible to alter the normal course of sex differentiation in fish towards the desired gonadal phenotype [74]. For instance, the addition of exogenous androgens, such as 17-α-methyltestosterone, has proven effective in the masculinization of genetically female fish in approximately 35 species [74,75]. Furthermore, hormonal treatments are not only effective in species which do not appear to possess GSD (e.g., the honeycomb grouper (*Epinephelus merra*) [74,76], red-spotted grouper (*E. akaara*) [77], barramundi (*Lates calcarifer)* [78], but has also proven an effective treatment in species with clear chromosomal sex-determination systems (e.g., Atlantic salmon [79], Nile tilapia [7], rainbow trout [80] and Atlantic cod (*Gadus morhua*) [6]). The effect of introducing exogenous hormones prior to and during the course of sex differentiation in fish is considered successful if their addition can override normal (genetic or environmental) sex determination pathways to achieve the desired phenotypic sex.

3.1.1. Principles of Exogenous Hormones and Other Chemicals Application for Sex Control

Control of sex differentiation through the administration of hormones is generally achieved through the exposure of sexually undifferentiated fish to an appropriate dosage of exogenous sex steroids. Yamamoto [71] highlighted that for effective sex reversal sex steroids should be given prior to any sign of gonadal differentiation, and at a dosage relevant to both the species treated and the nature and/or potency of the steroid. Yamamoto [71] further recommended that hormone administration be maintained until after the time when normal sex differentiation occurs. In general, fish are much more sensitive to the effects of steroid treatment during the labile period, when the gonads are undifferentiated [81,82]. The timing of this period is variable both within and between species. For example, the labile period for estrogen treatment in the masu salmon (*Oncorhynchus masou*) is from 5 to 22 days post-hatch (dph) [83], while in coho salmon (*Oncorhynchus kisutch)* it is between 8 days pre-hatch and 13 dph [84]. Furthermore, the differentiation period of fish destined to be female often precedes that of males (because the onset of ovarian development happens earlier than testicular differentiation), such as eight days prior to hatching and 6 dph in coho salmon [84], 20 dph and 42 dph in the African catfish (*Clarias gariepinus*) [85], and 19 and 90 dph in channel catfish, (*Ictalurus punctatus*) [86] for ovarian and testicular development, respectively. Whilst hormonal treatments performed during the labile period require the minimum steroid dosage and treatment duration to produce the desired phenotypic gender [87], this does not necessarily imply that fish cannot be successfully sex-reversed by hormonal treatment outside this period [83,88]. In the case of European seabass [89] and the chum salmon (*O. keta*), the gonad is responsive to exogenous steroids outside the period of gonadal differentiation, and well beyond the labile period [83]. Nevertheless, such treatments require higher doses of hormones and longer treatment times than those that are administered during the labile period in order to successfully reverse gonadal sex.

In addition to the use of sex steroids to feminize or masculinise fish, other chemically induced sex reversal strategies are available. Among these, the use of aromatase inhibitors, such as Fadrozole, are commonly applied to induce sex change in fish. Aromatase inhibitors work by either irreversibility deactivating the aromatase enzyme (which converts androgens to estrogens), or through competitive exclusion of aromatase to receptors in estrogen producing cells. Administration of Fadrozole has proven successful in inducing complete sex reversal of genetic females into phenotypic males in Nile tilapia [8,90], olive flounder (*Paralichthys olivaceus*) [91] and honeycomb grouper [9]. In adult female honeycomb grouper, Fadrozole implanted at 10 mg·kg^{-1} body weight resulted in reversion of the ovaries into testes, which produced sperm capable of fertilization after two and a half months treatment [9].

Administration of aromatase inhibitors or steroid hormones to control sex in fish can occur through several routes, including direct injection of the agent into the muscle or body cavity of the fish, or more commonly, direct immersion into culture water containing the agent, or as a dietary supplement [92]. The advantages and limitations of these methods have been discussed previously by Pandian and Sheela [74] and Beardmore *et al.* [75]. Due to simplicity of administration and applicability at a commercial scale, adding exogenous hormones via dietary supplementation is one of the most common techniques for sex reversal in aquaculture. Here, the hormone is dissolved into an evaporative solvent and added to the feed, which fish consume as part of their normal diet. The addition of hormones and chemicals to the feed has proven an effective strategy to produce 100% mono-sex populations [82,93,94], however, there are some limiting factors to this approach. These include degradation of hormone during storage or as part of the fishes normal digestive progress, variability in dosage among individuals due to non-uniformity and concentration of the hormone in the feed and behavioural hierarchies among fish that influence feeding rate and lead to differences in how much of the hormone in the diet fish may actually ingest. Whilst administration of sex-changing agents in the feed is routinely used for many species, for those species in which the gonadal labile period occurs before first feed (such as during embryogenesis, or in yolk sac larvae like most of the egg-laying salmonids [30]) immersion in water containing sex steroids offers an effective alternative. Immersion of coho salmon larvae with 400 μg/L of estradiol-17P for a period of 2 h results in clutches comprising 97% females [95]. For species in which the labile period is prolonged from embryogenesis to post-hatch, such as rainbow trout, multiple doses of steroids have been demonstrated more effective than a single dose [84,96].

3.1.2. Challenges to the Use of Hormones and Other Chemicals in Controlling Sex

Despite successful application in numerous aquaculture species, sex reversal of fish by hormonal administration manipulation should be undertaken with caution to prevent any adverse effects in the fish produced, to the farmers themselves and consumers, or to the environment. For example, hormonal overdoses or unnecessary long duration of treatments may induce deformities, or even skew sex ratios toward the non-target sex [75,93]. Within hatcheries, the use of hormones requires careful handling to avoid adverse effects on human health and hormone-laden water needs to be disposed of properly to limit environmental impacts. There are also consumer health concerns, particularly in cases where hormones are directly applied to fish destined for harvest (*i.e.*, not broodfish). In Europe, the application of hormones to commercial grow-out food fish is prohibited (with Directive 96/22/EC and Directive 2003/74/EC [97,98]), although the use of hormones to sex reverse broodfish is legal. To circumvent some of the legal, environmental and consumer issues associated with the use of sex steroids in fish, new chemicals to induce sex inversion that are regarded as safer (such as trenbolone acetate, 17α-methyldihydrotestosreone or neuroendocrine controlling substances) are being investigated [99].

3.2. Chromosome Ploidy Manipulation and Sex Control

Manipulation of chromosome-set ploidy levels (also referred to as whole genome manipulation) has been one of the most researched approaches to control sex in aquaculture species and, in several species, is routinely applied to commercial production (reviewed in [3,100,101,102,103]). Chromosome ploidy manipulation relies on the application of physical or chemical "shocks" to interfere with the normal developmental processes of gametogenesis and early post-fertilization. Depending on when these external shocks are applied, individuals that are haploid (n), triploid (3n), or tetraploid (4n) can be produced, as compared to the normal diploid (2n) somatic cell chromosome set number. Whilst there has been a body of research on producing mono-sex individuals with only a maternally (gynogens) or paternally (androgens) derived set of chromosomes, these techniques have had little direct impact on the production of commercial aquaculture stocks. Rather, most efforts have been focused towards the production of triploids and tetraploids. Triploids are desired in aquaculture due to an inability of homologous chromosomes to pair equally in gametogenesis and the organism effectively becomes sterile. Other benefits often associated with this gonadal sterility include delays in secondary sexual maturation characteristics, along with improved flesh quality and post-pubertal body growth [3]. In contrast, tetraploids are not sterile and can produce viable 2n gametes. Tetraploids are produced for sole use as broodstock to cross with normal diploids to produce 100% interploid triploids, as is common in tilapia (*Oreochromis* spp.) [104].

3.2.1. Principles of Triploidy Induction in Fish

When fish oocytes are released from the ovary they are arrested at the metaphase stage of meiosis II. Here, homologous chromosomes have been replicated, but the second process of chromosomal partitioning to bring the gamete to the haploid state has not occurred. Meiosis then remains arrested until after the egg is fertilized by the entry of a spermatozoon [105]. Therefore, at the time of fertilization the oocyte still contains two pairs (2n) of homologous maternal chromosomes. Immediately post-fertilization, entry of the sperm causes osmotic and ionic changes leading to the resumption of meiosis II, whereby the oocyte packages up the second set of maternal chromosomes into a vesicle called a polar body which carries the additional set of maternal chromosomes out of the nucleus. This means, that for a few minutes, the now fertilized fish zygote contains three sets of chromosomes (3n), one paternal and two maternal. Physical (e.g., temperature, pressure) or chemical (e.g., 6-dimethylaminopurine, cytochalasin B, caffeine) extrinsic shocks applied after fertilization and at the time of recommencement of meiosis II can disrupt this process of polar body extrusion resulting in retention of the second set of maternal chromosomes. From this point the embryo develops through the normal process of mitotic division and embryogenesis with all its cells containing three sets of chromosomes (3n). Triploids can also be produced in some species through the initial production of tetraploid broodstock and then subsequent crossing with diploids (see [3,106] for examples).

3.2.2. Uses of Triploids in Aquaculture Production

The production and use of triploids in aquaculture has been extensively summarised in [3], where triploids from at least 49 commercial species have been produced. However, despite this large number of species the routine commercial use of this approach to control sex and achieve other production benefits is a lot lower, with the technique largely commercially restricted to use for various salmonids (*S. salmo, Oncorhyncuus rhodurus, O. masou*) and trout (*O. mykiss, S. trutta, Salvelinus fontinalis*), Artic charr (*Salvelinus alpinus*), grass carp (*Ctenopharyngodon idella*), loach (*Misgurnus anguillicaudatus*), ayu (*Plecoglossus altivelis*), edible oysters (*Crassostrea gigas, C. virginica, Saccostrea commercialis*) and tilapia [88].

3.2.3. Advantages of Triploidy Induction in Controlling Sex

The major advantage to farmers in the culture of triploids is their sterility. This is because during meiosis the trivalent homologous chromosomes cannot pair correctly during prophase I [107]. Associated with this chromosomal sterility is a severe impairment of pubertal gonad development and subsequent reproductive maturation. In females this retardation of ovarian development and maturation is usually greater than in the male testes due to the post-meiotic process of vitellogenesis in oocytes. Consequently, ovaries remain quiescent and generally are smaller in triploid individuals [108]. One example of where triploidy is routinely applied to induce sterility is that of grass carp (*C. idella*) in North America. Here, meiotic triploidy induction is used as a reproductive containment approach to produce fish that are sterile and can therefore be safely stocked into canals for weed control without the concern that this non-native species will establish feral populations [109]. Triploidy to prevent precocious and uncontrolled reproduction was also investigated initially in blue (*O. aureus*) and Nile tilapia (*O. niloticus*) [110,111,112,113,114]. However, due to technical difficulties in producing 100% triploids in the field, the use of sex-reversal through addition of hormones (see Section 3.1) to produce YY "supermales" and their subsequent use to produce mono-sex XY males has become a more reliable solution to the problem of uncontrolled reproduction in tilapia.

Triploidy may also delay the onset of puberty, which has further advantages to aquaculture production. The first is that the onset of puberty will result in reallocation of energy resources away from somatic growth and towards gonad maturation, gamete production and spawning specific-sexual behaviours. In some species, like Atlantic cod, feed conversion efficiency and intake may decrease with the onset of puberty. The result of these two processes is that specific growth rates may significantly decrease once puberty is reached and can be further reduced if spawning occurs (e.g., Atlantic cod may lose up to 25% of their somatic body weight through a single spawning season [115,116]). During immature reproductive phases, diploids and triploids often grow at the same rate, however, in many species once normal maturation age is reached triploids outgrow their diploid counterparts (but not always, see [3,101]). Triploid Atlantic salmon, rainbow trout, Nile tilapia, channel catfish, turbot (*Scophthamus maximus*), for example, all exhibit higher growth rates after the onset of normal puberty than diploids. An added advantage in those species where females are more desired and the sex ratio is skewed in this direction is that females outgrow triploid males [3]. Puberty and sexual maturation also may influence flesh composition through the depletion of lipids, proteins and pigments such as the carotenoid, astaxhanthin. For example, in Atlantic salmon, pre-pubertal fish exhibit higher lipid stores than in sexually mature fish, while the fillet is depleted for lipids, proteins and astaxhanthin post-maturation [117]. Finally, delaying and/or preventing sexual maturation may have an influence on reproductive behaviours that often lead to antagonistic and territorial encounters, although in an aquaculture context there is limited evidence for such alterations. Garner *et al.*, [118] found that triploid chinook salmon (*O. hynchus tshawytscha*) exhibited a slightly less aggressive feeding behaviour than diploids, although this did not result in any discernible differences in growth.

3.2.4. Challenges of Producing Triploids

Whilst farming of triploids is often seen as a panacea to control sex in fish, there are several challenges with commercial application of the approach, thus having limited adoption. Besides the obvious challenge of obtaining newly fertilized eggs within minutes of spawning and before extrusion of the second polar body (difficult in some mass spawning species like barramundi and snappers (*Sparus* spp.)), to ensure 100% triploidy it is essential to interfere with polar body extrusion in all oocytes. Inducing 100% triploids as routine practice is, therefore difficult at commercial scales because often there is intra-specific family variability in the timing of polar body extrusion and/or oocyte sensitivity to intensity and duration of the environmental shock applied [119]. The timing, intensity/duration and best mode of shock is species-specific and in fishes to achieve 100% triploidy treatments have to cover the entire duration of polar body II extrusion and be of a sufficient magnitude to disrupt extrusion of the polar body. Thus protocols need to be individually optimized and precise in their application, as variation often leads to decreased percentages of triploidy induction success [120,121]. For instance, Piferrer, *et al.* [122] investigated the effects of cold shock timing and duration on induction of triploidy in turbot. A cold shock commencing at 5 min post-fertilisation (PF) and a duration of 20 min produced 92% triploid turbot, whereas cold shock starting at 6.5 min PF with 25 min duration produced only 83% triploids.

A further challenge in the commercial use of triploids for sex control is that often triploids exhibit lower survival, higher rates of deformities and reduced pre-pubertal growth than diploids, although this again is species-specific. As examples, triploid common carp (*Cyrpinus carpio*) had only a 70% survival rate and 85% of the growth as that of diploid controls [123], while all-female triploid Atlantic salmon, when reared in saltwater, had a survival rate of 40% compared to 60% of all-female diploids between stocking and harvest [13].

3.3. Environmental Manipulations and Opportunities for Sex Control

The influence of genotype by environment interactions on fish sex determination and differentiation allows for increased opportunity to alter fish sex ratios via environmental manipulations such as social factors, temperature, pH, and density. These manipulations can be used to alter processes of sexual differentiation during early development, or encourage sex change in adult hermaphrodite fish.

3.3.1. The Use of Social Factors to Influence Sex Change in Hermaphrodite Fish

Social interactions are commonly implicated in the onset of adult sex change in hermaphrodite fish [124,125]. Marine groupers are a group of premium aquaculture species, many of which are protogynous hermaphrodites (*i.e.*, they mature first as females and later sex change to males). A major challenge in establishing viable grouper hatcheries and breeding programs is a shortage of male broodstock [15]. Manipulation of social factors provides the possibility to attain males at a younger age. For example, to encourage female-to-male sex change in orange-spotted grouper (*Epinephelus coioides*), fish are initially stocked as mature females (~4 months) at low densities, and later transferred to higher stocking densities [25]. Social manipulations in *E. coioides* have not only been shown to influence sex ratios in adult fish, but also sexual differentiation in juveniles. By manipulating the number of juveniles reared in a single tank, Liu and Sadovy de Mitcheson [15] were able to encourage primary male maturation, with 39% of experimental juveniles maturing first as males, compared with <5% obtained under mariculture conditions.

3.3.2. The Use of Temperature to Manipulate Sex Ratios

The effect of temperature on sex ratios in fish has been documented in over 60 species, many of which are of commercial importance [30,37,63,65,126]. One of the best-studied examples of this occurs in the European seabass. In this species, farmed populations exhibit strongly male-biased sex ratios whereas in wild populations, female biases appear more common [29]. Exposure to high water temperatures during the thermosensitive period (0–60 days post-fertilisation) has been shown to produce male-biased sex ratios through an epigenetic modification known as DNA-methylation. In this system, DNA methylation of the gene encoding gonadal aromatase, *cyp19a*, is linked to a down regulation in gene expression and subsequent masculinisation of genetically female fish [16]. Experimentally, treatments of 20 °C and 15 °C can be applied to achieve 73% male and 77% female populations, respectively [66]. Such knowledge of thermosensitivity during sexual development in juvenile European seabass offers farmers a practical tool to achieve more desirable sex ratios. DNA methylation has also been proposed as the molecular mechanism allowing for manipulations of sex ratio in farmed populations of half-smooth tongue sole (*Cynoglossus semilaevis*). This species exhibits a female heterogametic (ZW) chromosomal system, in which high water temperature is able to induce ZW males (termed, "pseudomales") [62]. It was recently observed that the promoter of the key testis-determining gene, *dmrt1*, was hypermethylated and downregulated in ZW females, but not in ZW pseudomales, and further, that these epigenetic modifications were heritable, as the epigenetic signatures of sex-reversed fish were evident in progeny reared at normal temperatures [62]. This new understanding of how temperature influences sex ratio and the underlying genetic mechanisms driving these changes has stimulated exciting and innovative research and will likely lead to new and/or optimised methods for sex control in cultured species with GSD with temperature effects (GSD + TE).

3.3.3. Other Environmental Factors Affecting Fish Sex Ratios

In addition to social factors and temperature, there are other factors known to affect sex ratios in fish that could be considered when culturing fish. For example, stocking density can strongly influence sexual differentiation pathways. This phenomenon is particularly prevalent in *Anguillidae* spp., a group of catadromous eels for which high stocking densities produce a greater number of males [127]. A reduction in stocking density is advantageous, however, where this is not economically viable, periodic transfer from high to low density can also result in higher proportions of females and increased growth [128]. Acidity of culture water may also influence sex ratios. In poeciliids, such as *Xiphophorus helleri* and *Poecilia melanogaster*, acidic waters resulted in 100% and 90% male populations, respectively [129]. Sensitivity to pH is also common in fish, such as cichlids like *Apistogramma caetei*, where 96% female progenies can be achieved at pH 6.5 [130]. More general farming practices must also be considered, for example separation of fish by size grade is often used to reduce competition and/or avoid cannibalism. However, this procedure can lead to skewed sex ratios because of the strong association between growth and sex in some fish [17,131,132].

3.3.4. Challenges for Environmental Control of Sex in Fish

There is great diversity in the types of environmental stimuli that influence fish sexual development and the effectiveness of these manipulations ranges from monosex population production (as in ESD or other highly sensitive species), to only marginal, or sometimes negligible changes in sex ratio (GxE or strict GSD). To add to this complexity, the sensitivity of sexual differentiation in a species to a given environmental cue is also highly specific [32]. For example, the European seabass is sensitive to temperature, but not pH or density [132], tilapia are sensitive to temperature, but not salinity, density or confinement [32], and although not an aquaculture species, zebrafish appear sensitive to a broad range of factors including temperature, density and hypoxia [133,134,135,136]. An important future challenge for species of commercial importance will be to document species-specific characterisations of the environmental stimuli able to influence sexual differentiation processes, and further to identify the sensitive period during development during which treatments are most effective. Where this is achievable, environmental manipulations provide a particularly attractive technique for aquaculture production, as they provide a relatively consumer-friendly approach to sex control.

3.4. Selection for Altered Sex Ratios

Powerful and ever more accessible genomic approaches have greatly contributed to the rapid discovery of genes, gene functions and pathways associated with sex differentiation and maintenance of gonadal state in commercially important fish species [34]. The use of next generation sequencing and high throughput SNP genotyping technologies have allowed for high-resolution linkage mapping and quantitative trait loci (QTL) studies and their potential use in marker assisted selection (MAS) breeding schemes [137]. Although identification of sex-specific markers in fishes has traditionally been performed with low throughput random amplified polymorphic DNA (RAPD) markers (e.g., sex-specific DNA markers #1 and #2 from African catfish, *Clarias gariepinus*, *CgaY1* and *CgaY2* respectivley [138,139]), high density marker genome screening using restriction site associated DNA (RAD) markers are enabling the rapid identification of sex-associated loci and the development of sex-specific markers in species with genetically determined systems (e.g., Nile tilapia X-linked *Oni*23063 and *Oni*28137 SNP markers [140] and *Oni3161* in linkage group 20 [141]). The recent discovery that the underlying mechanism determining sex in fish may not be exclusively related to sex-specific genes (or alleles) themselves, but in fact related to epigenetic modifications, has resulted in a broadened perspective of where and how we should be looking at the DNA [59]. Nonetheless, the three complimentary fields of genome mapping [142], epigenetics [35] and selection for altered sex ratios [143] are likely to deepen our understanding on how to effectively manipulate and control sex differentiation in fishes.

3.4.1. Heritability and Potential for Selection of Altered Sex Ratios

For selective breeding to be effective, biases in sex ratio in a population must be (i) a quantitative genetic trait and (ii) under additive genetic control (heritable). As yet, there is very little information on the heritability of sex ratios in most commercial species, and for those species for which data do exist, heritability for biased sex ratios has been shown to be nearly 2-fold that of the most selected trait in aquaculture, growth (e.g., Nile tilapia: h^2_{sex} = 0.77 *vs.* h^2_{weight} = 0.31 [19,144], rainbow trout: h^2_{sex} = 0.67 *vs.* h^2_{weight} = 0.37 [145,146], European seabass: h^2_{sex} = 0.62 *vs.* h^2_{weight} = 0.41 [43,147]). These data suggest that within a population, individuals are genetically inclined to produce families with either more male, or more female offspring. Indeed, trials with Nile tilapia have shown that sex ratio can be selected [20]. Male Nile tilapia originating from a family comprised of >90% males produced progeny with equally biased sex ratios (85%–100%), regardless of whether they were mated with females from families with balanced, or male biased sex ratios [20]. However, selected offspring did not always produce similarly biased sex ratios to parents, suggesting that in some cases, the trait was not inherited [148,149]. Wessels and Hörstgen-Schwark [144] later showed that sex ratios may be influenced not by the sex-determining genes themselves, but their thermo-sensitivity (because sex ratio in families exhibited changes dependent on whether 9 day old progeny were exposed to 36 °C for 10 days). Over three generations of divergent selection for a high male line (families with >80% male offspring) and a low male line (families with <60% male offspring), Wessels and Hörstgen-Schwark [143] observed changes in the proportion of males to 92.7% and 50.4%, respectively. The authors also showed that for sex ratios to be selected for progenies must be exposed to temperature treatments during the thermo-sensitive period, suggesting an epigenetic mechanism is also at play. Similar results to tilapia have since been found within a single generation of rainbow trout [145].

3.4.2. The Use of Quantitative Trait Loci to Improve Sex Control

When variable sex ratios within a species exist within a population, or among different families, due to both genetic and environmental effects, sex can be regarded as a quantitative threshold trait [43,150]. In such instances, QTL mapping can assist in determining the number and effect of chromosomal regions determining sex, as well as to narrow down regions of the genome where major sex determining genes are located [137]. QTLs have been identified for various phenotypic traits, including sex traits, in over 40 aquaculture species [137,142]. Most of these efforts have been directed towards finding sex, or early maturation QTLs in species such as tilapia, sea bream *Sparus aurata*, rainbow trout, turbot [21] and others [140,151,152,153,154,155,156]. In Nile tilapia, for instance, QTLs for sex-determination have been detected on linkage groups 1 and 23, with the key sex gene *amh* residing within the middle QTL of linkage group 23 [153]. In turbot the *Sma-USC30* microsatellite on LG5 has been identified and used in commercial hatcheries as a precocious sex marker and to establish expected sex ratios in progeny [157]. Ultimately, QTL mapping aims to provide useful sex-linked and temperature-dependent sex reversal markers as a tool to accelerate the genetic improvement of monosex populations through selective breeding programs [140,152,155,158,159]. Alternatives to such DNA markers include a number of transcriptomic approaches, which aim to screen for genes that are differentially expressed between testis and ovaries [35]. The use of RNA-Seq technology to identify sex-biased gene expression has greatly contributed to the identification of major coding sequences underlying sex-determination/differentiation pathways in an increasing number of aquaculture species such as the Asian sea bass [160], olive flounder [161], sharpsnout seabream (*Diplodus puntazzo*) [162] and African cichlids [163].

4. OTHER METHODS AND FUTURE DIRECTIONS

Recent developments in genetic technologies have introduced the opportunity to control sex through the direct targeting of genes involved in either sex determination and/or reproductive maturation. Two of these technologies, namely antisense RNA gene knock-down and insertion of transgenes to achieve transgenic sterilisation hold greatest potential for commercial application [164]. Despite this, there are few examples of their application to control sex at this time and this paucity of examples is likely due to an inefficiency in achieving complete and reliable outcomes [165]. Whilst this inefficacy may be due to technical issues, there also are regulatory hurdles that need to be overcome. Regardless of these issues, preliminary work on controlling sex via direct targeting of sex and reproductive genes has commenced in several species including tilapia, rainbow trout and common carp (*Cyprinus carpio*) [166,167,168,169]. Studies with these three species have targeted key reproductive hormone genes like gonadotropin-releasing hormone (GnRH), a key hormone released from the hypothalamus stimulating the release of pituitary gonadotropins, follicle-stimulating hormone (FSH) and luteinizing hormone (LH), which signal the gonads to differentiate and mature. In rainbow trout, for example, Uzbekova *et al.* [166] were one of the first to employ a single-stranded antisense RNA silencing approach using DNA from the Atlantic salmon GnRH3 gene. In this experiment, *GnRH* mRNA in transgenic rainbow trout was down regulated, however, the effect on FSH and LH levels was not significant. Despite this, Uzbekova *et al.* [166] showed that maturation among treated fish showed increased variability than those in the control group. Similarly in tilapia, females with a GnRH3 construct under the control of a constitutive common carp *β-actin* promoter were half as fertile as control females [167]. Whilst these studies using transgenic anti-sense technology were not able to induce complete sterility, they highlight the potential of this approach to control sexual processes in aquaculture fish species in the future.

5. CONCLUDING REMARKS

Sexual development is a complex, often species-specific process in fish. Therefore, technological innovations that successfully control sex in one species are certainly not guaranteed to work in another. Despite this complexity, there is an armory of approaches available to farmers to enable them to influence sex in their culture species, and many of these have now achieved commercial implementation. Our knowledge of the genes involved in sexual development and the mechanisms by which environmental modifiers can induce phenotypic changes is also rapidly increasing, due to the ease with which whole-genome sequencing and transcriptomic studies can now be conducted. Incorporation of current approaches to control sex with this new genetic (and epigenetic) understanding will undoubtedly lead to further advances in sex control of fish and will be a significant catalyst for selective breeding and the culture of more productive populations of farmed fish in the near future.

ACKNOWLEDGMENTS

The authors acknowledge the reviewers and the editor for their helpful suggestions and comments.

REFERENCES

1. Mires, D. Production of aquatic animals. In *The Tilapias*; Nash, C.E., Novotony, A.J., Eds.; Elsevier: New York, NY, USA, 1995; pp. 133–152.
2. Lawrence, C.S.; Morrissy, N.M.; Vercoe, P.E.; Williams, I.H. Hybridization in Australian freshwater crayfish—Production of all-male progeny. *J. World Aquac. Soc.* **2000**, *31*, 651–658.
3. Piferrer, F.; Beaumont, A.; Falguière, J.-C.; Flajšhans, M.; Haffray, P.; Colombo, L. Polyploid fish and shellfish: Production, biology and applications to aquaculture for performance improvement and genetic containment. *Aquaculture* **2009**, *293*, 125–156.
4. Ridha, M.T. Evaluation of monosex culture of GIFT and non-improved strains of Nile tilapia *Oreochromis niloticus* in recirculating tanks. *Int. Aquat. Res.* **2011**, *3*, 189.
5. Coman, F.E.; Sellars, M.J.; Norris, B.J.; Coman, G.J.; Preston, N.P. The effects of triploidy on *Penaeus (Marsupenaeus) japonicus* (Bate) survival, growth and gender when compared to diploid siblings. *Aquaculture* **2008**, *276*, 50–59.
6. Lin, S.; Benfey, T.J.; Martin-Robichaud, D.J. Hormonal sex reversal in Atlantic cod, *Gadus morhua*. *Aquaculture* **2012**, *364*, 192–197.
7. Tayamen, M.M.; Shelton, W.L. Inducement of sex reversal in *Sarotherodon niloticus* (Linnaeus). *Aquaculture* **1978**, *14*, 349–354.
8. Kwon, J.Y.; Haghpanah, V.; Kogson-Hurtado, L.M.; McAndrew, B.J.; Penman, D.J. Masculinization of genetic female Nile tilapia (*Oreochromis niloticus*) by dietary administration of an aromatase inhibitor during sexual differentiation. *J. Exp. Zool.* **2000**, *287*, 46–53.
9. Bhandari, R.K.; Higa, M.; Nakamura, S.; Nakamura, M. Aromatase inhibitor induces complete sex change in the protogynous honeycomb grouper (*Epinephelus merra*). *Mol. Reprod. Dev.* **2004**, *67*, 303–307.
10. Pruginin, Y.; Rothbard, S.; Wohlfarth, G.; Halevy, A.; Moav, R.; Hulata, G. All-male broods of *Tilapia nilotica* × *T. aurea* hybrids. *Aquaculture* **1975**, *6*, 11–21.
11. Wolters, W.R.; de May, R. Production characteristics of striped bass x white bass and striped bass x yellow bass hybrids. *J. World Aquac. Soc.* **1996**, *27*, 202–207.

12. Chourrout, D.; Quillet, E. Induced gynogenesis in the rainbow trout: Sex and survival of progenies production of all-triploid populations. *Theor. Appl. Genet.* **1982**, *63*, 201–205.
13. Galbreath, P.F.; Thorgaard, G.H. Saltwater performance of all-female triploid Atlantic salmon. *Aquaculture* **1995**, *138*, 77–85.
14. Benfey, T.J.; Sutterlin, A.M. Triploidy induced by heat shock and hydrostatic pressure in landlocked Atlantic salmon (Salmo salar L.). *Aquaculture* **1984**, *36*, 359–367.
15. Liu, M.; Sadovy de Mitcheson, Y. The influence of social factors on juvenile sexual differentiation in a diandric, protogynous grouper *Epinephelus coioides*. *Ichthyol. Res.* **2011**, *58*, 84–89.
16. Navarro-Martín, L.; Viñas, J.; Ribas, L.; Díaz, N.; Gutiérrez, A.; di Croce, L.; Piferrer, F. DNA methylation of the gonadal aromatase (*cyp19a*) promoter is involved in temperature-dependent sex ratio shifts in the European sea bass. *PLoS Genet.* **2011**, *7*, e1002447.
17. Koumoundouros, G.; Pavlidis, M.; Anezaki, L.; Kokkari, C.; Sterioti, A.; Divanach, P.; Kentouri, M. Temperature sex determination in the European sea bass, *Dicentrarchus labrax* (L., 1758) (Teleostei, Perciformes, Moronidae): Critical sensitive ontogenetic phase. *J. Exp. Zool.* **2002**, *292*, 573–579.
18. Saillant, E.; Fostier, A.; Haffray, P.; Menu, B.; Thimonier, J.; Chatain, B. Temperature effects and genotype-temperature interactions on sex determination in the European sea bass (*Dicentrarchus labrax* L.). *J. Exp. Zool.* **2002**, *292*, 494–505.
19. Bentsen, H.B.; Gjerde, B.; Nguyen, N.H.; Rye, M.; Ponzoni, R.W.; Palada de Vera, M.S.; Bolivar, H.L.; Velasco, R.R.; Danting, J.C.; Dionisio, E.E.; *et al.* Genetic improvement of farmed tilapias: Genetic parameters for body weight at harvest in Nile tilapia *(Oreochromis niloticus)* during five generations of testing in multiple environments. *Aquaculture***2012**, *338–341*, 56–65.
20. Wedekind, H. *Studies on the Inheritance of Sex with African Cichlids (Oreochromis Niloticus) (in German)*; Göttingen University: Göttingen, Germany, 1987.
21. Martínez, P.; Bouza, C.; Hermida, M.; Fernández, J.; Toro, M.A.; Vera, M.; Pardo, B.; Millán, A.; Fernández, C.; Vilas, R. Identification of the major sex-determining region of turbot (*Scophthalmus maximus*). *Genetics* **2009**, *183*, 1443–1452.
22. Evelyn, G.D.J.-A.; Felix, G.A.; Valentin, T. Early development and seed production of Asian Seabass, *Lates calcarifer*. In *Biology and Culture of Asian Seabass Lates Calcarifer*; Jerry, D.R., Ed.; CRC Press: Boca Raton, FL, USA, 2013; pp. 17–30.
23. Palmer, P.; Blackshaw, A.; Garrett, R. Successful fertility experiments with cryopreserved spermatozoa of barramundi, *Lates calcarifer* (Bloch), using dimethylsulfoxide and glycerol as cryoprotectants. *Reprod. Fertil. Dev.* **1993**, *5*, 285–293.
24. Jerry, D.R.; James Cook University, Townsville, Australia. Personal observation, 2014.
25. Quinitio, G.F.; Caberoy, N.B.; Reyes, D.M., Jr. Induction of sex change in female *Epinephelus coioides* by social control. *Isr. J. Aquac. Bamidgeh* **1997**, *49*, 77–83.
26. Mackie, M.C. Socially controlled sex-change in the half-moon grouper, *Epinephelus rivulatus*, at Ningaloo Reef, Western Australia. *Coral Reefs* **2003**, *22*, 133–142.
27. Knuckey, R.; Finfish Enterprise, Cairns, Australia. Personal communication, 2014.
28. Penman, D.J.; Piferrer, F. Fish gonadogenesis. Part I: Genetic and environmental mechanisms of sex determination. *Rev. Fish. Sci.* **2008**, *16*, 16–34.
29. Piferrer, F.; Blazquez, M.; Navarro, L.; Gonzalez, A. Genetic, endocrine, and environmental components of sex determination and differentiation in the European sea bass (*Dicentrarchus labrax* L.). *Gen. Comp. Endocrinol.* **2005**, *142*, 102–110.

30. Devlin, R.H.; Nagahama, Y. Sex determination and sex differentiation in fish: An overview of genetic, physiological, and environmental influences. *Aquaculture* **2002**, *208*, 191–364.
31. Guiguen, Y.; Fostier, A.; Piferrer, F.; Chang, C.-F. Ovarian aromatase and estrogens: A pivotal role for gonadal sex differentiation and sex change in fish. *Gen. Comp. Endocrinol.* **2010**, *165*, 352–366.
32. Baroiller, J.F.; D'Cotta, H.; Saillant, E. Environmental effects on fish sex determination and differentiation. *Sex. Dev.***2009**, *3*, 118–135.
33. Kobayashi, Y.; Nagahama, Y.; Nakamura, M. Diversity and plasticity of sex determination and differentiation in fishes. *Sex. Dev.* **2013**, *7*, 115–125.
34. Heule, C.; Salzburger, W.; Böhne, A. Genetics of sexual development: An evolutionary playground for fish. *Genetics***2014**, *196*, 579–591.
35. Piferrer, F.; Ribas, L.; Díaz, N. Genomic approaches to study genetic and environmental influences on fish sex determination and differentiation. *Mar. Biotechnol.* **2012**, *14*, 591–604.
36. Koopman, P.; Gubbay, J.; Vivian, N.; Goodfellow, P.; Lovell-Badge, R. Male development of chromosomally female mice transgenic for *sry*. *Nature* **1991**, *351*, 117–121.
37. Baroiller, J.F.; D'Cotta, H. Environment and sex determination in farmed fish. *Comp. Biochem. Physiol. Part C* **2001**, *130*, 399–409.
38. Campos, H.H. Karyology of three Galaxiid fishes *Galaxias maculatus, G. platei,* and *Brachygalaxias bullocki*. *Copeia* **1972**, *1972*, 368–370.
39. Bertollo, L.A.; Mestriner, C.A. The X1X2Y sex chromosome system in the fish *Hoplias malabaricus*. II. Meiotic analyses. *Chromosome Res.* **1998**, *6*, 141–147.
40. Oliveira, C.; Foresti, F.; Hilsdorf, A. Genetics of neotropical fish: From chromosomes to populations. *Fish Physiol. Biochem.* **2009**, *35*, 81–100.
41. Kosswig, C. Polygenic sex determination. *Experientia* **1964**, *20*, 190–199.
42. Liew, W.C.; Orban, L. Zebrafish sex: A complicated affair. *Brief. Funct. Genomics* **2014**, *13*, 172–187.
43. Vandeputte, M.; Dupont-Nivet, M.; Chavanne, H.; Chatain, B. A polygenic hypothesis for sex determination in the European seabass *Dicentrarchus labrax*. *Genetics* **2007**, *176*, 1049–1057.
44. Ross, J.A.; Peichel, C.L. Molecular Cytogenetic Evidence of rearrangements on the Y chromosome of the threespine stickleback fish. *Genetics* **2008**, *179*, 2173–2182.
45. Kamiya, T.; Kai, W.; Tasumi, S.; Oka, A.; Matsunaga, T.; Mizuno, N.; Fujita, M.; Suetake, H.; Suzuki, S.; Hosoya, S. A trans-species missense SNP in *amhr2* is associated with sex determination in the tiger pufferfish, *Takifugu rubripes*(fugu). *PLoS Genet.* **2012**, *8*, e1002798.
46. Wilson, C.A.; High, S.K.; McCluskey, B.M.; Amores, A.; Yan, Y.-l.; Titus, T.A.; Anderson, J.L.; Batzel, P.; Carvan, M.J.; Schartl, M. Wild sex in zebrafish: Loss of the natural sex determinant in domesticated strains. *Genetics* **2014**, *114*, 1291–1308.
47. Matsuda, M.; Nagahama, Y.; Shinomiya, A.; Sato, T.; Matsuda, C.; Kobayashi, T.; Morrey, C.E.; Shibata, N.; Asakawa, S.; Shimizu, N. *Dmy* is a Y-specific DM-domain gene required for male development in the medaka fish. *Nature* **2002**, *417*, 559–563.
48. Nanda, I.; Kondo, M.; Hornung, U.; Asakawa, S.; Winkler, C.; Shimizu, A.; Shan, Z.; Haaf, T.; Shimizu, N.; Shima, A. A duplicated copy of *dmrt1* in the sex-determining region of the Y chromosome of the medaka, *Oryzias latipes*. *Proc. Natl. Acad. Sci. USA* **2002**, *99*, 11778–11783.
49. Hattori, R.S.; Strussmann, C.A.; Fernandino, J.I.; Somoza, G.M. Genotypic sex determination in teleosts: Insights from the testis-determining *amhy* gene. *Gen. Comp. Endocrinol.* **2013**, *192*, 55–59.
50. Hattori, R.S.; Murai, Y.; Oura, M.; Masuda, S.; Majhi, S.K.; Sakamoto, T.; Fernandino, J.I.; Somoza, G.M.; Yokota, M.; Strüssmann, C.A. A Y-linked anti-Müllerian hormone duplication takes over a critical role in sex determination. *Proc. Natl. Acad. Sci. USA* **2012**, *109*, 2955–2959.

51. Myosho, T.; Otake, H.; Masuyama, H.; Matsuda, M.; Kuroki, Y.; Fujiyama, A.; Naruse, K.; Hamaguchi, S.; Sakaizumi, M. Tracing the emergence of a novel sex-determining gene in medaka, *Oryzias luzonensis*. *Genetics* **2012**, *191*, 163–170.
52. Yano, A.; Guyomard, R.; Nicol, B.; Jouanno, E.; Quillet, E.; Klopp, C.; Cabau, C.; Bouchez, O.; Fostier, A.; Guiguen, Y. An immune-related gene evolved into the master sex-determining gene in rainbow trout, *Oncorhynchus mykiss*. *Curr. Biol.* **2012**, *22*, 1423–1428.
53. Takehana, Y.; Matsuda, M.; Myosho, T.; Suster, M.L.; Kawakami, K.; Shin-I, T.; Kohara, Y.; Kuroki, Y.; Toyoda, A.; Fujiyama, A.; *et al.* Co-option of as the male-determining factor on the Y chromosome in the fish *Oryzias dancena*. *Nat. Commun.* **2014**, *5*, 4157.
54. Kent, J.; Wheatley, S.C.; Andrews, J.E.; Sinclair, A.H.; Koopman, P. A male-specific role for *sox9* in vertebrate sex determination. *Development* **1996**, *122*, 2813–2822.
55. Smith, C.A.; Roeszler, K.N.; Ohnesorg, T.; Cummins, D.M.; Farlie, P.G.; Doran, T.J.; Sinclair, A.H. The avian Z-linked gene *dmrt1* is required for male sex determination in the chicken. *Nature* **2009**, *461*, 267–271.
56. Wang, D.-S.; Kobayashi, T.; Zhou, L.-Y.; Paul-Prasanth, B.; Ijiri, S.; Sakai, F.; Okubo, K.; Morohashi, K.-I.; Nagahama, Y. *Foxl2* up-regulates aromatase gene transcription in a female-specific manner by binding to the promoter as well as interacting with Ad4 binding protein/steroidogenic factor 1. *Mol. Endocrinol.* **2007**, *21*, 712–725.
57. Cutting, A.; Chue, J.; Smith, C.A. Just how conserved is vertebrate sex determination? *Dev. Dyn.* **2013**, *242*, 380–387.
58. Trukhina, A.V.; Lukina, N.A.; Wackerow-Kouzova, N.D.; Smirnov, A.F. The variety of vertebrate mechanisms of sex determination. *BioMed Res. Int.* **2013**, *2013*, 58746.
59. Piferrer, F. Epigenetics of sex determination and gonadogenesis. *Dev. Dyn.* **2013**, *242*, 360–370.
60. Russo, V.E.; Martienssen, R.A.; Riggs, A.D. *Epigenetic Mechanisms of Gene Regulation*; Cold Spring Harbor Laboratory Press: Cold Spring Harbor, NY, USA, 1996.
61. Turner, B.M. Epigenetic responses to environmental change and their evolutionary implications. *Philos. Trans. R. Soc. Lond. B* **2009**, *364*, 3403–3418.
62. Shao, C.; Li, Q.; Chen, S.; Zhang, P.; Lian, J.; Hu, Q.; Sun, B.; Jin, L.; Liu, S.; Wang, Z. Epigenetic modification and inheritance in sexual reversal of fish. *Genome Res.* **2014**, *24*, 604–615.
63. Conover, D.; Kynard, B. Environmental sex determination: Interaction of temperature and genotype in a fish. *Science***1981**, *213*, 31.
64. Karube, M.; Fernandino, J.I.; Strobl-Mazzulla, P.; Strüssmann, C.A.; Yoshizaki, G.; Somoza, G.M.; Patiño, R. Characterization and expression profile of the ovarian cytochrome P450 aromatase (*cyp19A1*) gene during thermolabile sex determination in pejerrey, *Odontesthes bonariensis*. *J. Exp. Zool. Part A* **2007**, *307*, 625–636.
65. Ospina-Alvarez, N.; Piferrer, F. Temperature-dependent sex determination in fish revisited: Prevalence, a single sex ratio response pattern, and possible effects of climate change. *PLoS ONE* **2008**, *3*, e2837.
66. Socorro, S.; Martins, R.S.; Deloffre, L.; Mylonas, C.C.; Canario, A.V. A cDNA for European sea bass (*Dicentrachus labrax*) *11β-hydroxylase*: Gene expression during the thermosensitive period and gonadogenesis. *Gen. Comp. Endocrinol.* **2007**, *150*, 164–173.
67. Valenzuela, N.; Adams, D.C.; Janzen, F.J. Pattern does not equal process: Exactly when is sex environmentally determined? *Am. Nat.* **2003**, *161*, 676–683.
68. Nakamura, M.; Hourigan, T.; Yamauchi, K.; Nagahama, Y.; Grau, E.G. Histological and ultrastructural evidence for the role of gonadal steroid hormones in sex

change in the protogynous wrasse *Thalassoma duperrey*. *Environ. Biol. Fishes* **1989**, *24*, 117–136.
69. Godwin, J.; Luckenbach, J.A.; Borski, R.J. Ecology meets endocrinology: Environmental sex determination in fishes. *Evolut. Dev.* **2003**, *5*, 40–49.
70. Rahman, M.A.; Arshad, A.; Marimuthu, K.; Ara, R.; Amin, S. Inter-specific hybridization and its potential for aquaculture of fin fishes. *Asian J. Anim. Vet. Adv.* **2013**, *8*, 139–158.
71. Yamamoto, T.-O. Sex differentiation. In *Fish Physiology*; Hoar, W.S., Randall, D.J., Eds.; Academic Press: New York, NY, USA, 1969; Volume 3, pp. 117–175.
72. Yamamoto, T.-O. Artificially induced sex-reversal in genotypic males of the medaka (*Oryzias latipes*). *J. Exp. Zool.* **1953**, *123*, 571–594.
73. Yamamoto, T.-O. Artificial induction of functional sex-reversal in genotypic females of the medaka (*Oryzias latipes*). *J. Exp. Zool.* **1958**, *137*, 227–263.
74. Pandian, T.; Sheela, S. Hormonal induction of sex reversal in fish. *Aquaculture* **1995**, *138*, 1–22.
75. Beardmore, J.A.; Mair, G.C.; Lewis, R.I. Monosex male production in finfish as exemplified by tilapia: Applications, problems, and prospects. *Aquaculture* **2001**, *197*, 283–301.
76. Bhandari, R.K.; Alam, M.A.; Higa, M.; Soyano, K.; Nakamura, M. Evidence that estrogen regulates the sex change of honeycomb grouper (*Epinephelus merra*), a protogynous hermaphrodite fish. *J. Exp. Zool. Part A Ecol. Genet. Physiol.* **2005**, *303A*, 497–503.
77. Li, G.-L.; Liu, X.-C.; Lin, H.-R. Effects of aromatizable and nonaromatizable androgens on the sex inversion of red-spotted grouper (*Epinephelus akaara*). *Fish Physiol. Biochem.* **2006**, *32*, 25–33.
78. Anderson, T.A.; Forrester, J. Administration of oestradiol to barramundi, *Lates calcarifer*, induces protandrous sex change. In *Perspective in Comparative Endocrinology, Unity and Diversity*; Goos, H.I.T., Rastogi, R.K., Vaudry, H., Pierantoni, R., Eds.; Monduzzi Editore: Bologna, Napoli, Italy, 2001; pp. 155–164.
79. Sower, S.A.; Dickhoff, W.W.; Flagg, T.A.; Mighell, J.L.; Mahnken, C.V. Effects of estradiol and diethylstilbesterol on sex reversal and mortality in Atlantic salmon (*Salmo salar*). *Aquaculture* **1984**, *43*, 75–81.
80. Padoa, E. Further observations on the differentiation of sex, normal and modified by the administration of follicular hormone, in the iridée trout (*Salmo irideus*) (in French). *Biomorphosis* **1939**, *1*, 337–354.
81. Hackmann, E.; Reinboth, R. Delimitation of the critical stage of hormone-influenced sex differentiation in *Hemihaplochromis multicolor* (Hilgendorf) (Cichlidae). *Gen. Comp. Endocrinol.* **1974**, *22*, 42–53.
82. Nakamura, M.; Takahashi, H. Gonadal sex differentiation in *Tilapia mossambica*, with special regard to the time of estrogen treatment effective in inducing complete feminization of genetic males. *Bull. Fac. Fish. Hokkaido Univ.* **1973**, *24*, 1–13.
83. Nakamura, M. Effects of estradiol-17β on gonadal sex differentiation in two species of salmonids, the masu salmon, *Oncorhynchus masou*, and the chum salmon, *O. keta*. *Aquaculture* **1984**, *43*, 83–90.
84. Piferrer, F.; Donaldson, E.M. Gonadal differentiation in coho salmon, *Oncorhynchus kisutch*, after a single treatment with androgen or estrogen at different stages during ontogenesis. *Aquaculture* **1989**, *77*, 251–262.
85. Van den Hurk, R.; Slof, G.A. A morphological and experimental study of gonadal sex differentiation in the rainbow trout, *Salmo gairdneri*. *Cell Tissue Res.* **1981**, *218*, 487–497.
86. Patiño, R.; Davis, K.B.; Schoore, J.E.; Uguz, C.; Strüssmann, C.A.; Parker, N.C.; Simco, B.A.; Goudie, C.A. Sex differentiation of channel catfish gonads: Normal development and effects of temperature. *J. Exp. Zool.* **1996**, *276*, 209–218.
87. Piferrer, F.; Baker, I.J.; Donaldson, E.M. Effects of natural, synthetic, aromatizable, and nonaromatizable androgens in inducing male sex differentiation in genotypic

female Chinook salmon (*Oncorhynchus tshawytscha*). *Gen. Comp. Endocrinol.* **1993**, *91*, 59–65.
88. Piferrer, F. Endocrine sex control strategies for the feminization of teleost fish. *Aquaculture* **2001**, *197*, 229–281.
89. Blázquez, M.; Piferrer, F.; Zanuy, S.; Carrillo, M.; Donaldson, E.M. Development of sex control techniques for European sea bass (*Dicentrarchus labrax L.*) aquaculture: Effects of dietary 17 α-methyltestosterone prior to sex differentiation. *Aquaculture* **1995**, *135*, 329–342.
90. Kobayashi, T.; Kajiura-Kobayashi, H.; Nagahama, Y. Induction of XY sex reversal by estrogen involves altered gene expression in a teleost, tilapia. *Cytogenet. Genome Res.* **2003**, *101*, 289–294.
91. Kitano, T.; Takamune, K.; Nagahama, Y.; Abe, S.-I. Aromatase inhibitor and 17α-methyltestosterone cause sex-reversal from genetical females to phenotypic males and suppression of P450 aromatase gene expression in Japanese flounder (*Paralichthys olivaceus*). *Mol. Reprod. Dev.* **2000**, *56*, 1–5.
92. Crim, L.W. Methods for acute and chronic hormone administration in fish. In *Reproduction and Culture of Milkfish*; Lee, C.S., Liao, I.C., Eds.; Tungkang Marine Laboratory: Tungkang, Taiwan, 1985; pp. 1–13.
93. Chatain, B.; Saillant, E.; Peruzzi, S. Production of monosex male populations of European seabass, *Dicentrarchus labrax* L. by use of the synthetic androgen 17α-methyldehydrotestosterone. *Aquaculture* **1999**, *178*, 225–234.
94. Nakamura, M. Dosage-dependent changes in the effect of oral administration of methyltestosterone on gonadal sex differentiation in *Tilapia mossambica*. *Bull. Fac. Fish. Hokkaido Univ.* **1975**, *26*, 99–108.
95. Hunter, G.A.; Solar, I.I.; Baker, I.J.; Donaldson, E.M. Feminization of coho salmon (*Oncorhynchus kisutch*) and chinook salmon (*Oncorhynchus tshawytscha*) by immersion of alevins in a solution of estradiol-17β. *Aquaculture* **1986**, *53*, 295–302.
96. Feist, G.; Yeoh, C.-G.; Fitzpatrick, M.S.; Schreck, C.B. The production of functional sex-reversed male rainbow trout with 17α-methyltestosterone and 11 β-hydroxyandrostenedione. *Aquaculture* **1995**, *131*, 145–152.
97. Parliament, E. Directive 96/22/EC concerning the prohibition on the use in stockfarming of certain substances having a hormonal or thyrostatic action and of β-agonists, and repealing Directives 81/602/EEC, 88/146/EEC and 88/299/EEC. *Off. J. Eur. Union* **1996**, *L25*, 3–9.
98. Parliament, E. Directive 2003/74/EC of the European Parliament and of the Council amending Council Directive 96/22/EC concerning the prohibition on the use in stockfarming of certain substances having a hormonal or thyrostatic action and of beta-agonists. *Off. J. Eur. Union* **2003**, *L262*, 19–21.
99. Pandian, T.J.; Kirankumar, S. Recent advances in hormonal induction of sex-reversal in fish. *J. Appl. Aquac.* **2003**, *13*, 205–230.
100. Felip, A.; Zanuy, S.; Carrillo, M.; Piferrer, F. Induction of triploidy and gynogenesis in teleost fish with emphasis on marine species. *Genetica* **2001**, *111*, 175–195.
101. Hulata, G. Genetic manipulations in aquaculture: A review of stock improvement by classical and modern technologies. *Genetica* **2001**, *111*, 155–173.
102. Tiwary, B.K.; Kirubagaran, R.; Ray, A.K. The biology of triploid fish. *Rev. Fish Biol. Fish.* **2004**, *14*, 391–402.
103. Ihssen, P.; McKay, L.; McMillan, I.; Phillips, R. Ploidy manipulation and gynogenesis in fishes: Cytogenetic and fisheries applications. *Trans. Am. Fish. Soc.* **1990**, *119*, 698–717.
104. Myers, J.M. Tetraploid induction in *Oreochromis* spp. *Aquaculture* **1986**, *57*, 281–287.
105. Colas, P.; Dubé, F. Meiotic maturation in mollusc oocytes. *Seminars Cell Dev. Biol.* **1998**, *9*, 539–548.
106. Guo, X.; Allen, S.K. Reproductive potential and genetics of triploid Pacific oysters, *Crassostrea gigas* (Thunberg). *Biol. Bull.* **1994**, *187*, 309–318.

107. Cuñado, N.; Terrones, J.; Sánchez, L.; Martínez, P.; Santos, J. Sex-dependent synaptic behaviour in triploid turbot, *Scophthalmus maximus* (Pisces, Scophthalmidae). *Heredity* **2002**, *89*, 460–464.
108. Benfey, T.J. The physiology and behavior of triploid fishes. *Rev. Fish. Sci.* **1999**, *7*, 39–67.
109. Cassani, J.R. *Managing Aquatic Vegetation with Grass Carp. A Guide for Water Resource Managers*; Cassani, J.R., Ed.; American Fisheries Society: Bethesda, MD, USA, 1996.
110. Varadaraj, K.; Pandian, T. Production of all-female sterile-triploid *Oreochromis mossambicus*. *Aquaculture* **1990**, *84*, 117–123.
111. Hussain, M.G.; Chatterji, A.; McAndrew, B.; Johnstone, R. Triploidy induction in Nile tilapia, *Oreochromis niloticus* L. using pressure, heat and cold shocks. *Theor. Appl. Genet.* **1991**, *81*, 6–12.
112. Mair, G.C. Chromosome-set manipulation in tilapia—Techniques, problems and prospects. *Aquaculture* **1993**, *111*, 227–244.
113. Byamungu, N.; Darras, V.; Kühn, E. Growth of heat-shock induced triploids of blue tilapia, *Oreochromis aureus*, reared in tanks and in ponds in Eastern Congo: Feeding regimes and compensatory growth response of triploid females. *Aquaculture* **2001**, *198*, 109–122.
114. Valenti, R. Induced polyploidy in *Tilapia aurea* (*Steindachner*) by means of temperature shock treatment. *J. Fish Biol.* **1975**, *7*, 519–528.
115. Fordham, B.S.; Trippel, E. Feeding behaviour of cod (*Gadus morhua*) in relation to spawning. *J. Appl. Ichthyol.* **1999**, *15*, 1–9.
116. Karlsen, Ø.; Holm, J.C.; Kjesbu, O.S. Effects of periodic starvation on reproductive investment in first-time spawning Atlantic cod (*Gadus morhua* L.). *Aquaculture* **1995**, *133*, 159–170.
117. Taranger, G.L.; Carrillo, M.; Schulz, R.W.; Fontaine, P.; Zanuy, S.; Felip, A.; Weltzien, F.-A.; Dufour, S.; Karlsen, Ø.; Norberg, B. Control of puberty in farmed fish. *Gen. Comp. Endocrinol.* **2010**, *165*, 483–515.
118. Garner, S.; Madison, B.; Bernier, N.; Neff, B. Juvenile growth and aggression in diploid and triploid Chinook salmon *Oncorhynchus tshawytscha* (Walbaum). *J. Fish Biol.* **2008**, *73*, 169–185.
119. Felip, A.; Zanuy, S.; Carrillo, M.; Martínez, G.; Ramos, J.; Piferrer, F. Optimal conditions for the induction of triploidy in the sea bass (*Dicentrarchus labrax* L.). *Aquaculture* **1997**, *152*, 287–298.
120. Piferrer, F.; Cal, R.M.; Álvarez-Blázquez, B.; Sánchez, L.; Martinez, P. Induction of triploidy in the turbot (*Scophthalmus maximus*): I. Ploidy determination and the effects of cold shocks. *Aquaculture* **2000**, *188*, 79–90.
121. Peruzzi, S.; Chatain, B. Pressure and cold shock induction of meiotic gynogenesis and triploidy in the European sea bass, *Dicentrarchus labrax* L.: Relative efficiency of methods and parental variability. *Aquaculture* **2000**, *189*, 23–37.
122. Piferrer, F.; Cal, R.M.; Gómez, C.; Bouza, C.; Martí, P. Induction of triploidy in the turbot (*Scophthalmus maximus*): II. Effects of cold shock timing and induction of triploidy in a large volume of eggs. *Aquaculture* **2003**, *220*, 821–831.
123. Cherfas, N.B.; Gomelsky, B.; Ben-Dom, N.; Peretz, Y.; Hulata, G. Assessment of triploid common carp (*Cyprinus carpio*L.) for culture. *Aquaculture* **1994**, *127*, 11–18.
124. Munday, P.L.; Wilson White, J.; Warner, R.R. A social basis for the development of primary males in a sex-changing fish. *Proc. R. Soc. B* **2006**, *273*, 2845–2851.
125. De Mitcheson, Y.S.; Liu, M. Functional hermaphroditism in teleosts. *Fish Fish.* **2008**, *9*, 1–43.
126. Baroiller, J.F.; Guiguen, Y.; Fostier, A. Endocrine and environmental aspects of sex differentiation in fish. *Cell. Mol. Life Sci.* **1999**, *55*, 910–931.
127. Davey, A.J.; Jellyman, D.J. Sex determination in freshwater eels and management options for manipulation of sex. *Rev. Fish Biol. Fish.* **2005**, *15*, 37–52.

128. Wickström, H.; Westin, L.; Clevestam, P. The biological and economic yield from a long-term eel stocking experiment. *Ecol. Freshwater Fish* **1996**, *5*, 140–147.
129. Rubin, D.A. Effect of pH on sex ratio in cichlids and a *poecilliid* (Teleostei). *Copeia* **1985**, *1985*, 233–235.
130. Römer, U.; Beisenherz, W. Environmental determination of sex in *Apistogrammai* (Cichlidae) and two other freshwater fishes (Teleostei). *J. Fish Biol.* **1996**, *48*, 714–725.
131. Papadaki, M.; Piferrer, F.; Zanuy, S.; Maingot, E.; Divanach, P.; Mylonas, C. Growth, sex differentiation and gonad and plasma levels of sex steroids in male-and female-dominant populations of Dicentrarchus labrax obtained through repeated size grading. *J. Fish Biol.* **2005**, *66*, 938–956.
132. Saillant, E.; Chatain, B.; Menu, B.; Fauvel, C.; Vidal, M.O.; Fostier, A. Sexual differentiation and juvenile intersexuality in the European sea bass *(Dicentrarchus labrax)*. *J. Zool.* **2003**, *260*, 53–63.
133. Uchida, D.; Yamashita, M.; Kitano, T.; Iguchi, T. An aromatase inhibitor or high water temperature induce oocyte apoptosis and depletion of P450 aromatase activity in the gonads of genetic female zebrafish during sex-reversal. *Comp. Biochem. Physiol. Part A* **2004**, *137*, 11–20.
134. Nusslein-Volhard, C.; Dahm, R. *Zebrafish*; Oxford University Press: Oxford, UK, 2002.
135. Shang, E.H.; Yu, R.M.; Wu, R.S. Hypoxia affects sex differentiation and development, leading to a male-dominated population in zebrafish (*Danio rerio*). *Environ. Sci. Technol.* **2006**, *40*, 3118–3122.
136. Lawrence, C.; Ebersole, J.P.; Kesseli, R.V. Rapid growth and out-crossing promote female development in zebrafish (*Danio rerio*). *Environ. Biol. Fishes* **2008**, *81*, 239–246.
137. Yue, G.H. Recent advances of genome mapping and marker-assisted selection in aquaculture. *Fish Fish.* **2014**, *15*, 376–396.
138. Kovács, B.; Egedi, S.; Bártfai, R.; Orbán, L. Male-specific DNA markers from African catfish (*Clarias gariepinus*). *Genetica* **2000**, *110*, 267–276.
139. Pandian, T. *Sex Determination in Fish*; CRC Press: Boca Raton, FL, USA, 2011.
140. Palaiokostas, C.; Bekaert, M.; Khan, M.G.Q.; Taggart, J.B.; Gharbi, K.; McAndrew, B.J.; Penman, D.J. Mapping and validation of the major sex-determining region in Nile tilapia (*Oreochromis niloticus* L.) using RAD sequencing. *PLoS ONE* **2013**, *8*, e68389.
141. Palaiokostas, C.; Bekaert, M.; Khan, M.G.; Taggart, J.B.; Gharbi, K.; McAndrew, B.J.; Penman, D.J. A novel sex-determining QTL in Nile tilapia (*Oreochromis niloticus*). *BMC Genomics* **2015**, *16*, 171.
142. Kocher, T.D.; Kole, C. *Genome Mapping and Genomics in Fishes and Aquatic Animals*; Springer: New York, NY, USA, 2008.
143. Wessels, S.; Hörstgen-Schwark, G. Temperature dependent sex ratios in selected lines and crosses with a YY-male in Nile tilapia (*Oreochromis niloticus*). *Aquaculture* **2011**, *318*, 79–84.
144. Wessels, S.; Hörstgen-Schwark, G. Selection experiments to increase the proportion of males in Nile tilapia (*Oreochromis niloticus*) by means of temperature treatment. *Aquaculture* **2007**, *272*, S80–S87.
145. Magerhans, A.; Hörstgen-Schwark, G. Selection experiments to alter the sex ratio in rainbow trout (*Oncorhynchus mykiss*) by means of temperature treatment. *Aquaculture* **2010**, *306*, 63–67.
146. Sae-Lim, P.; Kause, A.; Mulder, H.A.; Martin, K.E.; Barfoot, A.J.; Parsons, J.E.; Davidson, J.; Rexroad, C.E.; van Arendonk, J.A.M.; Komen, H. Genotype-by-environment interaction of growth traits in rainbow trout (*Oncorhynchus mykiss*): A continental scale study. *J. Anim. Sci.* **2013**, *91*, 5572–5581.
147. Dupont-Nivet, M.; Vandeputte, M.; Vergnet, A.; Merdy, O.; Haffray, P.; Chavanne, H.; Chatain, B.A. Heritabilities and GxE interactions for growth in the European

sea bass (*Dicentrarchus labrax* L.) using a marker-based pedigree. *Aquaculture* **2008**, *275*, 81–87.

148. Wohlfarth, G.W.; Wedekind, H. The heredity of sex determination in tilapias. *Aquaculture* **1991**, *92*, 143–156.
149. Phelps, R.P.; Carpenter, R.H. *Monosex Tilapia Production through Androgenesis: Selection of Individuals for Sex Inheritance Characteristics for Use in Monosex Production*; Oregon State University: Corvallis, OR, USA, 2002.
150. Grossen, C.; Neuenschwander, S.; Perrin, N. Temperature dependent turnovers in sex-determination mechanisms: A quantitative model. *Evolution* **2011**, *65*, 64–78.
151. Martyniuk, C.J.; Perry, G.M.L.; Mogahadam, H.K.; Ferguson, M.M.; Danzmann, R.G. The genetic architecture of correlations among growth-related traits and male age at maturation in rainbow trout. *J. Fish Biol.* **2003**, *63*, 746–764.
152. Cnaani, A. The tilapias' chromosomes influencing sex determination. *Cytogenet. Genome Res.* **2013**, *141*, 195–205.
153. Eshel, O.; Shirak, A.; Weller, J.I.; Hulata, G.; Ron, M. Linkage and physical mapping of sex region on LG23 of Nile tilapia (*Oreochromis niloticus*). *G3: Genes/Genomes/Genet.* **2012**, *2*, 35–42.
154. Loukovitis, D.; Sarropoulou, E.; Tsigenopoulos, C.S.; Batargias, C.; Magoulas, A.; Apostolidis, A.P.; Chatziplis, D.; Kotoulas, G. Quantitative trait loci involved in sex determination and body growth in the gilthead sea bream (*Sparus aurata*) through targeted genome scan. *PLoS ONE* **2011**, *6*, e16599.
155. Palaiokostas, C.; Bekaert, M.; Davie, A.; Cowan, M.; Oral, M.; Taggart, J.; Gharbi, K.; McAndrew, B.; Penman, D.; Migaud, H. Mapping the sex determination locus in the Atlantic halibut (*Hippoglossus hippoglossus*) using RAD sequencing. *BMC Genomics* **2013**, *14*, 566.
156. Shirak, A.; Seroussi, E.; Cnaani, A.; Howe, A.E.; Domokhovsky, R.; Zilberman, N.; Kocher, T.D.; Hulata, G.; Ron, M. *Amh* and *dmrta2* genes map to tilapia (*Oreochromis* spp.) linkage group 23 within quantitative trait locus regions for sex determination. *Genetics* **2006**, *174*, 1573–1581.
157. Martínez, P.; Viñas, A.M.; Sánchez, L.; Díaz, N.; Ribas, L.; Piferrer, F. Genetic architecture of sex determination in fish: Applications to sex ratio control in aquaculture. *Front. Genet.* **2014**, *5*, 340.
158. Lühmann, L.M.; Knorr, C.; Hörstgen-Schwark, G.; Wessels, S. First evidence for family-specific QTL for temperature-dependent sex reversal in Nile Tilapia *(Oreochromis niloticus)*. *Sex. Dev.* **2012**, *6*, 247–256.
159. Song, W.; Li, Y.; Zhao, Y.; Liu, Y.; Niu, Y.; Pang, R.; Miao, G.; Liao, X.; Shao, C.; Gao, F.; *et al.* Construction of a high-density microsatellite genetic linkage map and mapping of sexual and growth-related traits in half-smooth tongue sole (*Cynoglossus semilaevis*). *PLoS ONE* **2012**, *7*, e52097.
160. Ravi, P.; Jiang, J.; Liew, W.; Orban, L. Small-scale transcriptomics reveals differences among gonadal stages in Asian seabass (*Lates calcarifer*). *Reprod. Biol. Endocrinol.* **2014**, *12*, 5.
161. Fan, Z.; You, F.; Wang, L.; Weng, S.; Wu, Z.; Hu, J.; Zou, Y.; Tan, X.; Zhang, P. Gonadal transcriptome analysis of male and female olive flounder (*Paralichthys olivaceus*). *BioMed Res. Int.* **2014**, *2014*, 10.
162. Manousaki, T.; Tsakogiannis, A.; Lagnel, J.; Sarropoulou, E.; Xiang, J.; Papandroulakis, N.; Mylonas, C.; Tsigenopoulos, C. The sex-specific transcriptome of the hermaphrodite sparid sharpsnout seabream (*Diplodus puntazzo*). *BMC Genomics* **2014**, *15*, 655.
163. Böhne, A.; Sengstag, T.; Salzburger, W. Comparative transcriptomics in east african cichlids reveals sex- and species-specific expression and new candidates for sex differentiation in fishes. *Genome Biol. Evolut.* **2014**, *6*, 2567–2585.
164. Wu, G.-C.; Chiu, P.-C.; Lin, C.-J.; Lyu, Y.-S.; Lan, D.-S.; Chang, C.-F. Testicular *dmrt1* is involved in the sexual fate of the ovotestis in the protandrous black porgy. *Biol. Reprod.* **2012**, *86*, 41.

165. Wong, A.C.; van Eenennaam, A.L. Transgenic approaches for the reproductive containment of genetically engineered fish. *Aquaculture* **2008**, *275*, 1–12.
166. Uzbekova, S.; Chyb, J.; Ferriere, F.; Bailhache, T.; Prunet, P.; Alestrom, P.; Breton, B. Transgenic rainbow trout expressed *sGnRH*-antisense RNA under the control of *sGnRH* promoter of Atlantic salmon. *J. Mol. Endocrinol.* **2000**, *25*, 337–350.
167. Maclean, N.; Hwang, G.; Molina, A.; Ashton, T.; Muller, M.; Rahman, M.A.; Iyengar, A. Reversibly-sterile fish via transgenesis. *Inf. Syst. Biotechnol. News Rep.* **2003**, *12*, 3–5.
168. Maclean, N.; Rahman, M.; Sohm, F.; Hwang, G.; Iyengar, A.; Ayad, H.; Smith, A.; Farahmand, H. Transgenic tilapia and the tilapia genome. *Gene* **2002**, *295*, 265–277.
169. Hu, W.; Wang, Y.; Zhu, Z. A perspective on fish gonad manipulation for biotechnical applications. *Chin. Sci. Bull.* **2006**, *51*, 1–6.

CHAPTER 10

Disease Resistant Fish and Shellfish Are within Reach: A Review

Trygve Gjedrem

Nofima AS, Osloveien 1, P.O. Box 210, NO-1431 Ås, Norway

ABSTRACT

Disease in fish and shellfish is one of the main problems facing aquaculture production. Therefore, all attempts should be made to increase the rate of survival and, thus, reduce economic losses. Much has been done to develop vaccines and medical treatments to reduce mortality; and however, farming of aquatic species has a long way to go to optimize the environmental conditions for the animals and, thus, reduce stress and improve animal welfare. However, the good news is that there is the potential to increase disease resistance by selective breeding. By challenge-testing fingerlings from a number of families per generation, and including the rate of survival in the breeding goal, the results so far are very promising. By focusing on one disease at a time it is possible to increase the rate of survival by at least 12.5% per generation for most diseases studied. Unfortunately, selective breeding is only used to a small degree in aquatic species. In 2010, it was estimated that only 8.2% of aquaculture production was based on genetically improved stocks.

Keywords: aquaculture; resistance; selection

1. INTRODUCTION

Aquaculture is the fastest growing primary industry for production of animal protein, with a growth rate of 6%–7% per year [1]. By 2017/2018, aquaculture production is expected to exceed fisheries in terms of volume, and if the present rate of increase continues, in a few more years it will also surpass production of poultry and pigs [1], and, thus, become the highest source of animal protein. One of the reasons for this development is that fish and shellfish are recommended as healthy food and, in addition, they are more efficient converters of energy and protein than terrestrial farm animals. Studies have shown that Atlantic salmon has 2.3 and 1.6 times higher rates of energy retention than poultry and pigs, respectively [2,3]. Further, retention of protein was also higher in salmon compared with poultry (1.5 times) and pigs (1.75 times) [2,3]. In addition, it has been shown that feed conversion rate (FER) has improved by 23% during five generations of selection in Atlantic salmon [4]. Under farming conditions, Atlantic salmon use 1.1 to 1.2 kg feed per kg of growth. Today, Asian countries account for 88% of global aquaculture production, with China as the leading producer [5]. The prospects for further expansion are promising and it has been shown that the potential for future increase of aquaculture production is very high both in fresh water and particularly in marine water [6]. However, the availability of feed resources may reduce the rapid expansion in coming years. Therefore, there needs to be more focus on production of shellfish and seaweeds, which feed on naturally occurring plants and algae in marine waters.

One of the major problems facing aquaculture production are the diseases caused by bacteria, viruses, fungi and parasites. Efficient medicines and vaccines have been developed in recent years to meet the challenges of some diseases. Today, technology allows vaccination of individual fish, which makes it possible to vaccinate a large number of fingerlings. As an example, Norwegian farmers vaccinate every year around 400 million smolts (fingerlings ready for seawater) against several diseases.

Nevertheless, the mortality due to diseases is still relatively high. A major challenge surrounds the medical treatment of disease outbreaks, since it is not possible to sort the infected animals from the non-infected. The medicine must therefore be given to all fish/shellfish in the pond or cage so both infected and non-infected animals are treated. The medicine is usually supplied in pelleted feed.

Mortality is also caused by different stressors such as: handling, high-density, low dissolved oxygen concentration, rapid variation in water temperature and quality, few and irregular feeding per day, among others. These problems occur particularly for non-domesticated strains newly brought in from the wild. To reduce mortality, the industry is continuously searching for methods to increase the welfare of the animals. The feed is of course an important factor and its content should cover all the nutritional needs for the animals. However, although efforts are made to optimize the environmental conditions for animals, still there will be mortality and it is not possible to fully control the spread of pathogens and parasites between and within farms.

The purpose of this paper is to discuss the possibility of increasing the defense mechanisms or resistance of the animals against diseases and parasites through selective breeding.

2. MEASUREMENT OF DISEASE RESISTANCE

A central question for including a trait in a breeding program is how accurately it can be recorded. A common method of measurement is to count the number of mortalities that occur in different families of fish and shellfish during a certain period of life. However, the heritability of percent survival of fingerlings is rather low, $h^2 = 0.10$ [7] and less than 0.1 for fry [8], which means that it is difficult in practice to standardize the numerous environmental conditions which affect the mortality of fish.

An alternative method is to use challenge test in which fingerlings that are individually tagged from a number of families are kept under standardized environmental conditions (in a tank) and infected with a pathogen at a time. Mortality is recorded until it ceases. The results from challenge testing are very promising and the heritability for rate of survival is usually high ($h^2 =$ 0.43–0.69) [9,10,11]. It has also been shown that the results from challenge tests have been highly correlated with mortality in the field [12,13].

3. SOME RESULTS FROM SELECTION EXPERIMENTS

The objective of some of the first selection experiments in aquatic species was to reduce mortality. The first selection experiment in brook trout increased the survival rate from 2% to 69% after three generations of selection [14], although family selection was not practiced. Selection response was also obtained for reduced mortality due to furunculosis in brook trout and brown trout [15]. Further, a response of 30%–40% was achieved by selecting four to five generations against dropsy disease in common carp [16,17].

4. BREEDING PROGRAMS TO INCREASE DISEASE RESISTANCE

Early selection programs to increase growth rate used mass selection or random mating, without collecting information about the relationship between animals. Since growth rate has a relatively high heritability, 0.2–0.4 several of the selected breeders would be closely related. As time went on the animals become inbred and the result would automatically be low selection response or no response. This was most likely the result in common carp [18] and in Nile tilapia [19,20,21]. These results show how important it is to avoid inbreeding, which can be obtained by producing families and apply tagging or use microsatellites for identification of relationship.

Efficient and sustainable breeding programs are, therefore, family based and selection is usually for the traits of highest economic importance. Today breeding companies will usually include some major diseases in the breeding goal together with traits like growth rate and meat quality traits [22]. The breeding company usually discusses with the farmers which traits should be included in the breeding goal and which traits should be given highest weight in the selection index. One should, however, be aware that as the number of traits is increased in the breeding goal, the genetic gains for each trait will be reduced.

It is encouraging that positive genetic correlations have been reported between growth rate and most diseases [23]. In this study, the average of eight positive estimates of genetic correlation was $r_G = 0.25$. However, there are also some negative genetic correlation between growth rate and diseases, that between growth rate and Taura Syndrome virus in shrimp, was $r_G = -0.12$ [23]; between growth rate and VHS disease in rainbow trout $r_G = -0.22$ [23,24] and between growth rate and WSSV in shrimp $r_G = -0.55$ to -0.64 [25]. This means that selection for increased growth rate will give a positive correlated response in disease resistance for most species while negative in other.

5. DATA COLLECTION

Challenge tests for particular diseases have been shown to give very reliable estimates of disease resistance. To get results with high repeatability, a sample of 30 to 50 tagged fry/fingerlings from each family should be sent to a specialized laboratory equipped for such tests. The families of fish/shellfish should be kept in one tank and infected by the pathogen/parasite in question. As mortality starts, each dead fingerling should be removed from the tank and the time of death and tag number recorded. The recording continues until the mortality has ceased [26]. Frequencies of survival will later be included in the selection index as one of the traits in the breeding goal.

The recording of other traits takes place during the grow-out period, as well as at the time for harvesting. At harvesting, some results from test stations will be available. The final selection of animal for breeding should take place when all data are available and selection indices have been calculated [6]. The selection index for each trait should be weighted by its economic importance and its heritability. In addition, genetic correlations between the traits should be taken into account.

6. GENETIC GAINS OBTAINED

A number of selection experiments and breeding programs have successfully reduced mortality. Some results from selection for disease resistance are given in Table 1. Genetic gain in survival is rather low (5.0%–8.4%) per generation [27,28,29]. The reason seems to be that it is difficult to standardize the environmental rearing conditions for fish and shellfish over a lifetime, and there will be a number of factors causing mortality. In addition, multitrait selection has been practiced, which has lowered the weight of survival in the selection index. However, by applying challenge tests under standardized environmental conditions, more reliable information can be collected about resistance to disease.

The genetic gain obtained by applying challenge tests for Infectious Pancreatic Necrosis (IPN) [30] and Vibrio salmonicida[31] is very high applying single trait selection reaching 19% per generation. The reason for the low genetic gain in White Spot Syndrome Virus (WSSV) resistance, varying from 1.7% [27] to 6.3% [32] is not obvious, it could partly be due to difficulty and inaccuracy of mortality recording. Multi trait selection for Taura syndrome in P. vannamei is also high [33,34] while it is of medium magnitude for M. Sydney parasite in oyster [35]. The amoebic gill disease in Atlantic salmon has been shown to have genetic variation ($h^2 = 0.09–0.56$) [36]. The selection strategy is to reduce number of treatments and it is predicted to increase the interval between treatments by 3% per year [36]. Genetic variation is also shown for salmon louse with a heritability of $h^2 = 0.26$ [37].

These genetic gains, obtained through breeding programs, show that it is possible to develop lines which are resistant to diseases and parasites. How long this will take depends on several factors: Magnitude of genetic variation or heritability of the trait, what weight the disease or diseases get in the selection index, and the magnitude of genetic correlation between the diseases and the other traits in the breeding goal. If, for example, the aim was to improve resistance against Taura syndrome only and with a mortality at the beginning of 50% and a genetic gain of 15.4% per generation, it would take five generations to develop a full resistant strain. This rapid gain is because the improvement is cumulative; the newly selected generation builds on those made in previous generations.

New biotechnological developments show promising results to improve disease resistance. A single QTL was found to explain more than 80% of the genetic variation for IPN (Infectious Pancreatic Necrosis) resistance in Atlantic salmon [38,39]. In addition, a QTL for PD (Pancreatic Disease) was found explaining a considerable part of the variation in this disease [40]. The possibility for genetic gain by applying marker-assisted selection has been discussed [41]. These findings when included as a tool in breeding programs will increase the genetic gain in future generations and the methods are already in use by breeding companies.

Table 1. Average genetic gains of selection for increased disease resistance. Multi trait selection is marked with *.

Traits	Genetic Gain per Generation, %	Number of Generations Selected	References
Survival:			
Whiteleg shrimp, *P. vannamei*	5.7 *	4	Gitterle, *et al.* 2007 [27]
Blue tilapia, *Oreochromis aureus*	8.4 *	4	Thodesen, *et al* 2012 [28]
Red tilapia, *Oreochromis* spp.	5.0 *	4	Thodesen, *et al.* 2013 [29]
IPN resistance:			
Atlantic salmon, *Salmo salar*	18.7	1	Storseth, *et al.* 2007 [30]
***Vibrio salmonicida*:**			
Rainbow trout, *Oncorhyncus mykis*	19.0	1	Leeds, *et al.* 2010 [31]
White spot, WSSV:			
Whiteleg shrimp, *P. vannamei*	1.7 *	4	Gitterle, *et al.* 2007 [27]
Whiteleg shrimp, *P. vannamei*	6.3	4	Huang, *et al.* 2012 [32]
Taurasyndrom, TVS:			
Whiteleg shrimp, *P. vannamei*	12.4 *	1	Fjalestad, *et al.* 1997 [33]
Whiteleg shrimp, *P. vannamei*	18.4 *	1	Argue, *et al.* 2002 [34]
***M.sydneyi*, parasite:**			
Oyster, *S. glomerata*	11.0	2	Nell and Hand, 2003 [35]

7. DISCUSSION

In farming aquatic species mortality can cause serious economic losses and reduce profits. In addition, it is discouraging for the farmer to harvest and dispose of the dead animals. Therefore every effort should be made to improve the welfare of the animals and increase their survival rate by applying challenge tests. As documented above, selective breeding offers good potential to increase the rate of survival. A genetic gain of 12.5% per generation will have a marked increase in survival rate, since the effect is cumulative. In addition, a breeding program will always select for improved growth rate and because of the positive genetic correlation between growth rate and resistance to most pathogens, selection for improved growth rate will increase the genetic gain for survival.

Biotechnological research can be beneficial by providing additional knowledge about the genetics of diseases. The information about the QTL for resistance to IPN and PD is very promising for increasing the rate of resistance to these diseases through selective breeding. Hopefully the future will bring more knowledge in this field, which will increase our possibility to reduce mortality for aquatic species farmed in aquaculture.

The fact that only a small part of aquaculture production is based on genetically improved stocks is discouraging. Recent estimates [42,43] show that only 8.2% of aquaculture production in 2010 was based on family breeding programs. This is a disadvantageous situation, particularly since it is well documented that it is possible to get a rapid improvement of growth rate, disease resistance and product quality through family based breeding programs.

ACKNOWLEDGMENTS

I am grateful to Nofima who supported this work and to Nick Robinson for reading the paper.

CONFLICTS OF INTEREST

The authors declare no conflict of interest.

REFERENCES

1. FAO. Food Outlook. Biannual Report on Global Food Markets; FAO: Rome, Italy, 2013; p. 133.
2. Bjørkli, J. Protein and Energy Account in Salmon, Chicken, Pig and Lamb (Protein og Energi Rekneskap Hos Laks, Gris, og Lam). Master Thesis, Agricultural University of Norway, Ås, Akershus, Norway, 2002; p. 41. (In Norwegian).
3. Ytrestøyl, T.; Aas, T.S.; Berge, G.M.; Hatlem, B.; Sørensen, M.; Ruyter, B.; Thomassen, M.; Hognes, E.S.; Zigler, E.S.; Sund, V.; et al. Resource Utilisation and Ecoefficiency of Norwegian Salmon Farming in 2010; Nofima Report 53/2011,ISBN: 978-82-7251-945-1; Nofima: Tromsø, Norway, 2011; p. 65.
4. Thodesen, J.; Grisdale-Helland, B.; Helland, S.J.; Gjerde, B. Feed intake, growth and feed utilization of offspring from wild and selected Atlantic salmon (Salmo salar). Aquaculture **1999**, 189, 237–246.
5. FAO. Fishery and AQUACULTURE Statistics. Aquaculture Production; FAO: Rome, Italy, 2012.
6. Gjedrem, T.; Robinson, N.; Rye, M. The importance of selective breeding in aquaculture to meet future demands for animal protein: A review. Aquaculture **2012**, 350–353, 117–129.
7. Rye, M.; Lillevik, K.M.; Gjerde, B. Survival in early life of Atlantic salmon and rainbow trout: Estimate of heritabilities and genetic correlations. Aquaculture **1990**, 89, 209–216.
8. Kanis, E.; Refstie, T.; Gjedrem, T. A genetic analysis of egg, alevin and fry mortakity in salmon (Salmo salar), sea trout (Salm trutta) and rainbow trout (Salmo gairdneri). Aquqculture **1976**, 8, 259–268.
9. Chevassus, B.; Dorson, M. Genetics of resistance to disease in fishes. Aquaculture **1990**, 85, 83–107.
10. Gjedrem, T.; Salte, R.; Gjøen, H.M. Genetic variation in susceptibility of Atlantic salmon to furunculosis. Aquaculture **1991**, 97, 1–6.
11. Ødegård, J.; Sommer, A.I.; Præbel, A.P. Heritability of resistance to nervous necrosis Atlantic cod (Gadus morhua L.). Aquaculture **2010**, 399, 59–64.
12. Gjøen, H.M.; Refstie, T.; Ulla, T.; Gjerde, B. Genetic correlation between survival of Atlantic salmon in challenge and field tests. Aquaculture **1997**, 157, 277–288.
13. Storset, A.; Strand, C.; Wetten, M.; Kjøglum, S.; Ramstad, A. Response to selection for resistance against infectious pancreatic necrosis in Atlantic salmon (Salmo salar L.). Aquaculture **2007**, 272, 62–68.
14. Embody, G.C.; Hyford, C.D. The advantage of rearing brook trout fingerlings from selected breeders. Trans. Am. Fish. Soc. **1925**, 55, 135–138.
15. Ehlinger, N.F. Selective breeding of trout for resistance to furunculosis. N. Y. Fish Game J. **1977**, 24, 25–36.
16. Schaperclaus, W. Trate de Pisciculture en Etang; Vigot Freres: Paris, France, 1962; Volume 208, pp. 208–227.
17. Ilyassov, Y.L. Genetic principles of fish selection for disease resistance. In Selection, Hybridization and Genetic Engineering in Aquaculture; Tiewes, K., Ed.; Heenemann Verlagsgesellschaft: Berlin, Germany, 1987; pp. 455–469.
18. Moav, R.; Wohlfarth, G.W. Two way selection for growth rate in the common carp (Cyrinus carpio L.). Genetics **1976**, 82, 83–101.
19. Hulata, G.; Wohlfarth, G.W.; Halevy, A. Mass selection for growth rate in the Nile tilapia (Oreochromis niloticus). Aquaculture **1986**, 57, 177–184.
20. Teichert-Coddington, D.R.; Smitherman, R.O. Lack of response by tilapia nilotica to mass selection for rapid early growth. Trans. Am. Fish. Soc. **1988**, 117, 297–300.
21. Huang, C.M.; Liao, I.C. Response to mass selection for growth rate in Oreochromis niloticus. Aquaculture **1990**, 85, 199–205.

22. Gjerde, B. Prediction of breeding values. In Selection and Breeding Programs in Aquaculture; Gjedrem, B., Ed.; Springer: Berlin, Germany, 2005; p. 364.
23. Gjedrem, T.; Olesen, I. Basic statistical parameters. In Selection and Breeding Programs in Aquaculture; Gjedrem, B., Ed.; Springer: Berlin, Germany, 2005; p. 364.
24. Henryon, M.; Jokumsen, A.; Berg, P.; Lund, I.; Pedersen, P.B.; Olesen, N.J.; Slierendrecht, W.J. Genetic variation for growth rate, feed conversion efficiency, and disease resistance exists within a farmed population of rainbow trout. Aquaculture **2002**, 209, 59–76.
25. Gitterle, T; Salte, R.; Gjerde, B.; Cock, J.; Johansen, H.; Salazar, M.; Lozano, C.; Rye, M. Genetic (co) variation in resistance to White Spot Syndrome Virus (WSSV) and harvest weight in Penaeus (Litopenaeus) vannamei). Aquaculture**2005**, 246, 139–149.
26. Ødegård, J.; Baranski, M.; Gjerde, B.; Gjedrem, T. Methodology for genetic evaluation of disease resistance in aquaculture species: Challenges and future prospects. Aquac. Res. **2011**, 42, 103–114.
27. Gitterle, T.; Johansen, H.; Erazo, C.; Erazo, C.; Lozano, C.; Cock, J.; Salazar, M.; Rye, M. Response to multi trait selection for harvest body weight, overall survival, and resistance to white spot syndrome virus (WSSV) in Penaeus (Litopenaeus). Aquaculture **2007**, 272 (Suppl. 1), S262.
28. Thodesen (Da-Yong Ma), J.; Rye, M.; Wang, Y.X.; Li, S.J.; Bentsen, H.B.; Gjrdrem, T. Genetic improvement of tilapia in China: Genetic parameters and selection responses in growth, pond survival and cold-water tolerance of Blue tilapia (Oreochromis aureus) after four generations of multi-trait selection. Aquaculture **2012**, 396–399, 32–42.
29. Thodesen (Da-Yong Ma), J.; Rye, M.; Wang, Y.X.; Li, S.J.; Bentsen, H.B.; Gjedrem, T. Genetic improvement of tilapia in China: Genetic parameters and seletion responses in growth, survival and external color traits of red tilapia (Oreochromis spp.) after four generations of multi-trait selection. Aquaculture **2013**, 416–417, 354–366.
30. Storset, A.; Strand, C.; Wetten, M.; Kjøglum, S.; Ramstad, A. Response to selection for resistance against infectious pancreatic necrosis in Atlantic salmon (Salmo salar L.). Aquaculture **2007**, 272, 62–68.
31. Leeds, T.D.; Silverstein, J.T.; Weber, G.M.; Vallejo, R.L.; Palti, Y.; Rexroad, C.N., III; Evenhuis, J.; Hadidi, S.; Welsh, T.J.; Wiens, G.D. Response to selection for bacterial cold water disease resistance in rainbow trout. J. Anim. Sci. **2010**, 88, 1936–1946.
32. Huang, Y.; Yin, Z.; Weng, S.; He, J.; Li, S. Selective breeding and preliminary commercial performance of Penaeus vannamei for resistance to white spot syndrome virus (WSSV). Aquaculture **2012**, 364–365, 111–117.
33. Fjalestad, K.T.; Gjedrem, T.; Carr, W.H.; Sweeney, J.N. The Shrimp Breeding Program. Selective Breeding of Penaeus Vannamei; Report no. 17/97; The Oceanic Institute: Waimanalo, HI, USA, 1997; p. 85.
34. Argue, B.J.; Arce, S.M.; Moss, S.M. Selective breeding of Pacific white shrimp (Litopenaeus vannamei) for growth and resistance to Taura Syndrome Virus. Aquaculture **2002**, 204, 447–460.
35. Nell, J.A.; Hand, R.E. Evaluation of the progeny of second-generation Sydney rock oyster Saccostrea glomerata (Gold, 1850) breeding lines for resistance to QX disease Marteilia sydneyi. Aquaculture **2003**, 228, 27–35.
36. Kube, P.D.; Taylor, R.S.; Elliot, N.G. Genetic variation in parasite resistance of Atlantic salmon to amoebic gill disease over multiple infections. Aquaculture **2012**, 364–365, 165–172.
37. Kolstad, K.; Heuch, P.A.; Gjerde, B.; Gjedrem, T.; Salte, R. Genetic variation in resistance of Atlantic salmon (Salmo salar) to the salmon louse Lepeophtheirus salmonis. Aquaculture **2005**, 247, 145–151.

38. Housten, R.D.; Hamilton, A.; Guy, A.D.R.; Tinch, A.E.; Taggart, J.B.; McAndrew, J.B.; Bishop, S.C. Major QTL affect resistance to infectious pancreatic necrosis in Atlantic salmon (Salmo salar). Genetics **2008**, 178, 1109–1115.
39. Moen, T.; Baranski, M.; Sonesson, A.; Kjøglum, A. Conformation and fine-mapping of a major QTL for resistance in infectious pancreatic necrosis in Atlantic salmon (Salmo salar): Population-level association between markers and trait. BMC Genomics **2009**, 10, 368.
40. Gonen, S.; Baranski, M.; Thorland, I.; Norris, A.; Grove, H.; Arnesen, P.; Bakke, H.; Lien, S.; Bishop, S.C.; Housten, R.D. Mapping and validating of a major QTL affecting the resistance of Atlantic salmon (Salmo salar) to pancreas disease. Heredity **2015**. accepted.
41. Yue, G.H. Recent advances of genome mapping and marker-assisted selection in aquaculture. Fish Fish. **2014**, 15, 376–396.
42. Neira, R. Breeding in aquaculture species: Genetic improvement program in developing countries. In Proceedings of the 9th World Congress on Genetics Applied to Livestock Production, Lipzig, Germany, 1–6 August 2010; p. 8.
43. Rye, M.; Gjerde, B.; Gjedrem, T. Genetic development programs for aquaculture species in developed countries. In Proceedings of the 9th World Congress on Genetics Applied to Livestock Production, Lipzig, Germany, 1–6 August 2010; p. 8.

CHAPTER 11

New Approach for Fish Breeding by Chemical Mutagenesis: Establishment of TILLING Method in Fugu (Takifugu Rubripes) with ENU Mutagenesis

Miwa Kuroyanagi[1†], Takashi Katayama[2], Tadashi Imai[2], Yoshihisa Yamamoto[2], Shin-ichi Chisada[3], Yasutoshi Yoshiura[3], Tomokazu Ushijima[4], Tomonao Matsushita[4], Masashi Fujita[1], Aoi Nozawa[1], Yuzuru Suzuki[1], Kiyoshi Kikuchi[1*] and Hiroyuki Okamoto[5*†]

[1]*Graduate School of Agricultural and Life Sciences, The University of Tokyo, 2971-4 Bentenjima, Maisaka, Hamamatsu, Shizuoka 431-0214, Japan*

[2]*Aquaculture Technology Division, National Research Institute of Aquaculture, Fisheries Research Agency, 422-1 Nakatsuhamaura, Minami-ise, Mie 516-0193, Japan*

ABSTRACT

Background

In fish breeding, it is essential to discover and generate fish exhibiting an effective phenotype for the aquaculture industry, but screening for natural mutants by only depending on natural spontaneous mutations is limited. Presently, reverse genetics has become an important tool to generate mutants, which exhibit the phenotype caused by inactivation of a gene.

TILLING (Targeting Induced Local Lesions IN Genomes) is a reverse genetics strategy that combines random chemical mutagenesis with high-throughput discovery technologies for screening the induced mutations in target genes. Although the chemical mutagenesis has been used widely in a variety of model species and also genetic breeding of microorganisms and crops, the application of the mutagenesis in fish breeding has been only rarely reported.

Results

In this study, we developed the TILLING method in fugu with ENU mutagenesis and high-resolution melting (HRM) analysis to detect base pair changes in target sequences. Fugu males were treated 3 times at weekly intervals with various ENU concentrations, and then the collected sperm after the treatment was used to fertilize normal female for generating the mutagenized population (F_1). The fertilization and the hatching ratios were similar to those of the control and did not reveal a dose dependency of ENU. Genomic DNA from the harvested F_1 offspring was used for the HRM analysis. To obtain a fish exhibiting a useful phenotype (e.g. high meat production and rapid growth), fugu myostatin (*Mstn*) gene was examined as a target gene, because it has been clarified that the *mstn* deficient medaka exhibited double-muscle phenotype in common with *MSTN* knockout mice and bovine *MSTN* mutant. As a result, ten types of ENU-induced mutations were identified including a nonsense mutation in the investigated region with HRM analysis. In addition, the average mutation frequency in fugu *Mstn* gene was 1 mutant per 297 kb, which is similar to values calculated for zebrafish and medaka TILLING libraries.

Conclusions

These results demonstrate that the TILLING method in fugu was established. We anticipate that this TILLING approach can be used to generate a wide range of mutant alleles, and be applicable to many farmed fish that can be chemically mutagenized.

Keywords: TILLING Fugu ENU HRM Myostatin Mutagenesis Fish breeding

BACKGROUND

Mutations are the basis of genetic variation and mutant populations are indispensable genetic resources in all organisms. This variation can be either naturally occurring or, in plants, animals and lower organisms, induced by chemical or physical treatments. Mutation induction, for example, has played an important role in the genetic improvement of crop species that are of economic significance.

In breeding of aquacultural fish species, it is a final goal to discover and generate the fish with various useful phenotypes for supplying to consumers. Currently genetic breeding is carried out with the search for natural variation in traits such as growth, productivity and disease resistance, and then improved strains are produced by genetic selection or marker-assisted breeding approaches [1]. Thus the success of fish breeding solely depends on the mutations pre-existing in the focal population. The rate of natural spontaneous mutations in fish species is generally lower than 1.0×10^{-6} at specific loci. On the other hand, mutation frequency caused by radiation and chemical mutagenesis range from 1.0×10^{-3} to 3.9×10^{-3} at specific loci when *in vivo*spermatogonial treatments were carried out in zebrafish and medaka [2, 3, 4]. Given the high frequency of the mutation ratio, it is reasonable to consider that the chemical mutagenesis will be an efficient way to produce new mutants for future genetic improvement in aquacultural species.

TILLING (Targeting Induced Local Lesions IN Genomes) is a reverse genetics strategy that identifies induced mutations in specific genes of interest in chemically mutagenized populations with high-throughput discovery technologies. TILLING, first described in 2000 for mutation detection in *Arabidopsis*[5], is now used in a wide range of plants including soybean [6], rice [7], barley [8] and maize [9] as well as for animal model systems, including *Drosophila*[10], *Caenorhabditis elegans*[11], rat [12], medaka [13], zebrafish [14] and for the discovery of naturally occurring polymorphisms in humans [15].

The first step in TILLING is chemical mutagenesis. In plants, mutations have been induced by using ethylmethane sulfonate (EMS) and methylnitrosourea (MNU) [5, 7]. In animals, EMS is used in many invertebrate species for mutagenesis, whereas *N*-ethyl-*N*-nitrosourea (ENU) is used for males of vertebrates, and is the most widely employed method [16]. ENU acts as an alkylating agent and transfers its ethyl group to nucleophilic nitrogen or oxygen sites on deoxyribonucleotides, leading to base mismatch during DNA replication. Single-base substitutions that resemble natural spontaneous mutations [17, 18], are mainly induced by ENU treatment and all genes are mutated at random [19]. Because this protocol induces mutations in germ-line stem cells, mutations are transmitted at the rate for at least several months after the treatment. Also, the mutations are induced with premeiotic ENU treatment, so that the initial mutation, usually an ethylation of a base on one DNA strand, is fixed by DNA replication prior to production of differentiated sperm, leading to non-mosaic offspring in the next generation [16]. On the other hand, it has been recently reported that some point mutations were induced by postmeiotic mutagenesis in *Ctenopharyngodon idellus*(grass carp) with ENU for mature sperm [20]. In zebrafish, the postmeiotic ENU mutagenesis resulted in mosaic progeny, but produced a 10 fold increase in the frequency of induced mutations [21]. This protocol in zebrafish has been also shown to induce mutations such as deletions and translocations triggered by chromosomal rearrangements [18]. These results indicate that postmeitotic ENU treatment of male germ cells can induce a variety of mutations from point mutations to deletions and translocations.

In zebrafish and medaka, which are model fish, mutations are introduced in spermatogonia by soaking founder fish in ENU solution [13, 22]. However, this protocol is difficult to apply to many aquaculture species including fugu, as the body size of these fish is markedly larger than that of medaka and zebrafish. Thus alternative protocols are required for fish with a large body size.

After the ENU treatment, a large F_1 population is generated to choose the mutations for purpose from many random heterozygous mutations in their genomes. As the next step in TILLING, the genomic DNA from these animals containing induced mutations is screened with discovery technologies. Sequencing is considered the standard for DNA-based mutation detection because it reveals the exact location and the type of mutations. Taniguchi et al. employed a direct sequencing approach even for the prescreening step [13]. CEL I nuclease-mediated screening of heteroduplexes are widely carried out to detect mutations in TILLING [23, 24]. Both methods are effective, but the costs are substantial and it takes a relatively long time for the screening. Recently, high resolution melting (HRM) analysis has succeeded to detect point mutations in plant populations [25], medaka TILLING library [26] and human disease [27]. HRM analysis makes use of characteristics in DNA as follows; the thermal stability of a DNA fragment is determined by its base sequence. When the DNA fragment contains an altered sequence, the duplex stability is changed, leading to different melting behavior. Thus changes in the sequence within the DNA fragment such as point mutation can be detected by the melting analysis.

To apply TILLING technology for breeding practices, it is critical to select a proper target gene in which mutations are likely to cause useful phenotypes such as rapid growth and disease resistance. One of the growth-related factors, Myostatin (Mstn) is a member of the transforming growth factor-β (TGF-β) superfamily, and functions as a negative regulator of skeletal muscle mass [28]. Naturally occurring mutations in the *Mstn*gene have been identified in double-muscling cattle breeds [29, 30], mice [28], dogs [31] and even humans [32]. Recently, we have demonstrated that the antagonistic role of Mstn against muscle growth is conserved from mammals to fish species [33]. Indeed *mstn*-deficient medaka (*mstnC315Y*), which were generated by the TILLING method, show a significant increase in body weight in comparison with wild-type medaka, due to over-growth of muscles. Therefore, *Mstn* is a promising target gene for mutation breeding in fishes.

Here, in order to initiate targeted mutation breeding in fish improvement, we searched for efficient methods for TILLING combining ENU mutagenesis with HRM analysis in *Takifugu rubripes*, tiger pufferfish (fugu). Fugu males were treated with ENU via intraperitoneal injection each week for three weeks, and the sperm collected from the founder fish was used for *in vitro* fertilization to obtain the progeny. We then isolated genomic DNA from the progeny, and carried out HRM analysis targeting the exon 3 of fugu *Mstn* gene. Our screening successfully identified ten mutations including one nonsense mutation in fugu *Mstn* gene. The mutation frequency in fugu *Mstn* gene was similar to that identified in the medaka TILLING library, demonstrating that we were able to establish an effective protocol of the TILLING method in fugu. This protocol is likely to be applicable to various species of aquacultured fish.

RESULTS

Development of the ENU mutagenesis protocol for fugu TILLING

In order to develop a protocol for ENU mutagenesis in fugu, we performed intraperitoneal injections of ENU, referring to the method used for mouse mutagenesis [34]. Previous study has shown that inheritable point mutations are induced by ENU treatment at the spermatogonia stage in mouse [35]. Based on the report, fugu males were injected with ENU at a dose of 70 mg/kg body weight, 3 times at weekly intervals at the stage before spermiation when early stages of spermatogenesis were progressing (Figure 1A). The injected fish began to die at 4 to 6 weeks after the last ENU treatment (average survival rate: 71%). The surviving fish were treated with the hormone, gonadotropin, to trigger sperm production and spermiation. Their sperm was collected for fertilization and cryopreservation during 6 to 7 weeks after the ENU treatment. No fish survived more than 9 weeks after the last ENU treatment. We then performed *in vitro* fertilization using the collected sperm to obtain mutagenized F_1 fry. The average fertilization ratio was 89%, which is comparable to the ratio of non-mutagenized sperm. Although the hatching ratio was low (average: 23%), most of the hatched fry had a normal appearance without teratogenesis. Therefore, we determined the method for the ENU mutagenesis in fugu as follows: the founder fish are treated with the 70 mg/kg of ENU concentration 3 times at weekly intervals during the early mature stage to the late maturation stage in spermatogenesis of the fish.

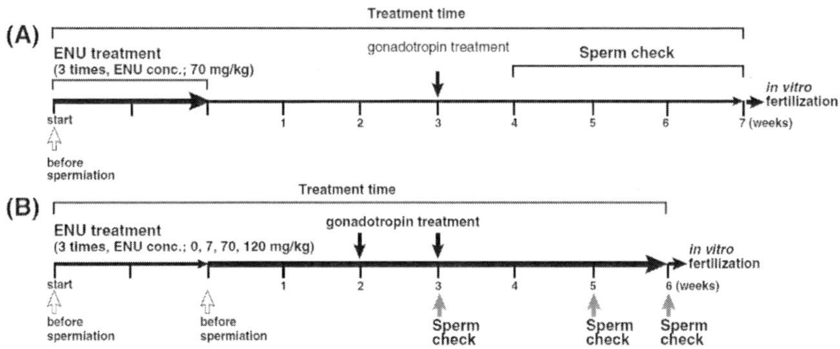

Figure 1 Scheme design of the ENU mutagenesis in fugu. A) The number of ENU treatments in fugu. The number and timing of ENU injections was examined. ENU (concentration: 70 mg/kg body weight of fugu) was injected 3 times at weekly intervals into the intraperitoneal of fugu males at the stage before spermiation. **B)** Treatment with various ENU concentrations, and timing of gonadotrophin (HCG/SPH) treatment and sperm collection. Founder fish were treated with three ENU concentrations (7, 70, 120 mg/kg body weight of fugu) at the stage before spermiation. As a control, the fish were treated with buffer (0 mg/kg body weight of fugu) instead of ENU. HCG/SPH was injected into the fugu intramuscularly at 2 or 3 weeks after the last ENU injection. Sperm was collected at 3, 5, and 6 weeks after the last ENU injection.

ENU treatment in TILLING method has been shown to effectively produce a lot of mutations. In mouse, ENU induces mutations at a relatively high rate with a dose dependency [36]. Next, we examined whether various ENU concentrations were effective on viability of the founder fish and their progeny (Figure 1B). The fish were injected intraperitoneally with ENU at a dose of 7, 70 or 120 mg/kg body weight, and with the buffer prepared for dissolving ENU as the control (0 mg/kg body weight) to remove the effect of injection on the fish. We examined the number of surviving founder fish after treatment with each ENU concentration. Although almost all fish with ENU in a dose of 7, 70 mg/kg (survival rates: 100% and 92%, respectively) survived at 4 weeks after the last ENU treatment, the fish with ENU in a dose of 120 mg/kg (survival rate: 67%) began to die at that time. The fish, which were collected sperm at 8 weeks after the last ENU treatment, survived till 5 months after the ENU experiments. Final survival rates of each ENU concentration were as follows: 7 mg/kg, 100%; 70 mg/kg, 50%;

120 mg/kg, 33%. After gonadotropin treatment in each family, the collected sperm was used for *in vitro*fertilization. The fertilization ratio (average: 72%) and the hatching ratio (average: 54%) in each cross were similar to those of control (67% and 53%, respectively; Table 1). More than ten thousand hatched fry of fugu were obtained in each family. They revealed that the fertilization and hatching ratios were unaffected by the dosage of ENU when the concentration of ENU was varied from 7 mg/kg to 120 mg/kg.

Table 1 Effects of various ENU concentrations on fertilization and hatching ratios of fugu

ENU conc. (mg/kg body weight)	Fertilization ratio (%)	Hatching ratio (%)
0	67	53
7	69	56
70	68	54
70	73	58
70	76	61
120	75	40

In vitro fertilization was carried out by using eggs from wild-type fugu and the sperm treated with ENU concentrations as described above. 100 of fertilized eggs in each family were counted at the 4 cell stage for fertilization ratio, while at the melanophore appearing stage for the hatching ratio. These ratios were calculated according to Methods. Founder fish treated with buffer was used as the control (ENU conc., 0 mg/kg).

In the course of the above experiment, we noted that the ENU-treated founder fish did not spermiate without the gonadotropin treatment injection. Then, we investigated the timing of the gonadotropin treatment to maximize the amount of sperm to collect (Figure 1B). Gonadotropin treatment was performed with the fish at 2 or 3 weeks after the last ENU treatment, which was the stage before the fish began to die as described above. To determine the time period for sperm collection after the gonadotropin treatment, we counted the number of spermiated fish at 3, 5 and 6 weeks after the last ENU treatment, and measured the volume of sperm (Table 2). In the gonadotropin treatment at 2 and 3 weeks after the last ENU treatment, sperm was collected at both 5 and 6 weeks, while fish did not spermiate at 3 weeks after the last ENU treatment. The frequency of spermiated fish was higher in the gonadotropin treatment at 3 weeks than at 2 weeks after the last ENU treatment. There was no noticeable difference between the volume of sperm in each week after the gonadotropin treatment. In the control without the gonadotropin treatment, the fish did not spermiate until 5 weeks after the last ENU treatment, and produce a very small amount of sperm at 6 weeks. The volumes of all sperm collected under these conditions were sufficient for cryopreservation, except for the fish without hormonal treatment. From the results, we fixed the method as follows: the gonadotropin treatment for the founder was carried out at 3 weeks after the last ENU treatment, and then the sperm from the fish was collected at 3 weeks after the gonadotropin treatment.

Table 2 Effects of gonadotropin treatment on timing of sperm collection and the sperm volume in fugu

		Period after the last ENU treatment					
		3 weeks		**5 weeks**		**6 weeks**	
	ENU conc. (mg/kg)	**No. of fish from which sperm was collected/No. of fish examined (%)***	**Sperm volume (ml)**	**No. of fish from which sperm was collected/No. of fish examined (%)***	**Sperm volume (ml)**	**No. of fish from which sperm was collected/No. of fish examined (%)***	**Sperm volume (ml)**
Gonadotropin treatment (2 weeks after the last ENU treatment)	70	0/4 (0)	0	2/4 (50)	5-15	3/4 (75)	4-10
Gonadotropin treatment (3 weeks after the last ENU treatment)	0	-	-	2/2 (100)	4-9	1/1 (100)	6
	7	-	-	3/3 (100)	7.5-30	3/3 (100)	8-10
	70	-	-	3/3 (100)	10-17.5	3/3 (100)	5-7
	120	-	-	0/2 (0)	0	1/1 (100)	3
Non-gonadotropin treatment (control)	70	0/3 (0)	0		0	3/3 (100)	0.2-2

(%)* reveals ratio of fish which were able to collect the ENU-mutagenized sperm to the examined fish.

Target gene for generation of a fugu mutant with TILLING

Our long-term goal is to generate a fugu line that exhibits a useful phenotype for the aquaculture industry by TILLING and to apply this technology to a variety of aquacultured species. Recently, we have reported that myostatin (*mstn*)-deficient medaka exhibited a double-muscling phenotype with hyperplasia and hypertrophy [33]. There is a high possibility that *Mstn*-deficient fugu exhibits such the phenotype, if a mutation is found in the same position or the significant region of *mstn* gene in medaka. Therefore, we amplified presumable full-length cDNA of fugu *Mstn* using cDNA from the muscle as a template. Sequencing of these products indicated that fugu *Mstn* comprises 3 exons, and encodes a protein consisting of 375 amino acids (Figure 2A). The fugu Mstn amino acid sequence has 83.8% identity to medaka and highly 88.1% identity in the C-terminal active peptide of Mstn (Figure 2B). Search in the fugu genome assembly suggested that fugu has a single copy of this gene.

Detection of ENU-induced mutations in fugu with HRM analysis and DNA sequencing

Since TILLING can identify mutations in the heterozygous condition, it is possible to screen using the F_1 offspring of ENU mutagenesis. We performed the HRM analysis to screen for ENU-induced mutations in the C-terminal region of fugu *Mstn* gene, since the region is highly conserved among vertebrates (Figure 2), and is responsible for the function of Mstn [37]. To screen correctly and quickly a large amount of genomic DNA, efficient protocols for HRM analysis were explored with a DNA automatic dispensing machine and a detection device for 384 samples. Genomic DNA was extracted from 1,056 fry obtained from in *vitro* fertilization with the ENU-treated sperm. These DNA were subjected to the nested PCR with some primers of the fugu Mstn (see Methods) and two amplicons Figure 2A (268 bp and 281 bp) were used for HRM analysis. Examples of the melting curves and the differentiation plots in fugu *Mstn* gene are shown in Figure 3. The mutation candidates were clearly indicated in each family (Figure 3, red lines). The amplicons selected as positive samples were subjected to direct sequencing in order to confirm whether they contain mutations. Overlapping peaks in the DNA sequence traces revealed that an F_1 individual carries a heterozygous point mutation (Figure 4), while this mutation was not observed in the rest of the siblings as well as in both parents. The results demonstrate that the mutations were induced by ENU treatment in fugu.

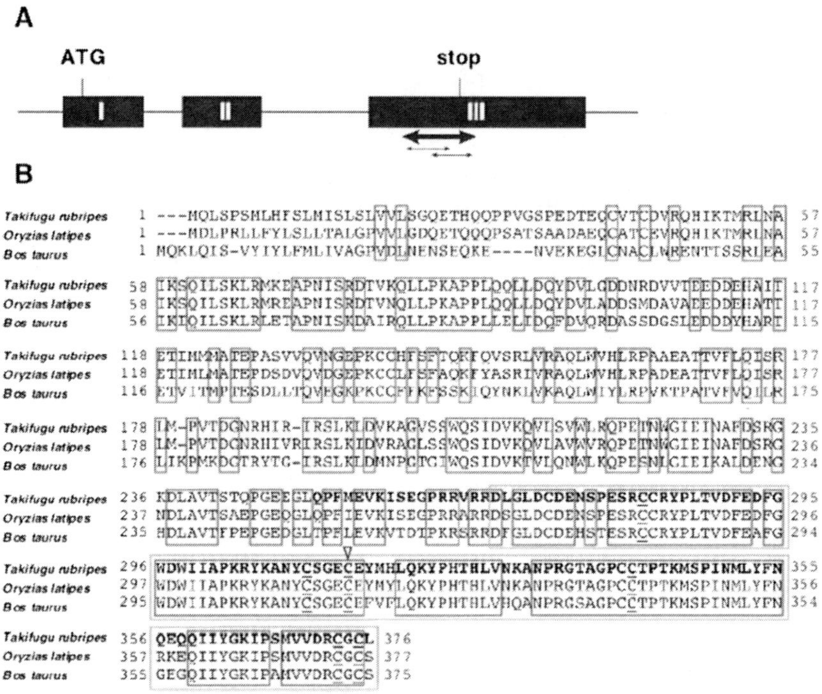

Figure 2 Fugu Mstn profiles. A) Genome structure of fugu *Mstn* gene (1905 bp). Black boxes, exons (with the Roman numeral representing the number of exons) and bars, introns. **B)** Sequence alignment of fugu [GenBank, ADT89782], medaka [GenBank, NP_001188428] and bovine [GenBank, NP_001001525] Mstn protein. Amino acids conserved across the three species are outlined in red. The amino-acid sequence of fugu Mstn shows 83.8% and 60.1% identity to that of medaka and bovine, respectively. The C-terminal peptide, which is considered to be significant for the function, is outlined in light blue. The underlined cysteines are essential for the activity of Mstn in vertebrate. The fugu C-terminal region has an 88.1% identity to medaka and bovine Mstn. Black triangle, the amino acid responsible for the loss-of-function phenotype in medaka and bovine. Bidirectional bold and thin arrows **(A)** and bold characters **(B)**, the region examined by the HRM analysis in this study.

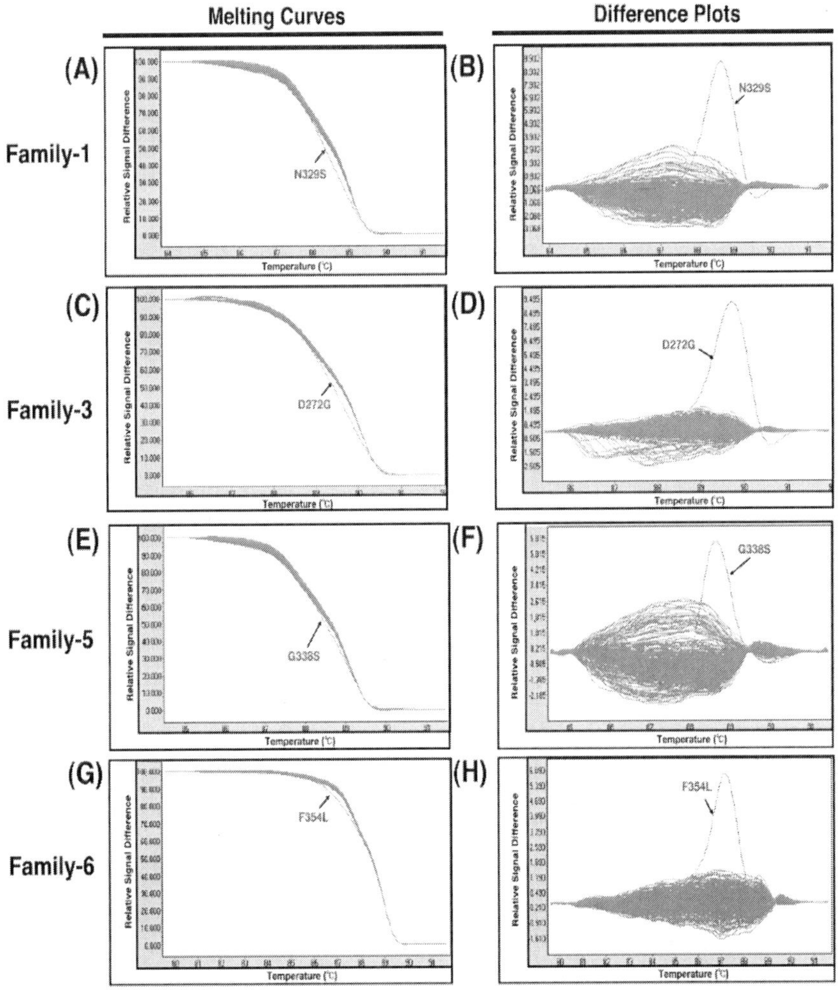

Figure 3 ENU-induced mutations in fugu *Mstn* gene detected by HRM analysis. The exon 3 of the *Mstn* gene was examined by HRM analysis with each family of fugu genomic DNA. The region contains the C-terminal peptide of the fugu Mstn protein. The family numbers mean the line of the hatched fry generated by *in vitro* fertilization (see Table 3). Left side of figures, the high-resolution melting curve **(A, C, E, G)**, and right side of the figures, the substitution plots **(B, D, F, H)**. X-axis and Y-axis, temperature (°C) and relative signal difference, respectively. Red lines and black arrows, the mutations screened by the HRM analysis.

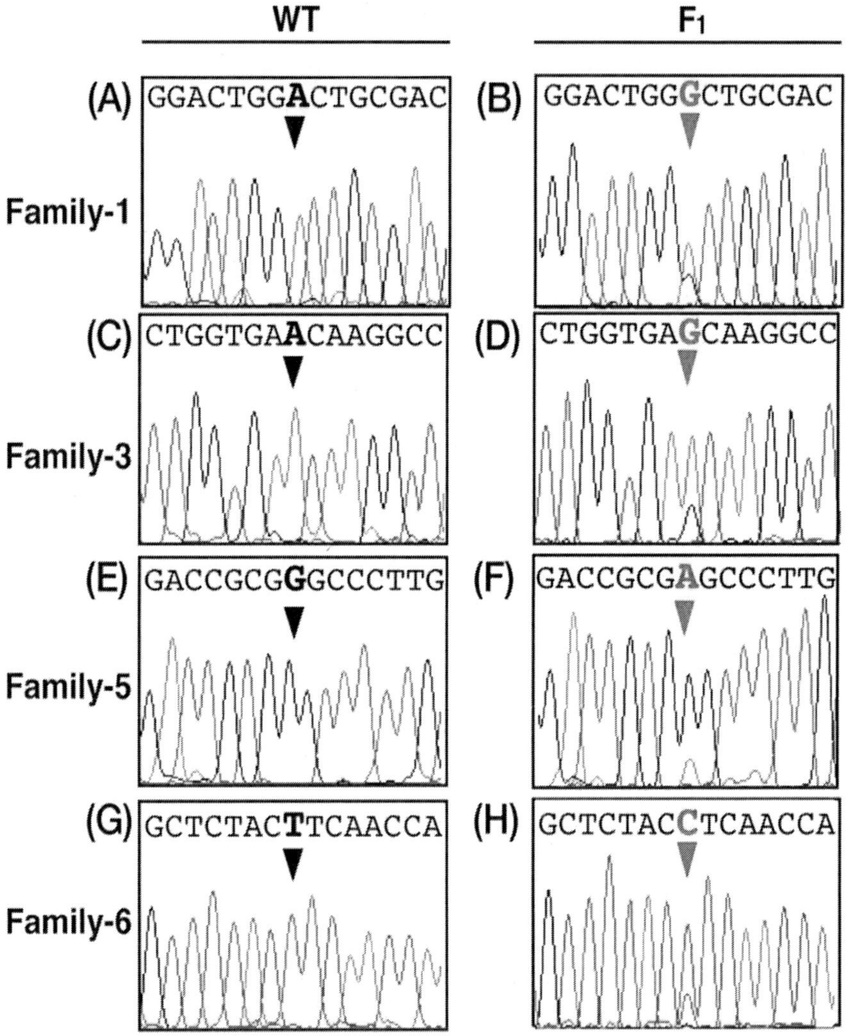

Figure 4 ENU-induced point mutations in fugu *Mstn* gene identified by direct sequencing. The PCR products from fugu genomic DNA screened by the HRM analysis were investigated by direct sequencing. Left side **(A, C, E, G)** and right side **(B, D, F, H)** of figures, the sequence traces of wild-type fugu (WT) and the fugu mutant (F_1). Lines in green, red, black and blue; adenine, thymine, guanine and cytosine, respectively. A nucleotide in the germ cell in the founder fish (black triangles) is replaced by another nucleotide due to the ENU-treatment (red triangles), resulting in an offspring that is heterozygous at the site. The numbers mean the line of the hatched fry generated by *in vitro*fertilization (see Table 3).

As shown in Table 3, ten types of mutations were detected by HRM analysis and then confirmed by direct sequencing (nonsense, 1; missense, 7; silent, 2). Also, it was found that all the ENU-induced mutations in fugu were point mutations. The number of the induced mutations was an average of 1 to 2 in each family treated with various ENU concentrations (7, 70, 120 mg/kg body weight). When the effect of injection into fugu on the mutations was examined in the control (0 mg/kg body weight), no mutations were recognized. This indicates that the number of induced mutations does not depend on the concentration of ENU within the range of 7 to 120 mg/kg. Consistent with this, the dose-dependent effects were not observed on the fertilization and the hatching ratios (Table 2).

Table 3 Summary of fugu MSTN mutations detected by HRM analysis

Family no.	ENU conc. (mg/kg body weight)	DNA sequence	Amino acid substitution
1	7	CTGGGACTGG(A > G)CTGCGACGAG	D272G
1	7	TTGCAGAAAT(A > G)CCCGCACACC	Y322C
2	70	gCAACCTTTC(A > G)TGGAGGTGAA	M254V
3	70	GAGACCTGGG(A > G)CTGGACTGCG	G270G
3	70	CACCTGGTGA(A > G)CAAGGCCAAC	N329S
4	70	AGTGTGAGTA(C > A)ATGCACTTGC	Y316X
5	70	CGCACACCCA(C > T)CTGGTGAACA	H326H
5	70	AGGGACCGCG(G > A)GCCCTTGCTG	G338S
6	120	CATGCTCTAC(T > C)TCAACCAAGA	F354L
6	120	CTCTACTTCA(A > G)CCAAGAACAG	N355S

(A > B): A is the nucleotide of wild-type and B is the nucleotide of the mutant. X indicates a stop codon. There were ten types of mutation in the investigated region with HRM analysis (nonsense, 1; missense, 7; silent, 2). The nonsense mutation is underlined.

Identification of Mstn-deficient mutation in fugu

The nonsense mutation we discovered changes a cytosine to an adenine in exon 3 of fugu *Mstn* converting a tyrosine codon to a stop codon in a fry from the family 4 (Figure 5A, B, Table 3). While the fry was heterozygous at this site, its parents and control fish were homozygous (Figure 5B), suggesting that the mutation was induced by the ENU treatment. DNA sequencing of the amplicon cloned into a plasmid vector confirmed the presence of this mutation (Figure 5C). The nonsense mutation was also examined with CEL I assay (Figure 5D). The cleavage product was detected in the CEL I reactions of the mutant DNA (lane 6), but they were not in the reactions of wild-type DNA (lanes 1–5). The mutation likely results in the loss-of-function of Mstn in fugu, as a critical part of the amino acid chain, the C-terminal peptide, are no longer created due to the stop codon.

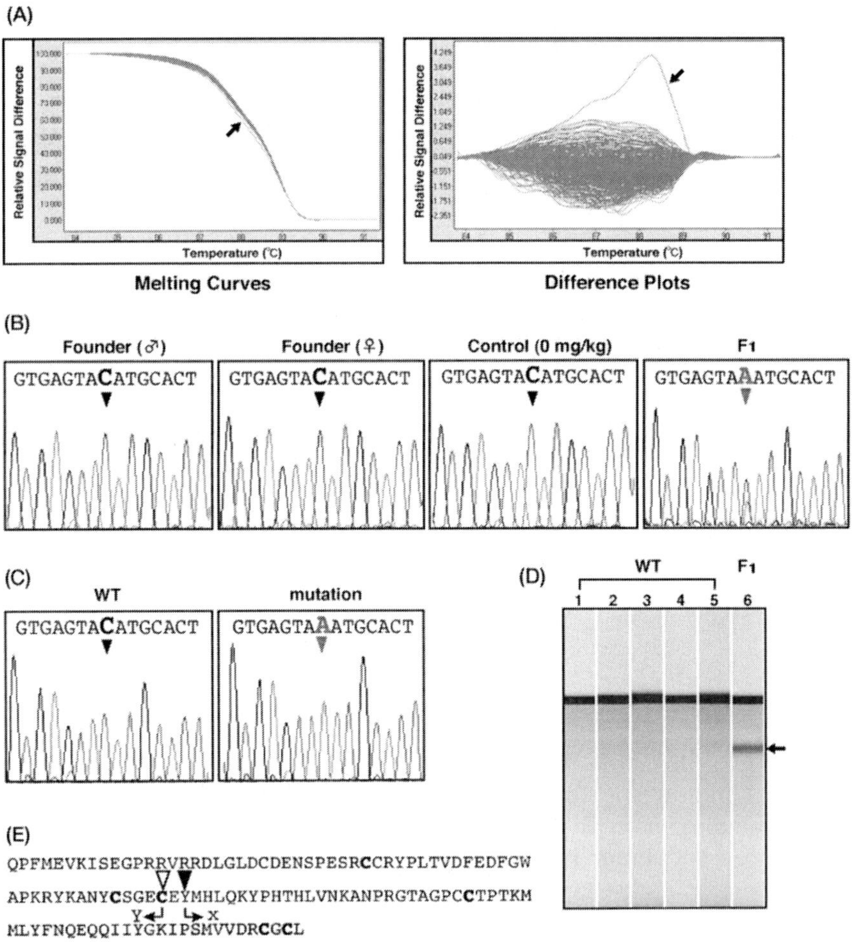

Figure 5 Identification of the ENU-induced nonsense mutation in fugu TILLING. A) The high-resolution melting curve (left) and the substitution plot (right). Exon 3 of the *Mstn* gene, which contains the C-terminal region of the Mstn protein, was examined by HRM analysis using the fugu family 4 (see Table 3). Red curve, the deviated curve due to mutation. **B)** Direct sequencing for the selected mutations. The selected genomic DNA with the

HRM analysis was investigated by direct sequencing. Lines in green, red, black and blue; adenine, thymine, guanine and cytosine, respectively. The sequence traces of the fugu parents (Founder-♂ and -♀), control (Control), and the ENU-induced mutant (F_1) are illustrated in order from left to right. Cytosine in the parents (black triangles) is substituted to adenine in the mutant (red triangle), resulting in the change from tyrosine to a stop codon at amino acid 316. **C, D)** Confirmation of the nonsense mutation in fugu TILLING. PCR amplicons were cloned and sequenced to confirm the substitution of the nucleotide **(C)**. The sequence traces of the cloned wild-type and the mutant DNA are illustrated in left (WT) and right (mutation), respectively. Cytosine to adenine nucleotide transition was recognized. The cleavage product was detected with CEL I assay in the mutant DNA (**D**, lane 6), but not in wild-type DNA (**D**, lanes 1–5). Black arrow, the cleavage product. **E)** Nonsense mutation in the C-terminal region of fugu Mstn. The ENU-induced point mutation in fugu TILLING caused the substitution to a stop codon in the C-terminal region. Bold characteristics, six cysteines which are essential for the Mstn conformation. X, a stop codon. Black and white triangles, amino-acid change from tyrosine to a stop codon in fugu Mstn; the change from cysteine to tyrosine in medaka Mstn responsible for the phenotype, respectively.

ENU-induced mutation ratio in fugu

The mutation ratio in the mutagenesis is a key point to produce the mutants effectively. Therefore, the ratio was calculated using data detected in fugu *Mstn* gene by HRM analysis. We screened about 3.0 Mb in total and identified 10 independent mutations (Table 3). Therefore, the ENU-induced mutation ratio for fugu *Mstn* gene was one substitution per 297 kb.

DISCUSSION

Establishment of fugu TILLING

Schematic model for fugu TILLING method is summarized in Figure 6 as follows: fugu males are treated with ENU 3 times at weekly intervals at the stage before spermiation and then treated with gonadotropin at 3 weeks after the last ENU treatment to facilitate the sperm production and spermiation. The sperm from the founder fish was collected at 3 weeks after the gonadotropin treatment and used for *in vitro* fertilization to obtain hatched fry. The genomic DNA extracted from the fry was subjected to PCR and the amplicons were used to screen ENU-induced mutations with HRM in a fugu gene of interest. Our study firstly demonstrated that TILLING technology is applicable to aquacultured fish. Interestingly, the TILLING approach in fugu could be used in a similar way as in plants [38]. We can treat them like plant seeds derived from mutagenized ancestor, because sufficient eggs and sperm to obtain a large amount of F_1 progeny are prepared for a single pair of fugu female and male. Furthermore, mutations induced by TILLING in fugu can be readily distinguished from endogenous SNP, since a single female and male pair is used for the cross. On the contrary, it has been pointed out that non-predictable frequency of endogenous SNP in a mutagenized bank derived from multiple founders is an obstacle for the screening in zebrafish and medaka [13, 22]. TILLING can be considered as an appropriate method for many aquacultured fish producing a large number of eggs.

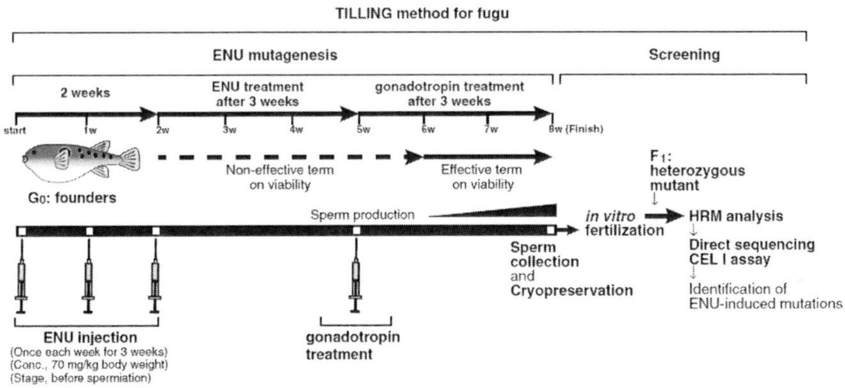

Figure 6 Establishment for the fugu TILLING method. Mutagenesis was carried out with injection of ENU (conc., 70 mg/kg body weight) into founder fugu 3 times at weekly intervals. Half of the founder fish died around 6 weeks after the last ENU treatment. Sperm was not obtained from the founder until 5 weeks after the last ENU treatment. To collect the sperm from the founder, gonadotrophin (HCG/SPH) was injected once 3 weeks after the last ENU treatment. The sperm was collected 3 weeks after the hormone treatment and then cryopreserved for a long-term TILLING resource. For screening strategy, the hatched fry were obtained *in vitro* fertilization with the collected sperm and eggs from wild-type female. Genomic DNA extracted from the fry was used to screen the induced mutations in *Mstn*gene with HRM analysis. The screened mutations were confirmed by direct sequencing and CEL I assay. These approaches demonstrate that the TILLING method was established in fugu.

Recently, *in vitro* ENU mutagenesis of postmeiotic sperm was performed in grass carp [20]. In their study, the induced mutations were investigated using growth-related genes (*Igf-2a, Igf-2b, Mstn-1, Mstn-2, Fst-1* and *Fst-2*) and the estimated mutation ratio was around 0.41%. However, the mutations are apparently mosaic in F_1individuals. Mosaicism of the induced mutation was also reported in zebrafish when postmeiotic sperm was mutagenized [21]. This situation makes recovery of the mutant line inefficient in contrast to the case of non-mosaic mutation induced by ENU-treatment of spermatogonia [21]. The mosaicism in F_1 individuals and resulting non-Mendelian transmission from F_1 to F_2 generations hinder some screening protocols. In addition, large-scale rearrangements of the genomic structure such as translocation of genomic segments have been reported following the mutagenesis of postmeiotic sperm [18]. Therefore, the suitability of the *in vitro* ENU mutagenesis of postmeiotic sperm for mutation breeding in fish improvement remains to be verified. Although the *in vitro* ENU treatment is more convenient than the *in vivo* methods used in fugu and other model animals, it will be required to examine whether the induced mutations will be inherited by the progeny in the mutants.

In TILLING, high efficiency for ENU-induced mutations is required to obtain various phenotypes in gene of interest. A previous report indicated that the induced mutation frequency increased with ENU dose dependently in mouse [36]. In this study, there was no difference in the number of surviving fish, fertilization and hatching ratios in ENU treatment with 7, 70 mg/kg (Table 2). ENU treatment with 120 mg/kg affected slightly on the number of surviving treated fugu. However, the number of the induced mutations did not tend to increase in each ENU concentration between 7 mg/kg and 120 mg/kg (Table 3). These results suggested that each ENU concentration in the method may become saturation in premeiotic sperm (spermatogonia and spermatocyte cells) of fugu.

Identification of fugu Mstn mutant

It has been reported that loss of the Mstn function in medaka and mice causes double-muscling phenotype with both hyperplasia and hypertrophy [28, 33]. In the present study, ENU-induced mutations in fugu Mstn were confirmed and in particular, the nonsense mutation (Y316X) was identified at C-terminal region of the peptide. The *mstn* mutant medaka (*mstnC315Y*) made by using the TILLING method [13] corresponded to a naturally occurring mutation (C313Y) in double-muscled Piedmontese cattle [29] (Figure 2B). The cysteine residue at the 315-position exists in C-terminal region of the Mstn peptide (Figures 2B and 5D) and is known to be one of the important amino acid residues involved in forming the cysteine knot structure in members of the TGF-β superfamily [37]. The structure functions as a homodimer to elicit a biological function [39]. Therefore, we speculate that the structural formation composed of six essential cysteines in fugu Mstn will be non functional by the truncated form of the peptide resulting from a point mutation at 317 of Tyr. Eventually, the mutation will be conceivable to cause loss of the Mstn function in fugu *Mstn* mutant. The double-muscling phenotype in medaka and bovine could be similarly expected for an *Mstn*-deficient mutant fugu.

Development of HRM analysis

For screening, there are a variety of mutation discovery technologies such as direct DNA sequencing, CEL I assay and HRM analysis [13, 23, 26]. We introduced the HRM analysis for screening system in fugu because there is no albino mutant in fugu for specific locus test (SLT) and the analysis has recently succeeded to detect point mutations as a screening step for the TILLING approach [25, 26]. But, there was room for improvement with respect to the cost and variability of output data. In this study, we have developed a system of HRM analysis as follows: cost-cutting with fluorescent intercalator; SYTO 9, introduction of the nested PCR for precision enhancement and avoidance of variability, speed-up with 384 samples treatment in a single experiment, and high uniformity of dispensed volume and removal of human error by using the automatic dispense device. As a result of the improvement, some mutations could be detected efficiently in the analysis (Figure 4 and Table 3). It is difficult for HRM analysis to investigate mutations with a long amplicon in comparison with the other methods, but the acquisition of melting profiles in HRM is not dependent on the location and number of mutations present in the analyzed fragment. It has been compared about the mutations detected by the screening approaches. The results indicate that HRM analysis is useful as resequencing in medaka [26] and more efficient than CEL I assay in mammals [40]. Taken together, these demonstrate that HRM analysis is a reliable high-throughput approach for screening ENU-mutagenized individuals in potentially various species.

Mutation ratios in fugu TILLING

The mutation ratio in ENU mutagenesis has been calculated in a SLT using the albino mutant in zebrafish and medaka [21, 41]. The characteristics of the albino mutant in fish reveal mainly red eyes and one of the responsible genes is tyrosinase. In some aquacultured fish with albino mutants, it may be easy for these ENU-treated fish to examine the mutation ratio with SLT. However, albino mutants are not found in nature in all animal species.

Instead of SLT, the mutation ratio was calculated using base-pairs screened and the number of mutations detected in fugu *Mstn* gene by HRM analysis (Table 3). The frequency in fugu (1 mutation per 297 kb) is similar to the mutation frequency in medaka (1 mutant per 345 kb) and in zebrafish (1 mutant per 235 kb), which were calculated by resequencing data [13, 42]. This suggests that ENU mutagenesis in fugu was efficiently performed in our way and the protocol detailed here will potentially be useful for constructing a TILLING library.

CG to AT (or reverse) changes are often detected in HRM analysis, but the endogenous SNPs are also detected as the same changes and give the most significant difference in this analysis. In our study, there were no endogenous SNPs in the C-terminal region of the *Mstn* gene in both the investigated family of fugu and their parents (data not shown). Although G > C, C > G, T > A, A > T mutations were not observed in the 10 mutations detected in the *Mstn* gene (Table 3), it was reasonable results since the appearance frequency of their mutations are also rare in human genome [43].

Generation of an Mstn-deficient mutant in fugu

Our next step will be to generate the *Mstn* mutant for the purpose of serving the aquaculture industry. One of the key components for a successful TILLING project is a near-perfect ability to recover valuable mutations once they have been found. For the recovery, TILLING libraries can either be 'living' or cryopreserved as suggested by Moens et al. [22]. A living library is one in which mutagenized F_1 fish are held in pools in tanks while their fin-tip genomic DNA is screened. A cryopreserved library is one in which each genomic DNA sample corresponds to vials of frozen sperm that can be used in an *in vitro* fertilization to generate F_2 heterozygotes. As a large amount of sperm can be stored from a single adult male of aquacultured fish, in general, the cryopreserved library is more desirable than 'living' library. However, generation time of many aquacultured fish including fugu is more than one year, making the construction of the cryopreserved library time-consuming. We are currently generating an *Mstn*-deficient fugu with 'living' library as the next step.

Conclusions

In the present study, the fugu TILLING method was established from the results in both ENU mutagenesis and a mutation search technology, HRM analysis. This is a first report that TILLING is an applicable method for aquacultured fish. The approach for the fish species might play an important role in producing a fish, which exhibits a valuable phenotype (e.g., rapid growth, disease resistance) for the aquaculture industry. We hope that the TILLING technique will be an alternative to the classic selection approach for improving traits in aquacultured fish species.

METHODS

Ethics statement

Fisheries Research Agency and The University of Tokyo rule that ethic approval is not necessary for study of fish. Nevertheless, this project was conducted in accordance with the Regulation for Animal Experiments in the Fisheries Research Agency, which stipulated in the Fundamental Guidelines for Proper Conduct of Animal Experiments and Related Activities in Research Institutions (Ministry of Agriculture, Forestry and Fisheries of Japan), and under the supervision of the Animal Care and Use Committee for the National Research Institute of Aquaculture. In this study, ENU and hormone injections for fugu were carried out under 0.05% anesthetic agent (2-phenoxy-ethanol) and hatching fry of fugu were harvested after they were soaked in sea water containing ice. All efforts were made to minimize suffering.

Experimental fish and eggs for *in vitro*fertilization

50 fugu males (body weight: around 1 kg) were purchased from a commercial supplier (Shihou Suisan, Minami-ise, Mie). The fish were kept at the Nansei station of National Research Institute until obtaining sperm from the ENU-treated fish. Three-year old fugu females (body weight: around 2–3 kg) were cultivated in 30–50 ton of seawater tanks at the Yashima station of National Center. Two independent fugu females were used to obtain eggs for *in vitro* fertilization. Mainly, the fertilization and harvest of the hatched fry were performed at the Yashima station. A part of the fertilized-eggs were packed into seawater without air and sent to Fisheries Laboratory attached to the University of Tokyo. The fertilized eggs were incubated at 18°C during 8 to 10 days until harvest of the hatched fry.

Mutagenesis with ENU

All working solutions of ENU (Conc.: 11 mg/ml, Sigma-Ardrich, St. Louis, MO USA) were freshly prepared following the methods described by Taniguchi et al. [13]. ENU was prepared at three concentrations (7, 70, 120 mg per kg, body weight of fugu). For ENU mutagenesis, fugu males (around 1 kg of body weight) were used and injected into the intraperitoneal with the ENU concentrations prepared for 10 mM sodium phosphate buffer (pH 6.3). The injections were repeated 3 times at weekly intervals to the founder fish. After the last ENU injection, the fish were injected intramuscularly with human chorionic gonadotropin (HCG, 500 IU/kg body weight) or salmon pituitary homogenate (SPH, 10 mg/kg body weight), consisting of acetone-dried chum salmon pituitary homogenized in Ringer, was added to HCG because of increasing the success rate of spermiation in ENU-treated males. After the hormone treatment, the sperm was collected from each fish.

Cryopreservation of sperm

The freshly-collected sperm from each fugu male was suspended at the rate of 1 to 4 in freezing diluents [10% dimethyl sulphoxide (Sigma-Aldrich Co., Tokyo, Japan) in 90% fetal bovine serum (Thermo Scientific Co., Tokyo, Japan)] and cryopreserved as described by Ohta et al. [44]. The amount of sperm held in each capillary (0.5 ml vol.) was enough to fertilize more than twenty thousand eggs.

*In vitro*fertilization

About one million unfertilized eggs were squeezed out from a fugu female, which was induced ovulation by administration of an LHRH analogue (des Gly10 [D-Ala6]-LHRH, 400 μg/kg body weight). Fresh sperm, obtained from each fugu family line, was used in this method, and the sperm activity was examined under a light microscope. 0.4 ml of the sperm was added to 12 g of the eggs (around twenty thousand eggs), and these were mixed well. After adding seawater to activate the sperm for fertilization, the eggs were immediately mixed and stood for 5 minutes. The eggs were washed 3 times with new seawater to remove the extra sperm after discarding the remaining seawater. The fertilized eggs were incubated for 4 to 6 hours at 18°C. The eggs were moved to a hatching tank with running water and kept at 18°C until they hatched. At 8 to 10 days after fertilization, around ten thousand hatched fry were obtained in each family. These fry were harvested in a 50 ml tube containing 99.5% ethanol (EtOH) and stored at −30°C.

The fertilization ratio was determined at the four-cell stage with Sera solution (EtOH 6: formalin 3: acetic acid 1), which makes the fertilized eggs transparent. In addition, one hundred eggs were counted at melanophore stage in the fertilized eggs of each cross, and then incubated at 18°C until hatching. The hatching ratio was calculated as the number of the hatched fry per 100 fertilized eggs.

Cloning of fugu *Mstn* gene

RNA was extracted from wild-type fugu skeletal muscle by RNeasy® Mini Kit (QIAGEN Sciences, MD, USA) according to the manufacturer's instructions. cDNA was synthesized using PowerScript (BD Biosciences Clontech, Palo Alto, CA, USA). To identify fugu *Mstn* gene orthologs, we used the basic local alignment search tool (BLAST) to search the fugu genome database (http://www.ensembl.org/Takifugu_rubripes/blastview). Fugu *Mstn* cDNA sequences was amplified using a reverse-transcription PCR by gene specific primer pairs flanking the ORF: 5′UTR-F1, 5′-TAGAACG CATTTCCACGTCTTC-3′, 3′UTR-R1: 5′-TGGTTTGAGGCCTCTACTCTGG-3′. Amplicons were cloned into a sequencing vector, and full-length ORF sequences were obtained.

Identification of mutations by HRM analysis

Extraction of genomic DNA

In this study, 1,056 hatched fry of fugu were prepared in each family obtained from *in vitro* fertilization. A single fry was placed into each well of 8-well tubes containing 45 μl of 50 mM NaOH, and incubated at 95°C for 10 min. After incubation, 5 μl of 1 M Tris–HCl (pH 8.0) was added to the lysate immediately for neutralization. These aliquots were centrifuged at 1,100 × g and room temperature for 20 min to remove tissue debris. Then, each 40 μl of supernatant was harvested into 96-deep well plates and stored at −30°C. Although the genomic DNA stocks were crude extract and freeze-thawed repeatedly, we could use the stocks for our assay during 10 months at least.

Primers

To find the loss-of-function mutations in the fugu *Mstn* gene, the third exon of the gene was screened using the primers designed by Primer3 programs on the web (source code available at http://bioinfo.ut.ee/primer3-0.4.0/primer3/). The primer sets and amplicon sizes were as follows, the forward and reverse primers #1: 5′-CAGCCATGATCTAACTTCTGCAAG-3′ and 5′-GTTTGAGGCCTCTACTCTGGTTCT-3′, yielding a fragment of 586 bp; the forward and reverse primers #2: 5′-CTCGCTCACATGCATTCTGTC -3′ and 5′-CTTGTTCACCAGGTGGGTGT-3′, yielding a fragment of 268 bp; the forward and reverse primers #3: 5′-AAGCGCTACAAGGCCAACTA -3′ and 5′-AAAGCGCTGGAACTGGTAGA-3′, yielding a fragment of 281 bp. These primers were used for the 1st-PCR (primer #1), and for the 2nd-PCR (primers #2 and #3) [see PCR].

PCR

PCR amplicons for HRM were obtained by using a nested-PCR approach. The genomic DNA stocks and all PCR reaction mixtures were dispensed by a laboratory automation system, Biomek®3000 (Beckman Coulter, CA, USA).

The 1st-PCR (PCR_1) was performed in 96-well PCR plates in 15 µl volumes. Reactions included 2 µl of the genomic DNA solution in 1x KOD-FX PCR buffer (Toyobo, Tokyo, Japan), with 400 µM of deoxynucleotide triphosphate, 0.1 U of KOD-FX polymerase (Toyobo), and 300 nM of each primer. The polymerase was able to increase precise amplicons with crude samples. The plates were centrifuged at 200 × g for 1 min and then PCR was performed in T-Professional Thermocycler (Biometra, Germany). The cycling conditions were set as follows: one cycle of 94°C for 2 min, followed by 40 cycles of 98°C for 10 sec; 60°C for 30 sec; 68°C for 40 sec. The PCR_1 products were firstly diluted 10 fold for DNA sequencing and then finally 140 fold with sterilized water by using Biomek® 3000. The 140 fold diluted samples were used as a template for second step of PCR reactions.

The 2nd-PCR (PCR_2) was performed in 384-well PCR plates (Roche Diagnostics, Basel, Switzerland) in 9 µl volumes. Reactions included 2 µl of the diluted PCR_1 products in 1x Hot-Start Gene Taq PCR buffer (Nippon Gene, Tokyo, Japan), with 200 µM of deoxynucleotide triphosphate, 1.25 µM of SYTO 9 (Molecular Probes, OR, USA) as a fluorescent substance, 0.16 U of Hot-Start Gene Taq polymerase (Nippon Gene), and 200 nM of each primer. The plates were centrifuged at 200 × g for 1 min and then PCR was performed in a 384-well Light Cycler 480 System (Roche Diagnostics). The cycling conditions were set as follows: one cycle of 95°C for 5 min, followed by 25 cycles of 95°C for 10 sec; 60°C for 30 sec; 72°C for 20 sec.

Several samples of the PCR_1 and PCR_2 products were examined with a microchip electrophoresis system, MultiNA (Shimadzu, Kyoto, Japan) whether there are the extra major fragments except for the proper fragment or not, because non-specific PCR products reduce the performance of HRM assay.

HRM assay

After PCR, the plates were imaged in the 384-well Light Cycler 480 System (Roche Diagnostics). The PCR products in the plates were heated at 0.1°C/second, and fluorescence measurements were collected from 65°C to 98°C. Briefly, background-subtracted fluorescence curves were normalized between 0% and 100%. The curves were viewed on the subtraction plots to magnify differences in the shape of the melting curves. The subtraction plots are useful to cluster the samples into groups. Clustering of the melting curves for genotype identification was carried out using the Light-Cycler software.

A single base-pair change within the probe region causes a significant shift in melting temperature (Tm) of the amplicon. In the TILLING method, all the induced mutations are heterozygous. PCR amplification of heterozygotes followed by heat denaturing and annealing result in the formation of 4 duplexes: two homoduplexes and two heteroduplexes. Each duplex has a characteristic melting temperature, and the sum of all transitions can be detected by melting curve analysis. The heterozygous samples are identified by differences in shape of the melting curve.

1,056 of the genomic DNA in each fugu family were subjected to PCR and two amplicons (268 bp and 281 bp) for screening were examined in HRM. The genomic DNA showing melting curves different from those of the wild-type allele were selected as mutant candidates. As previously reported in the medaka TILLING method [26], a substantial number of SNPs might exist in the fugu genomic DNA used. SNPs were discriminated from true mutations by defining the same type of base change detected in more than two sequencing reactions as an SNP. Moreover, the genomic DNA in the pair of fugu parents extracted from the cut fins was examined by DNA sequencing to confirm whether it was the true mutations or not.

Direct sequencing

The positive genomic DNA were screened as candidate mutations by HRM analysis. The PCR_1 products of the genomic DNA were already diluted 10 fold, and stocked at −30°C [see PCR]. These samples were used as a template for direct DNA sequencing. Sequencing reactions were carried out using a BigDye® terminator ver. 3.1 cycle sequencing kit (Applied Biosystems, CA, USA) and each forward and reverse primers. PCR cycling condition was set as follows: one cycle of 96°C for 60 seconds, followed by 25 cycles of 96°C for 10 seconds; 50°C for 5 seconds; 60°C for 120 seconds. Sequencing products cleaned up by magnetic bead-based purification system (Agencourt CleanSEQ, Beckman Coulter) were run on automated ABI 3130xl DNA analyzers (Applied Biosystems). Sequences were analyzed for the presence of heterozygous mutations using PolyPhred [45] and manual inspection of the mutated positions. All candidate mutations were confirmed with new PCR_1products and resequencing read. The PCR_1 products were treated with an ExoSAP-IT kit (Affymetrix, Ohio, USA) to remove unused primers and nucleotides.

CEL I assay

To confirm the presence of induced mutations in the target sequence, CEL I assay was performed with genomic DNA containing the mutations detected by direct sequencing. CEL I enzyme was purified from celery according to Anai et al. [46]. The enzyme activity for each batch of CEL I was determined experimentally using control samples.

PCR was performed in 9 µl volumes, and the primer sets and amplicon size were as follows, the forward and reverse primers: 5′-CTCGCTCACATGCATTCTGTC-3′ and 5′-AAAGCGCTGGAACTGGTAGA-3′, yielding a fragment of 465 bp. Reactions included 2 µl of the diluted PCR_1 products in 1x Hot-Start Gene Taq PCR buffer (Nippon Gene), with 200 µM of deoxynucleotide triphosphate, 0.16 U of Hot-Start Gene Taq polymerase (Nippon Gene), and 200 nM of each primer. The cycling condition was set as follows: one cycle of 95°C for 5 min, followed by 25 cycles of 95°C for 10 sec; 60°C for 30 sec; 72°C for 40 sec.

Directly following the PCR, heteroduplex was formed by incubating at 98°C for 8 min, 80°C for 20 sec, 70 cycles of 80°C for 7 sec with a decrement of 0.3 sec per cycle. Next, PCR reactions were incubated with 1 µl CEL I enzyme solution in a total volume of 10 µl at 37°C for 60 min. CEL I reactions were stopped by adding 10 mM EDTA. The cleavage products were detected with MultiNA (Shimadzu).

NOTES

Miwa Kuroyanagi and Hiroyuki Okamoto contributed equally to this work.

ABBREVIATIONS

TILLING: Targeting Induced Local Lesions In Genomes **ENU:** *N*-ethyl-*N*-nitrosourea **HRM:** High-resolution melting **Mstn:** Myostatin.

ACKNOWLEDGEMENTS

We would like to thank Miwa Osa (NRIA) for the excellent technical assistance during the experiment. This work was supported by the Program for Promotion of Basic and Applied Researches for Innovations in Bio-oriented Industry (BRAIN) and Grant-Scientific Research from Japanese Society for Promotion of Science.

REFERENCES

1. Moen T, Baranshi M, Sonesson AK, Kjøglum S: Confirmation and fine–mapping of a major QTL for resistance to infectious pancreatic necrosis in Atlantic salmon (*Salmo salar*): population-level associations between markers and trait. BMC Genomics. 2009, 10: 368-381. 10.1186/1471-2164-10-368.
2. Driever W, Solnica-Krezel L, Schier AF, Neuhauss SC, Malicki J, Stemple DL, Stainier DY, Zwartkruis F, Abdelilah S, Rangini Z, et al: A genetic screen for mutations affecting embryogenesis in zebrafish. Development. 1996, 123: 37-46.
3. Shima A, Shimada A: Development of a possible nonmammalian test system for radiation-induced germ-cell mutagenesis using a fish, the Japanese medaka (*Oryzias latipes*). Proc Natl Acad Sci U S A . 1991, 88 (6): 2545-2549. 10.1073/pnas.88.6.2545.
4. Loosli F, Koster RW, Carl M, Kuhnlein R, Henrich T, Mucke M, Krone A, Wittbrodt J: A genetic screen for mutations affecting embryonic development in medaka fish (*Oryzias latipes*). Mech Dev. 2000, 97 (1–2): 133-139.
5. McCallum CM, Comai L, Greene EA, Henikoff S: Targeted screening for induced mutations. Nat Biotechnol. 2000, 18 (4): 455-457. 10.1038/74542.
6. Cooper JL, Till BJ, Laport RG, Darlow MC, Kleffner JM, Jamai A, El-Mellouki T, Liu S, Ritchie R, Nielsen N, et al: TILLING to detect induced mutations in soybean. BMC Plant Biol. 2008, 8: 9-10.1186/1471-2229-8-9.
7. Suzuki T, Eiguchi M, Kumamaru T, Satoh H, Matsusaka H, Moriguchi K, Nagato Y, Kurata N: MNU-induced mutant pools and high performance TILLING enable finding of any gene mutation in rice. Mol Genet Genomics. 2008, 279 (3): 213-223. 10.1007/s00438-007-0293-2.
8. Slade AJ, Fuerstenberg SI, Loeffler D, Steine MN, Facciotti D: A reverse genetic, nontransgenic approach to wheat crop improvement by TILLING. Nat Biotechnol. 2005, 23 (1): 75-81. 10.1038/nbt1043.
9. Till BJ, Reynolds SH, Weil C, Springer N, Burtner C, Young K, Bowers E, Codomo CA, Enns LC, Odden AR, et al: Discovery of induced point mutations in maize genes by TILLING. BMC Plant Biol. 2004, 4: 12-10.1186/1471-2229-4-12.
10. Bentley A, MacLennan B, Calvo J, Dearolf CR: Targeted recovery of mutations in Drosophila. Genetics. 2000, 156 (3): 1169-1173.
11. Gilchrist EJ, O'Neil NJ, Rose AM, Zetka MC, Haughn GW: TILLING is an effective reverse genetics technique for *Caenorhabditis elegans*. BMC Genomics. 2006, 7: 262-10.1186/1471-2164-7-262.
12. Smits BM, Mudde J, Plasterk RH, Cuppen E: Target-selected mutagenesis of the rat. Genomics. 2004, 83 (2): 332-334. 10.1016/j.ygeno.2003.08.010.
13. Taniguchi Y, Takeda S, Furutani-Seiki M, Kamei Y, Todo T, Sasado T, Deguchi T, Kondoh H, Mudde J, Yamazoe M, et al: Generation of medaka gene knockout models by target-selected mutagenesis. Genome Biol. 2006, 7 (12): R116-10.1186/gb-2006-7-12-r116.
14. Sood R, English MA, Jones M, Mullikin J, Wang DM, Anderson M, Wu D, Chandrasekharappa SC, Yu J, Zhang J, et al: Methods for reverse genetic screening in zebrafish by resequencing and TILLING. Methods. 2006, 39 (3): 220-227. 10.1016/j.ymeth.2006.04.012.
15. Till BJ, Zerr T, Bowers E, Greene EA, Comai L, Henikoff S: High-throughput discovery of rare human nucleotide polymorphisms by Ecotilling. Nucleic Acids Res. 2006, 34 (13): e99-10.1093/nar/gkl479.
16. Kennedy CL, O'Bryan MK: N-ethyl-N-nitrosourea (ENU) mutagenesis and male fertility research. Hum Reprod Update. 2006, 12 (3): 293-301. 10.1093/humupd/dmk004.

17. Mullins MC, Hammerschmidt M, Haffter P, Nusslein-Volhard C: Large-scale mutagenesis in the zebrafish: in search of genes controlling development in a vertebrate. Curr Biol. 1994, 4 (3): 189-202. 10.1016/S0960-9822(00)00048-8.
18. Imai Y, Feldman B, Schier AF, Talbot WS: Analysis of chromosomal rearrangements induced by postmeiotic mutagenesis with ethylnitrosourea in zebrafish. Genetics. 2000, 155 (1): 261-272.
19. Knapik EW: ENU mutagenesis in zebrafish-from genes to complex diseases. Mamm Genome. 2000, 11 (7): 511-519. 10.1007/s003350010098.
20. Jiang XY, Sun CF, Zhang QG, Zou SM: ENU-induced mutagenesis in grass carp (*Ctenopharyngodon idellus*) by treating mature sperm. PLoS One. 2011, 6 (10): e26475-10.1371/journal.pone.0026475.
21. Riley BB, Grunwald DJ: Efficient induction of point mutations allowing recovery of specific locus mutations in zebrafish. Proc Natl Acad Sci U S A . 1995, 92 (13): 5997-6001. 10.1073/pnas.92.13.5997.
22. Solnica-Krezel L, Schier AF, Driever W: Efficient recovery of ENU-induced mutations from the zebrafish germline. Genetics. 1994, 136 (4): 1401-1420.
23. Colbert T, Till BJ, Tompa R, Reynolds S, Steine MN, Yeung AT, McCallum CM, Comai L, Henikoff S: High-throughput screening for induced point mutations. Plant Physiol. 2001, 126 (2): 480-484. 10.1104/pp.126.2.480.
24. Moens CB, Donn TM, Wolf-Saxon ER, Ma TP: Reverse genetics in zebrafish by TILLING. Brief Funct Genomic Proteomic. 2008, 7 (6): 454-459. 10.1093/bfgp/eln046.
25. Gady AL, Hermans FW, Van de Wal MH, van Loo EN, Visser RG, Bachem CW: Implementation of two high through-put techniques in a novel application: detecting point mutations in large EMS mutated plant populations. Plant Methods. 2009, 5: 13-10.1186/1746-4811-5-13.
26. Ishikawa T, Kamei Y, Otozai S, Kim J, Sato A, Kuwahara Y, Tanaka M, Deguchi T, Inohara H, Tsujimura T, et al: High-resolution melting curve analysis for rapid detection of mutations in a Medaka TILLING library. BMC Mol Biol. 2010, 11: 70-10.1186/1471-2199-11-70.
27. Li F, Niu B, Huang Y, Meng Z: Application of high-resolution DNA melting for genotyping in lepidopteran non-model species: *Ostrinia furnacalis* (Crambidae). PLoS One. 2012, 7 (1): e29664-10.1371/journal.pone.0029664.
28. McPherron AC, Lawler AM, Lee SJ: Regulation of skeletal muscle mass in mice by a new TGF-beta superfamily member. Nature. 1997, 387 (6628): 83-90. 10.1038/387083a0.
29. Kambadur R, Sharma M, Smith TP, Bass JJ: Mutations in myostatin (GDF8) in double-muscled Belgian Blue and Piedmontese cattle. Genome Res. 1997, 7 (9): 910-916.
30. McPherron AC, Lee SJ: Double muscling in cattle due to mutations in the myostatin gene. Proc Natl Acad Sci U S A . 1997, 94 (23): 12457-12461. 10.1073/pnas.94.23.12457.
31. Mosher DS, Quignon P, Bustamante CD, Sutter NB, Mellersh CS, Parker HG, Ostrander EA: A mutation in the myostatin gene increases muscle mass and enhances racing performance in heterozygote dogs. PLoS Genet. 2007, 3 (5): e79-10.1371/journal.pgen.0030079.
32. Schuelke M, Wagner KR, Stolz LE, Hubner C, Riebel T, Komen W, Braun T, Tobin JF, Lee SJ: Myostatin mutation associated with gross muscle hypertrophy in a child. N Engl J Med. 2004, 350 (26): 2682-2688. 10.1056/NEJMoa040933.
33. Chisada S, Okamoto H, Taniguchi Y, Kimori Y, Toyoda A, Sakaki Y, Takeda S, Yoshiura Y: Myostatin-deficient medaka exhibit a double-muscling phenotype with hyperplasia and hypertrophy, which occur sequentially during post-hatch development. Dev Biol. 2011, 359 (1): 82-94. 10.1016/j.ydbio.2011.08.027.
34. Wansleeben C, van Gurp L, Feitsma H, Kroon C, Rieter E, Verberne M, Guryev V, Cuppen E, Meijlink F: An ENU-mutagenesis screen in the mouse: identification of novel developmental gene functions. PLoS One. 2011, 6 (4): e19357-10.1371/journal.pone.0019357.

35. Russell WL, Kelly EM, Hunsicker PR, Bangham JW, Maddux SC, Phipps EL: Specific-locus test shows ethylnitrosourea to be the most potent mutagen in the mouse. Proc Natl Acad Sci U S A . 1979, 76 (11): 5818-5819. 10.1073/pnas.76.11.5818.
36. Shibuya T, Horiya N, Matsuda H, Sakamoto K, Hara T: Dose-dependent induction of recessive mutations with N-ethyl-N-nitrosourea in primordial germ cells of male mice. Mutat Res. 1996, 357 (1–2): 219-224.
37. Berry C, Thomas M, Langley B, Sharma M, Kambadur R: Single cysteine to tyrosine transition inactivates the growth inhibitory function of piedmontese myostatin. Am J Physiol Cell Physiol. 2002, 283 (1): C135-141. 10.1152/ajpcell.00458.2001.
38. Henikoff S, Till BJ, Comai L: TILLING. Traditional mutagenesis meets functional genomics. Plant Physiol. 2004, 135 (2): 630-636. 10.1104/pp.104.041061.
39. Lee SJ, McPherron AC: Regulation of myostatin activity and muscle growth. Proc Natl Acad Sci U S A . 2001, 98 (16): 9306-9311. 10.1073/pnas.151270098.
40. Bono C, Nuzzo D, Albeggiani G, Zizzo C, Francofonte D, Iemolo F, Sanzaro E, Duro G: Genetic screening of Fabry patients with EcoTILLING and HRM technology. BMC Res Notes. 2011, 4: 323-10.1186/1756-0500-4-323.
41. Shima A, Shimada A: The Japanese medaka, *Oryzias latipes*, as a new model organism for studying environmental germ-cell mutagenesis. Environ Health Perspect. 1994, 102 (Suppl 12): 33-35. 10.1289/ehp.94102s1233.
42. Wienholds E, Schulte-Merker S, Walderich B, Plasterk RH: Target-selected inactivation of the zebrafish rag1 gene. Science. 2002, 297 (5578): 99-102. 10.1126/science.1071762.
43. Venter JC, Adams MD, Myers EW, Li PW, Mural RJ, Sutton GG, Smith HO, Yandell M, Evans CA, et al: The sequence of the human genome. Science. 2001, 291: 1304-10.1126/science.1058040.
44. Ohta H: Cryopreservation of teleost spermatozoa. Cryobiology and Cryotechnology. 2008, 54 (1): 29-32.
45. Nickerson DA, Tobe VO, Taylor SL: PolyPhred: automating the detection and genotyping of single nucleotide substitutions using fluorescence-based resequencing. Nucleic Acids Res. 1997, 25 (14): 2745-2751. 10.1093/nar/25.14.2745.
46. Anai T, Yamada T, Hideshima R, Kinoshita T, Rahman SM, Takagi Y: Two high-oleic-acid soybean mutants, M23 and KK21, have disrupted microsomal omega-6 fatty acid desaturase, encoded by *GmFAD2-1a*. Breed Sci. 2008, 58: 447-452. 10.1270/jsbbs.58.447.

CHAPTER 12

Testing Strategies for Genomic Selection in Aquaculture Breeding Programs

Anna K Sonesson*†1 and Theo HE Meuwissen†2

1Nofima Marine AS, P.O. Box 5010, 1432 Ås, Norway and 2University of Life Sciences, P.O. Box 5003, 1432 Ås, Norway

ABSTRACT

Background

Genomic selection is a selection method where effects of dense genetic markers are first estimated in a test population and later used to predict breeding values of selection candidates. The aim of this paper was to investigate genetic gains, inbreeding and the accuracy of selection in a general genomic selection scheme for aquaculture, where the test population consists of sibs of the candidates.

Methods

The selection scheme started after simulating 4000 generations in a Fisher-Wright population with a size of 1000 to create a founder population. The basic scheme had 3000 selection candidates, 3000 tested sibs of the candidates, 100 full-sib families, a trait heritability of 0.4 and a marker density of $0.5N_e/M$. Variants of this scheme were also analysed.

Results

The accuracy of selection in generation 5 was 0.823 for the basic scheme when the sib-testing was performed every generation. The accuracy was hardly reduced by selection, probably because the increased frequency of favourable alleles compensated for the Bulmer effect. When sib-testing was performed only in the first generation, in order to reduce costs, accuracy of selection in generation 5 dropped to 0.304, the main reduction occurring in the first generation. The genetic level in generation 5 was $6.35\sigma_a$ when sib-testing was performed every generation, which was 72%, 12% and 9% higher than when sib-testing was performed only in the first generation, only in the first three generations or every second generation, respectively. A marker density above $0.5N_e/M$ hardly increased accuracy of selection further. For the basic scheme, rates of inbreeding were reduced by 81% in these schemes compared to traditional selection schemes, due to within-family selection. Increasing the number of sibs to 6000 hardly affected the accuracy of selection, and increasing the number of candidates to 6000 increased genetic gain by 10%, mainly because of increased selection intensity.

Conclusion

Various strategies were evaluated to reduce the amount of sib-testing and genotyping, but all resulted in loss of selection accuracy and thus of genetic gain. Rates of inbreeding were reduced by 81% in genomic selection schemes compared to traditional selection schemes for the parameters of the basic scheme, due to within-family selection.

BACKGROUND

In current family-based aquaculture breeding schemes, many important traits are tested on the sibs of the candidates. The test information is used to calculate breeding values for the selection of parents [1]. Only 50% of the genetic variation *i.e.* the between family variation, is used in these schemes, which are employed for traits that cannot be measured on the selection candidates such as disease challenge testing and slaughter quality traits.

Dense marker maps and high-throughput genotyping have become increasingly available in aquaculture species. Genomic selection is a selection method where the effects of dense genetic markers are first estimated in a test population and later used to predict breeding values of selection candidates [2]. The sib-test design can be used for marker-assisted selection [3] or genomic selection, where the association between markers and phenotypes is estimated in the sibs of the candidates, and the candidates are selected on breeding values that result from summing the estimates of the effects of their marker alleles. For aquaculture breeding schemes, genomic selection may prove very useful, because the breeding goals include many traits that are based on information from the sibs and not from the candidates, and therefore genomic selection can result in increased accuracy of selection for those traits by using both between- and within-family genetic variances.

The aim of this paper was to investigate genetic gains, inbreeding and the accuracy of selection in a general genomic selection scheme for aquaculture, where selection is based on information from the sibs of the candidates. Schemes with different numbers of candidates and test animals (sibs of the candidates) and with different heritabilities of the trait under selection were compared by computer simulation. In addition, the importance of performing the sib-test every generation, which is costly for the breeding program, was assessed. Finally, the effect of selection on the accuracy of selection was evaluated.

METHODS

Population

A population with an effective population size (N_e) of 1000 was simulated for 4000 generations according to the Fisher-Wright population model [4, 5]. Five hundred males and 500 females were randomly selected and mated using sampling with replacement.

Among the individuals of the last of these 4000 generations, 100 males and 100 females were randomly selected to create 100 full-sib families, which each produced 30 or 60 progeny to form generation Gen0. These progeny were selection candidates and were not performance tested. However, in addition to these selection candidates, every family also produced 30 or 60 full-sibs, which entered into a sib-test where they were performance tested. One hundred sires and 100 dams were selected from the candidates to produce generation Gen1 by either (a) random selection, whereby a sire and a dam were randomly sampled with replacement (RAND) or (b) directional selection, whereby sires and dams with the highest genome-wide breeding values (see Calculation of phenotypic values and true and estimated breeding values) were selected without any restriction on the number of parents selected from each family. Again each of the 100 sires was mated to one of the 100 dams, using sampling without replacement, to produce 30 or 60 full-sib selection candidates and 30 or 60 sib-test progeny in generation Gen1. This scheme was repeated until generation Gen10. Hence, with the number of families, Nfamilies, being 100, the total number of candidates, Ncand, was 3000 or 6000 and the total number of sib-test progeny, Ntested, was 3000 or 6000. In one scheme, Nfamilies = 200, Ncand = 3000 and Ntested = 3000.

Reduction of the number of sib-tests

The idea here was to reduce the number of sib-tests by not performing a sib-test every generation. Four different strategies of sib-testing were compared:

EVERY GENERATION (EVERY-GEN): sib-testing was performed in every generation Gen0-Gen10 as described above in 2.1.

FIRST GENERATION (FIRST-GEN): sib-testing was performed in generation Gen0.

EVERY SECOND GENERATION (EVERY-2GEN): sib-testing was performed in the odd generation numbers Gen1, Gen3, .., Gen9.

FIRST 3 GENERATIONS (FIRST-3GEN): sib-testing was performed only during the first three generations Gen0-Gen2.

Genome

Individuals had a diploid genome with ten 100 cM chromosomes. Recombinations were sampled at random positions on the chromosome assuming the Haldane mapping function. All polymorphisms were generated during the 4000 generations of the Fisher-Wright population model, where a mutation rate of 10^{-9} per nucleotide was assumed and the number of nucleotides per cM was 1000000. This effectively resulted in the infinite sites mutation model [6], *i.e.* every mutation occurred at a unique position and created a bi-allelic SNP. This mutation process generated numerous SNP, among which 100 per chromosome were sampled randomly as a QTL (sampling without replacement from the SNP with minor allele frequency (MAF) >0.05), and among the remaining SNP, the 1000 with the highest MAF were chosen as genetic markers. The latter resulted in a total of markers of Nmarkers = 10000 spread over 1000 cM. Reduced numbers of markers were obtained by taking every 10th marker and every 2nd marker, resulting in a total of Nmarkers = 1000 and 5000 markers, respectively.

The allelic effects of the QTL alleles were sampled from the gamma distribution with a shape parameter of 0.4 and a scale parameter of 1.66

[7]. The QTL effects were assumed to be either positive or negative with a probability of 0.5, because the gamma distribution only gives positive values. After sampling, these QTL allelic effects were standardised so that the total genetic variance was 1.0 in Gen0.

The expected average linkage disequilibrium (R^2) between two adjacent markers can be approximated by Hill [8]:

$$R^2 = \frac{5+2N_e c}{11+26N_e c+8N_e^2 c^2}$$

where c is the distance between adjacent loci, here on average 0.001 Morgan, which resulted in $R^2 = 0.342$. In generation Gen0, the realised average R^2 between adjacent markers was 0.374, which is slightly higher than Hill's approximation, probably due to the selection of the markers on their MAF.

Calculation of phenotypic values and true and estimated breeding values

The true breeding value of an individual was calculated as:

$$TBV_i = \sum_{j=1}^{1000} x_{ij1} g_{j1} + x_{ij2} g_{j2}$$

where x_{ijk} is the number of copies that individual i has at the jth QTL position and k^{th} QTL allele, and g_{jk} is the effect of the kth QTL allele at the jth position. The phenotypic values of the individuals in the sib-test were simulated by adding an error term sampled from a normal distribution to the true breeding value (TBV_i):

$$P_i = TBV_i + \varepsilon_i$$

where ε_i is an error term for animal i, which was normally distributed (0, σ^2_e) and σ^2_e was adjusted so the heritability was 0.1 or 0.4.

Marker effects were predicted using the BLUP method described in [2]. The statistical model used to estimate the marker effects was:

$$y_i = \mu + \sum_j^n X_{ij} a_j + e_i$$

where y_i is the record of test individual i; μ is the overall mean; X_{ij} denotes the marker genotype: 0 denotes that the individual is homozygous for the first allele; $1/\sqrt{H_j}$ denotes that it is heterozygous; and $2/\sqrt{H_j}$ denotes that it is homozygous for the second allele, where H_j is the marker heterozygosity and thus dividing by $\sqrt{H_j}$standardises the variance of the X_{ij} to 1; a_j is the random effect of the jth marker and Var(a_j) is assumed 1/Nmarkers (total genetic variance was standardised to 1.0); e_i is a random residual.

Genome-wide breeding values were estimated by summing the effects of the markers:

$$EBV_i = \sum_j^n X_{ij} a_j.$$

The accuracy of selection (acc) was calculated as the correlation between true and estimated breeding values. The acc was calculated for all schemes, also for the RAND scheme, although the EBV_i were not used for selection in RAND.

Statistics

Selection schemes were run for ten generations (Gen1-Gen10) and summary statistics for each of the schemes are based on 50 replicated simulations. The breeding schemes were compared for the genetic level (G, expressed in genetic standard deviation units of generation Gen0 (σ_a)), genetic gain (ΔG), accuracy of selection (acc), genetic variance, and level and rate of inbreeding (ΔF). Inbreeding was calculated based on pedigree, assuming that the Gen0 individuals are unrelated base parents. The values of these variables were either shown in figures over generations Gen1-Gen10 or in tables with the values in generation 5, when the Bulmer effect had stabilised, and inbreeding had started to build up in the population.

RESULTS

Basic scheme

Accuracy of selection was the highest for EVERY-GEN and increased from 0.647 to approximately 0.820 over generations (Figure 1a), due to the increased amount of information on marker effects that becomes available. When phenotypic and genotypic testing was only in the first generation, as for FIRST-GEN, accuracy of selection decreased rapidly over generations and was only 0.304 in generation Gen5. This reduction in accuracy of selection is mainly because of changes in the linkage disequilibrium between marker and QTL. Especially, spurious LD [9], which is not due to linkage, changes quickly over generations. When the sib-test was performed every second generation, as for EVERY-2GEN, the accuracy of selection fluctuated, being almost as high as for EVERY-GEN in the generations with sib-testing and lower in the other generations. However, fluctuations decreased over generations, probably because the selection pressure increased favourable marker alleles, and then, their effects became increasingly accurately estimated. For FIRST-3GEN, the accuracy of selection was as high as for EVERY-GEN until generation 3, as expected, and thereafter, the accuracy of selection decreased. However, the reduction during the first generation after selection had stopped was not as large as for EVERY-2GEN, probably because more generations of information on marker effects had built up. In general, the reduction in accuracy of selection after sib-testing had stopped was larger when there were fewer generations of information *i.e.* the reduction in accuracy of selection was the largest for FIRST-GEN, thereafter EVERY-2GEN, and FIRST-3GEN.

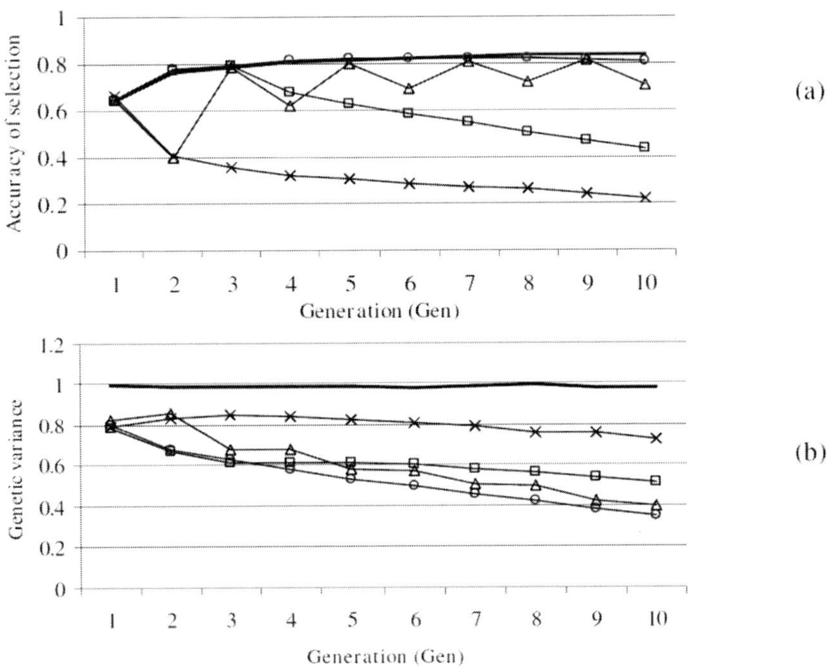

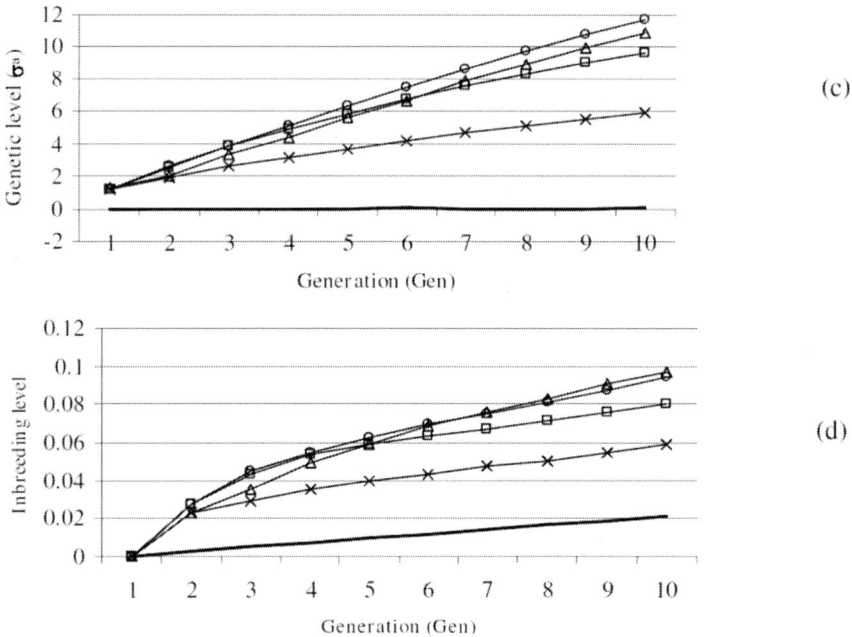

Figure 1 Accuracy of selection, genetic variance, genetic level and inbreeding level for the basic scheme. Accuracy of selection (a), genetic variance (b), genetic level (c) and inbreeding level (d) for schemes when sib-testing was every year (EVERY-GEN; circles), every second year (EVERY-2GEN; triangles), only the first year (FIRST-GEN; cross) or the first three years (FIRST-3GEN; squares); Random selection (solid line); Ncand = 3000, h^2 = 0.4, Nfamilies = 100, Ntested = 3000.

The RAND scheme had an accuracy of selection very similar to EVERY-GEN.

The genetic variance was reduced over generations of selection, except for the RAND schemes, as expected (Figure 1b). The genetic variance decreased most for the EVERY-GEN scheme, with the highest accuracy of selection, and least for the FIRST-GEN scheme. For FIRST-GEN, the largest reduction was between generations Gen1 and Gen2.

The accuracy of selection and genetic variance resulted in the highest genetic level for EVERY-GEN and the lowest for FIRST-GEN, as expected (Figure 1c). In generation Gen5, the genetic level was 6.35σ_a for EVERY-GEN and 3.69 σ_a for FIRST-GEN. For FIRST-3GEN, the genetic level was 5.82 σ_a. However, the genetic level of FIRST-3GEN became over time increasingly lower than that of EVERY-GEN. The genetic level of EVERY-2GEN in generation Gen5 was 5.66 σ_a and the difference in genetic level between EVERY-GEN and EVERY-2GEN increased over generations.

Inbreeding levels were also the highest for EVERY-GEN, probably because it had the highest accuracy of selection (Figure 1d). For FIRST-3GEN, the inbreeding level was as high as for EVERY-GEN until generation Gen4, and decreased thereafter, also because of lower accuracy of selection in the generations after the sib-testing had stopped.

Effect of marker density

The effect of different marker densities on the accuracy of selection was large *i.e.* when increasing Nmarkers from 1000 to 5000, the accuracy of selection increased from 0.661 to 0.823 using EVERY-GEN (Figure 2). However, when increasing Nmarkers from 5000 to 10000, the accuracy of selection hardly increased and was 0.842 in Gen5 with Nmarkers = 10000. For FIRST-GEN, the accuracy of selection decreased relatively much faster for the lowest number of markers *i.e.* 1000 (-51%) than for 5000 (-39%) or 10000 (-35%) markers. This suggests that if the markers are sufficiently close to the QTL, the linkage disequilibrium changes less and thus the estimates of the marker effects remain more accurate.

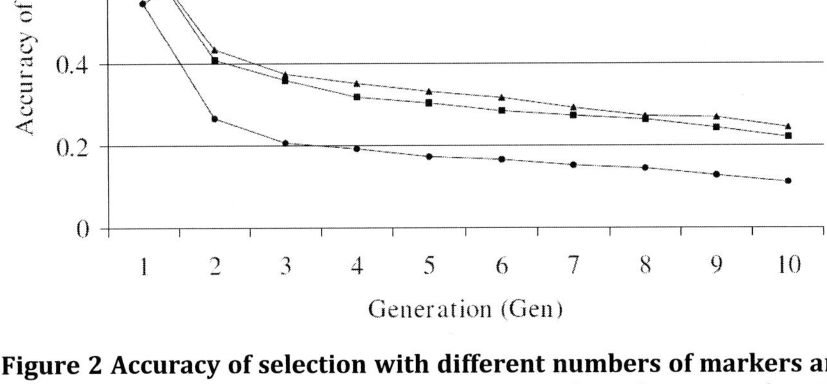

Figure 2 Accuracy of selection with different numbers of markers and sib-testing strategies. Accuracy of selection for schemes with 1000 (circles), 5000 (squares) or 10000 (triangles) markers when sib-testing was every year (EVERY-GEN; open) or only the first year (FIRST-GEN; filled); Ncand = 3000, h^2 = 0.4, Nfamilies = 100, Ntested = 3000.

Effect of heritability and numbers of candidates, tested sibs and families

Table 1 shows the sensitivity of the results to lower heritability of the trait, higher number of candidates and tested sibs, and higher number of families. Decreasing the heritability from $h^2 = 0.4$ to 0.1, decreased ΔG_5 by 7%. Similarly, acc_5 was 10% lower for the scheme with low heritability, but ΔF_5 was 50% higher for the scheme with low heritability, because of the increased between-family selection. The vg_5 was 5% higher for the scheme with a low heritability, because of lower selection accuracy.

Table 1 Effect of heritability and numbers of candidates, tested sibs and families

	ΔG_5 (se)	acc_5 (se)	ΔF_5 (se)	vg_5 (se)
Basic scheme[1]	1.22 (0.009)	0.823 (0.003)	0.008 (0.002)	0.532 (0.006)
Lower heritability, $h^2 = 0.1$		0.742 (0.004)	0.012 (0.002)	0.561 (0.009)
Higher number of candidates, Ncand = 6000	1.34 (0.019)	1.13 (0.012)	0.009 (0.002)	0.468 (0.010)
Higher number of tested sibs, Ntested = 6000	1.24 (0.009)	0.845 (0.002)	0.007 (0.002)	0.540 (0.007)
Higher number of families, Nfamilies = 200	0.99 (0.010)	0.749 (0.004)	0.004 (0.002)	0.642 (0.007)

Genetic gain (ΔG_5), accuracy (acc_5), rate of inbreeding (ΔF_5) and genetic variance (vg_5) in generation 5 for schemes with different numbers of candidates (Ncand = 3000 or 6000), tested sibs (Ntested = 3000 or 6000) and families (Nfamilies = 100 or 200) and heritability levels (0.1 or 0.4). EVERY-GEN testing strategy was used.
[1] $h^2 = 0.4$, Ncand = 3000, Ntested = 3000, Nfamilies = 100

Increasing the number of selection candidates from Ncand = 3000 to 6000 increased ΔG_5 by 10%. The increased genetic gain was mainly the result of higher selection intensity. However, this also resulted in larger increases of allele frequencies of favorable alleles, which in turn resulted in more accurate estimates of these alleles. Thus acc_5 was somewhat higher *i.e.* 0.837 for the scheme with 6000 selection candidates compared to 0.823 for the scheme with 3000 selection candidates. Increasing the number of tested sibs from Ntested = 3000 to 6000 increased acc_5 by only 2.7%, which hardly affected neither ΔG_5, vg_5 nor ΔF_5. This suggests that Ntested = 3000 is sufficient when $h^2 = 0.4$.

Increasing the number of families from Nfamilies = 100 to 200, reduced ΔF_5 from 0.008 to 0.004, because of the increased numbers of sires and dams selected. The latter resulted in higher genetic variance, which was 21% higher for the scheme with Nfamilies = 200 than with Nfamilies = 100. The acc_5 was 9% and ΔG_5 19% lower for the scheme with Nfamilies = 200 than with Nfamilies = 100, because selection intensity was reduced.

DISCUSSION

This simulation study examines the accuracies of selection and genetic gain that can be attained with genomic selection sib-testing schemes in aquaculture. Various strategies were evaluated to reduce the amount of sib-testing, but all resulted in a loss of accuracy of selection and thus of genetic gain. Whether these reductions in genetic gains are acceptable will depend on the relative sizes of the reduced benefits and the savings due to less sib-testing. In schemes where selection is for an index combining growth and sib-testing traits, the reduction in accuracy of selection and thus genetic gain will be less than that which was found here due to the reduced importance of the sib-testing traits. How much the accuracy of such an index will be reduced can be assessed using the results in Table 1 and selection index calculations. The general picture that emerges from Figure 1a is, however, that continuous phenotypic and genotypic testing is important to maintain the accuracy of the genome-wide breeding values over generations. The results showed that the number of previous generations of sib-testing affected the level of accuracy proportionally, such that the reduction in accuracy of selection over generations was the largest for FIRST-GEN, thereafter EVERY-2GEN, FIRST-3GEN and finally EVERY-GEN. It may be noted that the reduction in accuracy of selection was substantially larger for the first generation after sib-testing had stopped compared to later generations. This may be because within a generation, markers merely explaining family effects can be used for the prediction of breeding values, whereas across generations, the family effects decay. These results agreed with those of [10]. (Solberg, T.R., Sonesson, A.K., J. A. Woolliams, Ødegård, J., Meuwissen, T.H.E: Persistence of estimates of genome-wide markers over generations when including a polygenic effect, submitted) found a much smaller reduction in the accuracy over generations than we saw here. However, their study did not include the effect of selection. Also, the BayesB method of [2] was used instead of BLUP, which may have increased the weight given to markers that are in close LD with the QTL. Habier *et al.* [11] found that the accuracy of selection did not decrease as much with BayesB as with BLUP, because BayesB gives more weight to the LD when estimating genome-wide breeding values. Hence, the reduction observed in our study may be smaller if the BayesB method was used.

During later generations (say generations Gen6-10), the accuracy of selection was based on LD and its reduction was much smaller, indicating that the LD decayed slowly. Still, the low value of the selection accuracy suggests that in genomic selection breeding schemes also, the prediction of family effects is important and that breeding designs where family effects can be accurately predicted are important. This implies that the relationship between the individuals in the test group and the selection candidates should be as high as possible.

When decreasing heritability from 0.4 to 0.1, acc_5 decreased from 0.823 to 0.742 for the basic scheme with 30 sibs per family, *i.e.* a reduction of 0.08. The expected accuracy of selection for a traditional BLUP scheme is 0.66 for a heritability of 0.4 and 0.55 for a heritability of 0.1, *i.e.* a reduction of 0.11[12]. Hence, the effect of heritability on the accuracy of selection is somewhat larger for traditional BLUP schemes than for a scheme with genome-wide breeding values, which is in agreement with the literature on genomic selection [13] and marker assisted selection ([14] and others). The results also showed that the accuracy of selection increased by 25 and 35% when using genomic selection compared to traditional BLUP for the scheme with a heritability of 0.4 and 0.1, respectively. See [13] for a more detailed comparison of traditional and genomic selection schemes.

Truncation selection for traditional BLUP breeding values [15] resulted in a ΔF of 0.043 for the parameters of the basic scheme. Thus, ΔF was dramatically reduced by 81% for the genomic selection schemes compared to the traditional BLUP breeding schemes, although in practical breeding schemes, a way of reducing rates of inbreeding would probably be used, *e.g.* reducing the use of number of parents from single families or using optimum contribution selection. The reduced ΔF for genome-wide breeding values was probably due to the increased possibilities for within-family selection in genomic selection schemes, whilst this is not possible in traditional sib-testing schemes. It was also due to a stronger Bulmer effect since a more accurate selection increases the Bulmer effect, which leads to less between-family variance and thus more within-family selection [16].

In this study, we used 1000, 5000 and 10000 markers on a genome size of 1000 cM and an effective population size (N_e) of 1000. Hence, the marker density was 0.1, 0.5 and $1N_e/M$, which is not very high, but probably in accordance with what a first-generation SNP chip would include for most aquaculture species. For example, with a total genome size of 30M for salmon [17] and an assumed N_e of 1000 (see discussion [18]), $1N_e/M$ implies a SNP chip of 30000 markers. Furthermore, Figure 2 shows that increasing the marker density from 0.5 to $1N_e/M$ produces little extra gain. The low number of records relative to the N_e explains this plateau and such a plateau has also been reported by [19] with a SNP marker density of around $4N_e/M$.

The effect of selection on its accuracy was rather small here (EVERY-GEN versus RAND). This result is contrary to that of Muir [10], but he did not update the marker effects every generation, which made the accuracy decay rapidly. In the EVERY-GEN scheme, marker effects were updated every generation, but still one might expect that the Bulmer effect, which reduces the genetic variance (Figure 1b), would also reduce the accuracy of selection, because the signal to noise ratio is reduced. However, a selective genotyping effect occurs, due to selection, resulting in increased frequency and higher accuracy of marker alleles with positive effects. The Bulmer and selective genotyping effects seem to balance each other out approximately, resulting in an accuracy that is hardly reduced by selection.

For practical aquaculture breeding schemes, the results of this study have the following implications, which could lead to a new design of the breeding programs. (i) Genotyping with genomic selection generates extra costs, which can be partly recovered by higher ΔG due to the increased accuracy by genomic selection compared with conventional BLUP breeding values. Here, we found an accuracy of around 0.82 for genome-wide breeding values compared with 0.66 for conventional BLUP breeding values, for a sib-test with 30 sibs tested per family and a heritability of 0.4. (ii) The breeding companies could reduce phenotyping costs if they cancelled all sib-tests of the candidates in one generation, like in EVERY-2GEN. In many salmon breeding schemes, most traits are actually measured on the sibs of the candidates, and phenotypic testing for all these disease and slaughter tests constitute a large part of the cost of the breeding program. However, the reduction of the sib-testing reduces ΔG_5 by 16% from 3.80 σ_a for EVERY-GEN to 3.19 σ_a for EVERY-2GEN. FIRST-GEN reduces ΔG_5 even more. An alternative is to use field data, *e.g.* slaughter house data or practical disease outbreaks, to estimate the association between markers and phenotypes. The data must come from the population under selection and represent animals closely related to the selection candidates, because Figure 1 shows that since the relationship between test individuals and individuals whose EBV are estimated decreases, the accuracy of selection decreases. (iii) Genotyping all selection candidates, *e.g.* 100 families times 30 candidates per family = 3000 individuals, is probably not feasible in practice. Instead, pre-selection on *e.g.* growth and maybe other traits could be a way to reduce the number of candidates to be genotyped. However, this pre-selection step needs to be optimised, in order to get the desired weight on the traits in the pre-selection step relative to the other traits. Also the sib-tested individuals need to be genotyped in the breeding scheme presented. Here the number of genotypes could probably be reduced by a factor of 5 without a substantial loss of accuracy of selection using a selective genotyping strategy [20].

The results show that the accuracy of selection is mainly affected by the following parameters of the breeding scheme:

1. In general, accuracies are highly affected by the number of generations with information on marker effects. For the basic scheme, with 3000 selection candidates, 3000 tested sibs of the candidates, 100 full-sib families and a trait heritability of 0.4,

accuracy of selection increased from 0.647 to approximately 0.823 over generations for EVERY-GEN. When sib-testing was only in generation 1, as in the FIRST-GEN in order to reduce costs, accuracy of selection dropped rapidly and was only 0.304 in generation 5 for the basic scheme.

2. Genomic selection, using EVERY-GEN, showed a higher accuracy of selection than the theoretical maximum of 0.71 of a conventional sib-testing scheme.

3. After generation 1, the Bulmer effect will reduce the genetic variance, which indirectly reduces the accuracy of selection. The accuracy of selection increases when more information on the marker effects becomes available over generations, and this effect is larger in the selection scheme than in the random selection scheme, probably because the favourable alleles become more abundant and are thus more accurately estimated. Therefore, the accuracy of selection in generation 5 for genomic selection (0.823) is similar to the accuracy of selection for random selection (0.811) in generation 5 for the basic scheme.

4. Increasing Ntested from 3000 to 6000 increased accuracy of selection in generation 5 from 0.823 to 0.845. With a lower heritability of 0.1, the effect of increasing Ntested from 3000 to 6000 (not shown previously), increased the accuracy of selection in generation 5 from 0.742 to 0.763. Hence, the increase in accuracy of selection was similar.

5. 5.The reduction in accuracy of selection with FIRST-GEN was somewhat larger for the lowest marker density than for the two other marker densities.

CONCLUSION

The results using the current sib-breeding design of family-based aquaculture breeding schemes, which was not optimised in any way for genomic selection, show that genomic selection yields high genetic gain, accuracy of selection and very low rates of inbreeding, which makes it a promising selection scheme. Various strategies were evaluated to reduce the amount of sib-testing and genotyping, to reduce costs, but all resulted in loss of the accuracy of selection and thus of genetic gain. Genotyping costs may also remain high in a near future, and further research on strategies to reduce the number of fish to genotype is highly needed.

ACKNOWLEDGEMENTS

This study was supported by grant 159831/S40 from the Research Council of Norway.

Calculations were done on the TITAN computer cluster at University of Oslo, Norway.

REFERENCES

1. Gjedrem T: Improvement of productivity through breeding schemes. GeoJournal. 1985, 10: 233-241. 10.1007/BF00462124.
2. Meuwissen THE, Hayes BJ, Goddard ME: Prediction of total genetic value using genome-wide dense marker maps. Genetics. 2001, 157: 1819-1829.
3. Sonesson AK: Within-family marker-assisted selection for aquaculture species. Genet Sel Evol. 2007, 39: 301-317.
4. Fisher RA: The genetical theory of natural selection. 1930, Oxford: Clarendon Press
5. Wright S: Evolution in Mendelian populations. Genetics. 1931, 16: 97-159.
6. Kimura M: Number of Heterozygous Nucleotide Sites Maintained in A Finite Population Due to Steady Flux of Mutations. Genetics. 1969, 61: 893-903.
7. Hayes B, Goddard ME: The distribution of the effects of genes affecting quantitative traits in livestock. Genet Sel Evol. 2001, 33: 209-229.
8. Hill WG: Linkage Disequilibrium Among Multiple Neutral Alleles Produced by Mutation in Finite Population. Theor Popul Biol. 1975, 8: 117-126. 10.1016/0040-5809(75)90028-3.
9. Goddard ME, Meuwissen THE: The use of linkage disequilibrium to map quantitative trait loci. Aust J Exp Agric. 2005, 45: 837-845. 10.1071/EA05066.
10. Muir WM: Comparison of genomic and traditional BLUP-estimated breeding value accuracy and selection response under alternative trait and genomic parameters. J Anim Breed Genet. 2007, 124: 342-355.
11. Habier D, Fernando RL, Dekkers JCM: The impact of genetic relationship information on genome-assisted breeding values. Genetics. 2007, 177: 2389-2397.
12. Cameron ND: Selection indices and prediction of genetic merit in animal breeding. 2007, Wallingford: CAB International
13. Nielsen HM, Sonesson AK, Yazdi H, Meuwissen THE: Comparison of accuracy of genome-wide and BLUP breeding value estimates in sib based aquaculture breeding schemes. Aquaculture. 2009, 289: 259-264. 10.1016/j.aquaculture.2009.01.027.
14. Lande R, Thompson R: Efficiency of Marker-Assisted Selection in the Improvement of Quantitative Traits. Genetics. 1990, 124: 743-756.
15. Henderson C: Applications of Linear Models in Animal Breeding. 1984, Guelph, Canada: Guelph University Press
16. Daetwyler HD, Villanueva B, Bijma P, Woolliams JA: Inbreeding in genome-wide selection. J Anim Breed Genet. 2007, 124: 369-376.
17. Ng SHS, Artieri CG, Bosdet IE, Chiu R, Danzmann RG, Davidson WS, Ferguson MM, Fjell CD, Hoyheim B, Jones SJM, de Jong PJ, Koop BF, Krzywinski MI, Lubieniecki K, Marra MA, Mitchell LA, Mathewson C, Osoegawa K, Parisotto SE, Phillips RB, Rise ML, von Schalburg KR, Schein JE, Shin H, Siddiqui A, Thorsen J, Wye N, Yang G, Zhu B: A physical map of the genome of Atlantic salmon, Salmo salar. Genomics. 2005, 86: 396-404. 10.1016/j.ygeno.2005.06.001.
18. Hayes BJ, Gjuvsland A, Omholt S: Power of QTL mapping experiments in commercial Atlantic salmon populations, exploiting linkage and linkage disequilibrium and

effect of limited recombination in males. Heredity. 2006, 97: 19-26. 10.1038/sj.hdy.6800827.

19. Solberg TR, Sonesson AK, Woolliams JA, Meuwissen THE: Genomic selection using different marker types and densities. J Anim Sci. 2008, 86: 2447-2454. 10.2527/jas.2007-0010.

20. Darvasi A, Soller M: Selective Genotyping for Determination of Linkage Between A Marker Locus and A Quantitative Trait Locus. Theor Appl Genet. 1992, 85: 353-359. 10.1007/BF00222881.

CHAPTER 13

Can the Global Adoption of Genetically Improved Farmed Fish Increase Beyond 10%, and How?

Ingrid Olesen [1,*], Hans B. Bentsen [1], Michael Phillips [2] and Raul W. Ponzoni [3]

[1] *Nofima, P.O. Box 210, NO-1431 Ås, Norway*

[2] *WorldFish, Jalan Batu Maung, 11960 Batu Maung, Bayan Lepas, Penang, Malaysia*

[3] *Facultad de Agronomia, Universidad de la República, Avenida Garzon 780, Sayago 12900 Montevideo, Uruguay*

ABSTRACT

The annual production from global aquaculture has increased rapidly from 2.6 million tons or 3.9% of the total supply of fish, shellfish and mollusks in 1970, to 66.7 million tons or 42.2% in 2012, while capture fisheries have more or less leveled out at about 90 million tons per year since the turn of the century. Consequently, the future seafood supply is likely to depend on a further increase of aquaculture production. Unlike terrestrial animal farming, less than 10% of the aquaculture production comes from domesticated and selectively bred farm stocks. This situation has substantial consequences in terms of poorer resource efficiency, poorer product quality and poorer animal welfare. The history of biological and technical challenges when establishing selective breeding programs for aquaculture is discussed, and it is concluded that most aquaculture species may now be domesticated and improved by selection. However, the adoption of selective breeding in aquaculture is progressing slowly. This paper reports on a study carried out in 2012 to identify key issues to address in promoting the development of genetically improved aquaculture stocks. The study involved semi structured interviews of 34 respondents from different sectors of the aquaculture society in East and Southeast Asia, where 76% of the global aquaculture production is located. Based on the interviews and literature review, three key factors are identified: (i) long-term public commitment is often needed for financial support of the breeding nucleus operation (at least during the first five to ten generations of selection); (ii) training at all levels (from government officers and university staff to breeding nucleus and hatchery operators, as well as farmers); and (iii) development of appropriate business models for benefit sharing between the breeding, multiplier and grow-out operators (whether being public, cooperative or private operations). The public support should be invested in efforts of selective breeding on the most important and highest volume species, which may not be a priority for investment by private breeders due to, for instance, long generation intervals and delays in return to investment.

Keywords: aquaculture; breeding programs; selection; business models; technology adoption; Asia

1. INTRODUCTION

The annual production from global aquaculture has increased rapidly from 2.6 million tons or 3.9% of the total supply of fish, shellfish and mollusks in 1970, to 66.7 million tons or 42.2% in 2012 [1]. Capture fisheries have more or less leveled out at about 90 million tons per year since the turn of the century, partly because the exploitation has reached or surpassed the carrying capacity of many major fishery stocks [1,2]. Consequently, future increase of (or perhaps even maintaining) the global fish supply depends on the further growth of aquaculture production. For instance, China, the world's largest producer, consumer, processor, and exporter of finfish and shellfish, produces mainly from its highly expanding aquaculture sector, accounting for approximately 72% of its reported domestic fish production. China alone contributes more than 60% of global aquaculture production volume and roughly half of global aquaculture value [1,3,4]. Aquaculture is the country's fastest growing food sector (5% to 6% annual growth in volume from 2000 to 2012), and its output reached 40 million metric tons (including mollusks, excluding algae) in 2012, four times the production volume in 1990 [1,4]. Vietnam is another Southeast Asian country with a tremendous growth of aquaculture production during the past decade (tripled since 2003), exceeding its capture fisheries [1], and with a need and potential for further expansion [5,6].

Terrestrial farm animal production is almost entirely based on populations or breeds that have been genetically domesticated through millennia and are systematically and selectively bred for improved performance often over centuries. This approach has improved the productivity and resource efficiency of the livestock populations many fold beyond the capacity of their wild ancestors. The development of modern animal husbandry would be impossible without domesticated and genetically improved animals.

Aquaculture probably dates back as far as about 2000 B.C. with the development of pond culture in e.g., East Asia, the Mediterranean and Central America, but aquatic animals have not been through a similar process of genetic domestication and selective breeding as terrestrial farm animal species. Less than 10% of the global aquaculture production is based on genetically improved material from modern breeding programs allowing for control of inbreeding as well as genetic selection [7]. As a result, the majority of aquaculture production is still based on wild type animals that are usually poorly adapted to life in captivity. This situation implies poor growth rates and animal welfare, high mortality, inefficient use of resources such as feeds, water and energy, and higher cost per kg of fish produced. Farming of genetically improved fish has the potential to contribute significantly to increased supply of fish, including in those regions and countries with significant food security and nutrition challenges.

The two major aquaculture species under domestication worldwide are Atlantic salmon and Nile tilapia (3.1% and 4.8% of the global production) [1], which have been subjected to more than 40 years (10 generations) and 20 years (15–20 generations) of selective breeding, respectively [8,9]. However, most of the production of other important aquaculture species such as carps and other cyprinids (38% of the global aquaculture production [1]) is based on stocks that are not expected to differ largely from wild type animals.

Still, the feed conversion efficiency of many farmed fish species is higher than that of domesticated terrestrial farm animals [7], partly because of lower energy requirements for body thermoregulation. Furthermore, an increasing number of studies show considerable genetic variation and significant heritability for a wide range of performance traits in aquaculture species under farming conditions [10], laying the foundation for rapid domestication and performance improvement of specialized aquaculture stocks through modern selection programs [11,12]. Gjedrem and Thodesen [13] listed 21 estimates of selection response for growth rate in 10 aquatic species, averaging 14% per generation, which is more than double the response normally achieved in breeding programs with already highly domesticated terrestrial farm animals. Highly favourable benefit/cost ratios ranging from 8 to 400 are reported for investments in fish selective breeding programs [14,15,16].

Consequently, the prospects are good for developing high performing and resource efficient domesticated populations and breeds of aquaculture species that may secure the future supply of aquatic animal products and even outperform traditional farm animals in terms of feed conversion efficiency. The question is why a similar domestication process to that in traditional farm animals has not happened during the history of aquaculture, and what can be done to overcome the lack of progress even today in many farmed species. In the following, this will be discussed in terms of biological and technical constraints as well as social and economic conditions, since technology advances alone seem to be insufficient to promote the necessary progress. The discussion is based on literature studies and on results from interviews with stakeholders, including farmers, breeders and experts from the fish farming industry as well as aquaculture scientists and governmental representatives in Eastern and Southeastern Asia, where most of global aquaculture is located today. Following an established interdisciplinary approach applied earlier [17,18], the paper is structured with an initial section reviewing literature and reports on biological and technical constraints followed by a chapter on the perceived needs and interests of actors in Chinese, Malaysian and Vietnamese aquaculture, which is based on the interviews. In Section 4, we discuss the empirical data and semi-structured key-actor interviews to explore the reasons for the slow adoption of selective breeding technology in aquaculture and what can be done to overcome the challenges identified.

2. BIOLOGICAL AND TECHNICAL CONSTRAINTS

2.1. Domestication

Genetic domestication is defined as the result of natural and artificial selection of an animal species to adapt to life in a new environment in captivity. Genetic domestication may involve changes of traits like behavior, mortality, disease resistance, feed requirements and others. A prerequisite for genetic changes from domestication of a species is the closing of the life cycle in captivity and genetic separation from wild progenitor stocks. The captive animals must be able to tolerate the new environment, to mature sexually, to reproduce and to give rise to viable progeny in captivity. This will form the base for the accumulation of both natural and artificial selection effects from generation to generation in the new environment. Unlike terrestrial farm animals, most aquatic species reproduce by spawning gametes into the external aquatic environment where the egg is fertilized, followed by external fetal development. Consequently, the environmental conditions (e.g., water temperature and quality) required for successful reproduction are usually more critical than for mammals and birds. Historically, reproduction in captivity of many farmed aquatic species has been quite unpredictable or not occurred at all, and the use of wild-caught seed for stocking in culture has been widespread. This reliance of wild-caught animals excludes, of course, the closing of the life cycle in captivity and accumulation of domestication and selection effects. Even today, culture systems based on wild-caught animals are not uncommon (e.g., *Penaeus monodon*culture, which relies on brood stock collected from the wild, and, occasionally, ponds filled with water containing natural seed). However, in more recent times, techniques have been developed to induce or synchronize sexual maturation (e.g., hormone treatment, temperature and light control), to strip and fertilize eggs and milt, and to incubate and hatch the eggs in a controlled environment of many aquaculture species [19]. This has provided a basis for domestication and selective breeding in many farmed aquatic animals.

2.2. Inbreeding

A second prerequisite for long-term domestication is to prevent the rapid accumulation of inbreeding in farmed stocks. The great fecundity in most aquatic species [19] makes it possible to recruit an entire new generation of production animals from very few parents, which can result in minimal effective population sizes. On-farm reproduction, without careful management and control of selection of brood fish and mating structure is then likely to result in high rates of inbreeding and deterioration of performance because of inbreeding depression [20]. In many cases, inbreeding has created a need for and a tradition of "refreshing" the stocks with wild-caught breeders to reduce inbreeding depression. However, this breaks the life cycle in captivity and disrupts the domestication and selection processes.

Several measures may be used to control and avoid the accumulation of high inbreeding without sacrificing genetic progress. The simplest technique is to exchange breeders between not closely related farm stocks, as has been a historical practice in terrestrial farm animals in many situations. This approach requires some form of organized exchange between farmers or hatcheries as well as tracking of records of the genetic relationship between farmed stocks, practices that have not yet been developed for most aquaculture species. A similar approach may be applied on-farm, provided that sufficient resources are available to maintain several cohorts of physically separated and not closely related samples of brood stock that may be used in a rotational mating scheme [21,22]. Another procedure to limit the increase of inbreeding is to carry out single pair mating of a sufficiently large number of outbred parents, (e.g., 50) followed by separate hatching and early juvenile rearing of the progeny full-sib families until survival has stabilized. A restricted number of full-sibs from each family (e.g., 30–50) may then be communally stocked without tagging for rearing of a new generation of breeders before repeating the procedure [23]. Finally, techniques for individual identification of breeding candidates (e.g., external branding or tagging, internal electronic tags, DNA fingerprinting) are now available at reasonable costs, making it possible to maintain ancestry records for each breeding candidate. Appropriate mating plans may then be applied to avoid rapid accumulation of inbreeding [24].

2.3. Selective Breeding

In terrestrial farm animals, long-term on-farm natural genetic domestication and genetic separation from their wild progenitor stocks have formed a base for accumulation of response to artificial selection in the farmed stocks. Based on information about individual performance and pedigree, the best performing individuals have been selected as parents of new generations to gradually improve desired traits for farming purposes, and inbreeding could be controlled at a low level. During the latest century, statistical methods have been developed for more accurate ranking of the breeding candidates according to their additive genetic performance. This has significantly improved the prospects of response to selection.

In aquaculture species with incomplete genetic separation between cultured and wild stocks, it is expected that the gene flow from wild breeders will have counteracted the accumulation of possible response to selection in the cultured stocks. Furthermore, the effective population sizes have probably been too low to avoid rapid accumulation of inbreeding and loss of genetic variation in many genetically closed aquaculture populations. Since additive genetic variation is a major prerequisite for response to selection, inbred populations are expected to respond poorly to selection. Several early studies of mass selection (*i.e.*, selection based on individual phenotypes without control of inbreeding) for increased growth performance in closed fish populations showed no significant response [25,26,27,28]. As a result, aquaculture breeding has a history of attempts to identify and maintain well performing 'pure strains' and to apply crossing between strains to counteract inbreeding and utilize possible heterosis, more similar to the testing of plant varieties and hybrids in plant breeding than to the process of selection within heterogeneous populations practiced in livestock [29].

Consequently, the establishment of a base population for long-term selection in aquaculture should aim at identifying populations with high levels of heterogeneity as well as optimizing the performance in culture. Using genetic material collected from diverse wild stocks has also proven successful in the major selection programs with Atlantic salmon [30] and in the GIFT (Genetic Improvement of Farmed Tilapias) project with Nile tilapia [31]. Several of the tested wild stocks of Nile tilapia even outperformed material collected from cultured stocks [32]. The further process of selection should then be designed to maintain the genetic variation. This may be done for example by combining mass selection with measures to control inbreeding (see above). However, mass selection may only be used for traits that can be measured on the live breeding candidates (e.g., growth performance), and is less efficient for traits with low heritabilities. Mass selection will often require facilities for synchronized mating of a considerable number of breeding pairs to produce progeny of a similar age for testing, and will also require management involving facilities for separate hatching and early juvenile rearing of full-sib groups [23] or separate stocking of other genetic groups [21,22] that are normally not available at ordinary grow-out farms.

Furthermore, statistical techniques based on pedigree records that have been developed for estimating breeding values to rank and select livestock have also proven to be effective in aquaculture species (e.g., in Atlantic salmon [8], and in tilapias [9]). The techniques depend on keeping parentage records for all breeding candidates, but they provide more accurate estimates of breeding values and consequently increased response to selection, particularly for traits with low heritability. Furthermore, parentage records increase the possibilities to control inbreeding accumulation. Information about the performance of relatives may also be also used to select for traits that may not be recorded in the live breeding candidates (e.g., carcass quality traits). Large-scale parentage assignment routines have been complicated or impossible in most aquaculture species until recent decades. However, the introduction of affordable procedures for individual external or electronic tagging or DNA fingerprinting has made parentage assignment an option in most aquaculture species as well. External and electronic tagging require separate hatching and rearing of full-sib progeny groups until the body sizes are sufficiently large for application of tags.

2.4. Breeding Programs

With these technological advances in reproduction, control of inbreeding and pedigree recording, it is now possible to run commercial scale selection programs for most aquaculture species. However, the facilities and skills needed to control reproduction and maintain pedigree records will make it difficult for average fish farmers to establish on-farm selection programs, as has been the case in the history of terrestrial farm animals. Most aquaculture selection programs today are carried out by specialized nucleus operations that focus on brood stock management and repeated testing and selection of new generations of pedigreed breeders. The best non-selected breeders may then be used by farmers or by dedicated hatcheries/fry producers to supply the grow-out farmers with quality seed. If properly organized, a single breeding nucleus may provide seed or brood fish from the recent generation of selection to a number of such multiplier farms. Because of the very high fecundity in most aquaculture species, the multipliers may then offer genetically improved seed in large quantities to grow-out farmers with only one generation delay compared to the genetic progress in the breeding nucleus.

This three-step model (breeding nucleus, multipliers and grow-out farms) has proven successful in salmonids and tilapias [8,9]. Each step may be operated independently, they may be coordinated by collaborative or cooperative organizations, or they may be fully integrated in a closed, vertical business operation. However, the model has proven to require substantial long-term investments without immediate financial returns to the crucial breeding nucleus part of the operation. Still, Ponzoni *et al.* [15,16] estimated benefit cost ratios (BCR) in the range of 8.5–60 for selective breeding programs for Nile tilapia and 22–420 for common carp when assuming initial investment cost of USD 50,000–100,000 and annual recurrent costs of USD 30,000–90,000. Furthermore, the level of initial investment and annual costs showed a relatively small effect on estimated benefit and BCR in these studies, whereas the effect of the price of fish was substantial.

3. PERCEIVED NEEDS AND INTERESTS OF AQUACULTURE STAKEHOLDERS IN CHINA, VIETNAM AND MALAYSIA

One of the objectives of the present study was to gain knowledge and a greater understanding of the specific causes of the limited use of domesticated and genetically improved animals in aquaculture in Southeast and East Asia [7], where 76% of the global fish, shellfish and mollusk aquaculture production takes place [1].

More specific objectives were to improve understanding about the causes of deficiencies in skill development and training in aquaculture breeding and genetics, specific reasons for aquaculture stakeholders' reluctance to invest in selective breeding programs, the reasons for hatcheries and fish farmers' apparent low demand for genetically improved animals, and finally to make recommendations for overcoming challenges of establishing and managing selective breeding programmes for more productive, efficient and animal friendly aquaculture.

3.1. Material and Methods

This interdisciplinary study combined biological and social sciences using the approach described and applied by Olesen *et al.* [17] and Rosendal *et al.* [18]. This implies a methodology of descriptive and explorative case study within the qualitative domain, which allows more in-depth understanding of factors and mechanisms through multiple observations [33]. Interdisciplinary collaboration, involving both biological and social sciences, was needed to address the complex research questions here. For this, the collaboration with scientists at WorldFish allowed for combining competence on genetics and aquaculture with competence on social and economic analysis, including value chain analyses of the aquaculture sector, policy and institutional analysis for the improved governance of aquatic resources as well as methods for breeding improved farm fish for small-scale and household farmers.

The first author worked at WorldFish for six months in January–June 2012, and met with public and private sector stakeholders in Malaysia, Vietnam and China during this period. The stakeholders were granted the right to be anonymous. The three countries were chosen because they all had established tilapia breeding programs with successful adoptions of genetically improved material in the aquaculture industry, whereas there were few or no breeding programs for other important aquaculture species such as carps, African catfish and marine fish species in these countries [7]. Respondents were recruited for interviewing based on existing networks of the authors and using contact persons in research institutions in Malaysia, Vietnam and China. Stakeholders with different background and nationalities (both Western and Asian) were recruited for the interviews. Note that the method used purposely sampled respondents representing the main sectors of the aquaculture industry in the three countries studied, covering breeding, hatcheries, dissemination, middle men and farmers as well as research, development, education and governmental bodies. A total of 34 persons, including seven women, representing the different stakeholder groups in aquaculture were interviewed about fish farming and breeding with focus on tilapias, carps and catfish. The sample included four policy makers from governmental bodies, 20 persons from research institutions and universities, nine from the hatchery sector, 10 involved in selective breeding companies and seven from the farming sector, with the interviews covering various challenges and obstacles to implementing selective breeding programs for farm fish. Breeding and farming enterprises of different scales were represented including both small farmers and hatcheries as well as relatively large multinational breeding and farming companies, included an integrated company. Based on the limited number of respondents, this material did not qualify for any quantitative analyses and general conclusions. Hence, it was rather used for exploring and gaining understanding of factors and mechanisms that may affect the interests and needs of the actors in aquaculture breeding and farming.

In depth semi-structured interviews using mainly open-ended questions presented in different questionnaires or interview guides were designed for three different categories of actors, *i.e.*, farmers, hatchery operators and academic staff, were applied. Questions were asked about: personal background and organization; farm, hatchery and breeding characteristics; strategies; practices; market; costs; willingness to pay (WTP) for genetically improved seed and perceived limitations, challenges and interests as well as recommendations. Native interpreters with professional background assisted in the interviewing of non-English speaking persons.

3.2. Results

3.2.1. Public Policies

It was clear from the interviews that seed quality is a prioritized area of development in aquaculture by Government in China, Malaysia and Vietnam. In China, long-term funding of breeding programs has been an implemented policy since 2008 [34] through, for instance, approval of annual reports rather than demanding processes with proposals for breeding projects lasting for only three to five years. Governments in China and Vietnam also recognize the need for public funding of fish breeding programs according to the governmental officers and researchers interviewed. As a result, several selective breeding programs have been established in these two countries in recent years, including carps, shrimp, river catfish and tilapia. Respondents in China commented on the resulting significant improvements for research institutions in receiving funding for fish breeding projects in recent years. Still, a higher number of breeding programs were identified in tilapia (>9) than in carps (4) and catfish (1) in these countries. The higher number of breeding programs for tilapia may be due to the successful development and worldwide dissemination of GIFT tilapia [9,21]. However, attempts to mimic the GIFT success by international projects on selective breeding in common carp during the last decade have not yet resulted in the same positive development. Public support of hatcheries for disseminating subsidized improved seed is further carried out in all three countries studied here. This support reflects governmental policies to increase access of farmers to improved seed quality in aquaculture in both Vietnam and Malaysia. In addition, new high-value, often marine species more suitable for export markets are given priority by the governments for development in these countries, whereas lower value species serving domestic markets, such as carps in China and African catfish in Malaysia, get less attention and support according to some respondents. In Malaysia, a tilapia breeding program has been operated through the public support of Malaysia and EU, via WorldFish, since 2001. Apart from that, a public research and development program aims to support private sector initiatives in seed genetic improvement, and optimism about the private actors' possibility to establish and manage aquaculture breeding programs was expressed by a governmental representative. Here, public-private cooperation through support of private industry developments is the main strategy. However, to our knowledge no private operation of selective breeding program has taken place in Malaysia so far, but some farming relies on imports of seed originating from such family based selective programs abroad, which are run by multinational companies.

3.2.2. Human Resources

In all countries, shortage of human resources, competence and skills in selective breeding was noted as a major challenge to develop breeding programs by the respondents. Insufficient training and competence in selective breeding and its benefits at all levels from farmers' education and training programs to universities was emphasized. This is considered to be a general feature in agriculture in developing countries, and may not be special to aquaculture breeding. However, in both China and Vietnam, funding programs for students to carry out M.Sc. and Ph.D. studies abroad are available. Asian students' lack of skills in the English language, partly due to limited language education in primary school, was, however, mentioned as a relevant challenge, and motivated students often failed to pass English language tests when applying for such fellowships. As we know, Asian students and scientists may be very competitive in other fields such as Information and Communication Technology (ICT), and this result may be a consequence of weaker recruitment to agriculture and aquaculture universities and study programs. It was also pointed out by a respondent that agricultural universities are less popular among students than other universities and study programs.

To our knowledge, textbooks on selective aquaculture breeding are also not available in, for instance, Mandarin and Vietnamese. In Malaysia, English is a lesser constraint at the university level.

In China, collaboration in breeding projects was reported between agriculture university staff and scientists in research institutes serving the aquaculture industry. In both Vietnam and Malaysia, such collaboration was absent and missed by respondents representing both governmental body and research. Competition among actors seemed to reduce the motivation for cooperation and joining efforts to achieve synergies and support aquaculture development. This challenge might explain failures to succeed in establishing breeding and dissemination in more than one example reported by respondents.

3.2.3. Business Strategies and Challenges

The farmers interviewed, including tilapia, carp and catfish farmers, expressed high demand for genetically improved seed and wanted to see the same improvements in carps and catfish as obtained in Nile tilapia. However, it was also noted by farmers, breeders and scientists that there will be a need for more than three years or generations of selection to obtain the significant genetic gain needed to demonstrate and convince more farmers and hence hatcheries to pay a price premium (or royalty) to the nucleus breeding program. The willingness to pay (WTP) a price premium for improved seed with respect to growth and survival among the very limited number of farmers in this study was in the order of 20%. These farmers' stated WTP, which we do not claim to be representative for all farmers in this region, varied substantially between 10% and 100%. It was suggested by respondents that providing farmers free test samples of genetically improved seed, or by other means demonstrate the fish superior performance directly to the farmers, may be necessary to increase farmers' demand and willingness to pay for genetically improved seed.

However, some respondents considered private actors and investments in fish breeding as realistic in future when breeding program, infrastructure and markets are more firmly established, as we can see in the salmon industry today.

It was noted by a respondent that it may be hard for private breeders to reap these economic benefits from carp breeding due to the common and traditional production system and lack of trust and confidence in the improved material and perhaps in new actors of breeding and seed supply.

DNA technology was not considered important for improving the marketing of seed or crucial for establishing breeding programs today by most respondents representing the breeding sector. It was, however, mentioned by some that it may become more important when the breeding sector is more developed as already documented for Atlantic salmon in some European countries.

4. DISCUSSION

4.1. Human Resources

Among fish farmers and aquaculture researchers in the major aquaculture countries, the basic skills needed to induce and synchronize sexual maturation and carry out single pair mating and separate hatching and rearing of progeny full-sib groups now seems to be well established for the most important freshwater aquaculture species and many marine species. Consequently, the life cycle in captivity may be closed, the farmed populations may be genetically isolated from wild stocks, and the process of accumulation of genetic domestication may be started. However, the technical knowledge needed to control the accumulation of inbreeding and to carry out efficient artificial selection for improved performance in culture still seems to be scarce in the aquaculture community. Still, within the three countries studied, there may be well-trained individuals or groups with expertise in breeding programs methodology for traditional farm animals, for example at agricultural universities. Most successful aquaculture breeding programs today are the result of combining skills from fish biology and farming, with farm animal breeding technology, both within and across countries [35].

In order to compensate for the shortage of human skills and competence, cooperation and efficient management of the qualified staff available becomes crucial and creates opportunities for improvements in efficiency. Synergies from cooperation within research, development and teaching at university level could be obtained through for example incentive programs for cooperation projects and by exchange of staff through, for instance, adjunct professor, senior scientist and expert positions (e.g., minor adjunct positions of 10%–20% position) at universities and research institutes, respectively. Further, benefits of the frequently seen assignment of the limited number of highly qualified scientists with doctoral degrees in genetics and breeding to administration positions rather than continued work in research and development can be questioned. This practice, which is not unique to the field of genetics and breeding, may, however, give less or slower progress in science, development, as well as in innovation and implementation of aquaculture breeding. Furthermore, training of farmers and hatchery operators to improve their understanding about selective breeding and the associated benefits should be increased to facilitate more dissemination and farming of genetically improved fish, as aimed at by all the countries studied through, for example, public support of breeding for improved seed quality and dissemination of seed.

4.2. Initiation of Breeding Programs

Aquaculture in Asia, and particularly in China, involves a large variety of species compared to terrestrial livestock production, where just a limited number of domesticated species are being farmed. Hence, the need for prioritizing a limited and manageable number of aquaculture species for domestication and selective breeding is crucial. This was considered as a challenge by respondents in China and also stressed by Gjerde and Su Xiaowen [36].

Starting a selective breeding program should then focus on expected benefits of e.g., faster growing fish with higher survival rates and the costs, by improving the aquaculture species in question. Because of the great fecundity of most aquaculture species, a single, well-organized breeding nucleus may supply a huge market with continuously upgraded seed by delivering selected brood stock to a number of multiplier farmers after each generation of selection. Consequently, the running costs of the breeding programs will be more or less constant and independent of the volume of the target industry. The benefit/cost ratio will then often depend more on the volume of the target industry than on the price of the end product, both from the consumers' point of view and from a private business point of view. The total impact of an aquaculture breeding program on food supply and resource efficiency for the society will usually be greater for a large volume species. Furthermore, a small premium on genetically improved seed for a large production volume of a low value species may easily generate more income for a breeding program aiming at seed sales than a higher premium on a limited volume of seed for production of an exclusive and highly priced species. For larger integrated enterprises this may not always be the case. For China and Vietnam, common carp was identified as a prioritized carp species for genetic improvement in 2001 [37]. Common carp constitutes 12.6% of the total volume of cultured freshwater fish in China [1]. Substantial response to selection for improved growth rate has been reported [38]. Selective breeding of common carp accompanied with a well-functioning dissemination program and infrastructure has been shown to be particularly profitable with benefit cost ratios (BCR) between 22 and 420, due to the high reproductive capacity of the species [16].

The start of a selective breeding program will require a heterogeneous base population, because genetic variation is a prerequisite for response to selection. If there is access to a sufficient number of breeding candidates (e.g., 1000 individuals or more) from a well performing and documented heterogeneous genetic source, performance testing and selection (pedigree based selection or mass selection modified to control inbreeding) may be started directly [24]. Otherwise, the genetic variation may be secured by collecting genetic material from a variety of sources. For previously undomesticated species, collecting samples of genetic material from viable and numerous wild stocks of diverse geographic origin has proven successful e.g., for Atlantic salmon [30] and Nile tilapia [31].

Techniques and facilities should then be established for a minimum of 100–150 synchronized single pair matings, followed by separate rearing of full-sib progeny groups until tagging and then individual tagging of all breeding candidates (alternatively parentage determination by DNA fingerprinting). A couple of generations of systematic crossing of breeders of different genetic origin before selection is applied is also recommended to preserve genetic variation within the new, mixed population [31]. Furthermore, a few generations of accumulated response to selection may often be needed to prove the benefits of genetically improved seed to the grow-out farmers and stimulate the willingness to pay a premium.

This entire start phase will evidently require substantial investments without immediate returns. Experiences from salmonids and tilapias suggest a minimum investment period of six to ten years depending on the generation interval of the species. Apart from large integrated companies, it is obvious that most Asian fish farmers do not have the necessary resources to establish private on-farm domestication and selection programs, as has happened during the domestication history of traditional farm animals. Fish farmers' cooperatives or associations may be a tool to initiate breeding programs for key species in the member farms, provided that they are given proper training and professional support. It has been shown that technology improvements with e.g., improved seed quality may contribute to tremendous benefits to a society of small scale farmers [39]. Important success criteria identified from aquaculture development projects involving collective action in Indonesia (Aceh), India and Bangladesh were: common burning issues among farmers (drivers such as poor yields, disease problems, environmental constraints or other constraints that would gain the most from the provision of consistent and relevant improvement); farmer-oriented services, comprising local teams with technical and organizational skills; and collective action through group approaches to reach large numbers of farmers [39]. Economies of scale and knowledge transfer are advantages of collective action, and experience does suggest a good business case for investing through cooperatives and other producer organizations although both group organization and capacity building take time [39].

So far, however, specialized research facilities and substantial national or international public funding has often been necessary during this early phase of domestication programs with a new species. Private investors tend to await the results of the initial efforts and the reaction of the seed market before investing in commercial selection programs. However, recent examples from China of privately funded and successful domestication and selection programs have been reported for new short generation species like red tilapia and blue tilapia [40,41]. Still, as indicated by some interview respondents, it is likely that access to public funding and support will be necessary in the future as well to initiate domestication and selection programs with new species in Asia, in particular for species with longer generation intervals like carps.

4.3. Accumulation and Dissemination of Genetic Progress

Once a selective breeding program is established and operative, plans should be made to obtain long-term sustainable continuation of generation-by-generation accumulation of selection response. Long-term biologically, ecologically and sociologically sound breeding goals are important for sustainable animal production [42]. Such breeding plans should therefore include strategies for generating income to cover the running and development costs of the breeding program and strategies for broadening the breeding goal beyond the common initial short-term goal of improving growth rate.

In the early days of modern selective breeding programs for terrestrial farm animals, it was anticipated that selection limits would be reached quite soon for the traits selected for (e.g., egg production [43]). However, after more than 50 years of intensive selection, substantial response is still maintained in farm animal selection programs, and the average performance of modern farm animals exceeds by far the performance of the most extreme animals in the ancestor populations. This may mainly be explained by the nature of polygenic inheritance [44] and in the longer term also by new mutations [45]. Consequently, aquaculture breeding programs that want to stay in business will have to market genetic material that is continuously upgraded by new generations of selection. During the early phase of Atlantic salmon farming in Norway, most of the grow-out smolt originated from a single, publicly supported selection program [8]. Today, several large scale, commercial selection programs are competing internationally with their latest generation of selected genetic material. A similar development may be emerging for Nile tilapia selection programs [9].

According to interview respondents, supporting dissemination of genetically improved seed is considered a governmental policy in all of the three countries. Nevertheless, the costs of maintaining and further developing aquaculture selection programs will probably in most cases eventually have to be paid by the grow-out industry, where the benefits of improved production efficiency and product quality are realized in terms of increased profitability on the sales to consumers. Given the very high benefit to cost ratios for investments in selection programs (see above), the premium needed on the price of seed will be marginal, in particular for species with large production volumes. The major problems seem to be to convince grow-out farmers that repeated supply of seed from a well-operated breeding program is worth a premium, and to ensure that the premium flows back through the multiplier farms to cover the costs of the breeding nucleus [17].

A proven method to stimulate the demand for genetically improved seed has often been to offer free test samples to be stocked on-farm in a separate pond or cage along with the genetic material that is commonly used by the farmer. The benefits of using seed from a selection program have then often been obvious without scientific comparisons and have caused a boost in the demand. More scientific approaches have also been used, for instance, when GIFT tilapia was distributed to several Asian countries for testing [46]. It may also be useful to offer training courses to targeted farmers to increase the awareness of the effects on productivity of selective breeding and the importance of always using the latest generation from the selection program.

It was also noted by several interview respondents that western investors in aquaculture selection programs have found it difficult to access Asian seed markets. This was perceived as 'a lack of sound business culture and practices' by one European respondent. This perception may however also be due to unfamiliarity with local customs. Guanxi refers to a Chinese system of doing business on the basis of personal relationships and is representative of the business culture throughout much of the non-western world. For example, business partners may be chosen based on personal relationships, and this approach may influence the partnerships available for enterprises in Southeast Asia [47,48] in developing selective breeding in aquaculture. Such challenges were indicated by respondents representing both research, innovation and private breeding. Informal or personal long-term relationships (rather than contractual) are particularly important in Asian countries [47,48,49]. Such relationships may also be nurtured in aquaculture to provide mutual benefits to both parties in the form of stability, predictability and access to information and technical assistance [50,51]. Such conditions include an information-rich business environment in which fear of losing one's reputation for trustworthiness is a strong deterrent to unfair business practices. Lovett *et al.* [47] pointed out how eastern and western business practices may already be converging toward systems based on more complete models of trust to deal with the conditions of progress coupled with uncertainty that form our new economic reality. According to Lovett *et al.* [47], dealings among acquaintances rather than relatives (as traditionally in rural China) are becoming more common among urban Asians. As acquaintances can be selected on the basis of ability, there may be a trend towards a more western way of choosing business partners in Asia as it grows [47]. A corresponding trend has been observed in the Western business practises towards a more complete trust including attachments and considerations of ability. Younger and female Asian respondents in the current study also requested more emphasize of abilities and competence than relationships in research and development in aquaculture, which may reflect a shift in the Asian culture of choosing partners.

A common problem in the early phase of many selection programs, in particular programs based on public funding, is that the funding is insufficient to develop the necessary network of professional and dedicated multiplier farms [8,21,52]. Instead, the programs may have to rely on agreements or contracts with existing hatcheries and seed producers to use selected brood stock from the selection program to produce and market genetically improved seed to the grow-out farmers [17]. The hatcheries may be public, cooperative or private, and often have their own agendas. Because the premium that the grow-out farmers pay for the improved seed has to be claimed by the multiplier farms as a part of the price of the seed, the selection programs may be left dependent on the multipliers' willingness to share their increased profits with the operators of the selection program in terms of royalties, extra price on the brood stock, or other means of reward. As the grow-out farmers gain positive experience with genetically improved seed, resulting in an increased demand and willingness to pay a premium, the multipliers have sometimes seen an opportunity to reap large, short term profits by avoiding paying their dues to the selection program. This situation will however leave the further operation of the selection program short of income. In some cases, this has resulted in lack of continuity of selection and (or) dissemination that has deprived the grow-out farmers of several generations of genetic progress [52]. It has also happened that multipliers that experience a rapidly increasing demand for the improved seed are tempted to use any available, unselected brood stock to scale up their seed production, thus harming the grow-out farmers' confidence and the reputation of the genetic quality of the seed from the selection program [53]. This has increased the awareness of selection program operators about the need for control and monitoring of collaborating multiplier farms. Hussain *et al.* [54] recommended introducing a quality certificate to regulate the relationship between a carp selection program and collaborating hatcheries in Bangladesh. In Ghana, all tilapia material used since 2011 should, according to government recommendations, come from the genetically improved Akosombo strain, to control the quality and health of tilapia seed as far as possible [53]. Here, a number of hatcheries were to be selected and approved by the Government for multiplying and disseminating the Akosombo strain. Anyhow, the early phase of new selection programs should be sufficiently funded to include the operation of dedicated, integrated multiplier farms for direct sales of genetically improved seed to the grow-out industry. For similar reasons, private selection program operators tend to move towards integrated selection and multiplier solutions, where the premium paid on the genetically improved seed will go directly to the integrated operation. In some cases, as also commented by respondents in this study, private breeding companies choose to go for full, vertical integration of selection, multiplying and grow-out farming operations, which implies that only the grow-out farms that are owned or contracted by the operator of breeding nucleus will gain access to the improved genetic material. This will often mean that the production potential of genetically improved seed is not fully exploited, and consequently results in a reduced impact of the selection program on the total grow-out industry. As also noted in these interviews, other selection programs, including public funded actors, have signed contracts with feed companies to increase their funding. By this, the feed companies can offer free or subsidized genetically improved seed as part of their marketing of the feed to farmers. One reason for private breeding companies to avoid selling their improved seed on the open market is that the seed may easily be reared and used as brood stock by unauthorized hatcheries that may compete with the breeding company on the seed market without having to pay the costs of a selection program. Possible mechanisms for introducing legal intellectual property rights protection of selectively bred aquaculture populations have been discussed and found difficult to implement [55]. However, Atlantic salmon breeding companies seem to be less occupied by the need for legal protection, other than material transfer agreements, than by the competitive advantage of offering the best quality genetic material and the best customer support [17]. Seed produced by unauthorized 'piracy' of commercial genetic material from a selection program will often lag behind with as much as two to three generations of selection compared to the material supplied directly from authorized multipliers, and this may easily result in 20%–30% loss in performance [13]. This should normally be a sufficiently large difference to encourage properly informed grow-out farmers to avoid unauthorized genetic material.

Finally, planning for long-term domestication and genetic improvement of performance by selective breeding of aquaculture stocks is not only a question of securing sustainable funding of the operation of the breeding programs. Until now, most breeding programs have focused their breeding goal on increasing the average growth rate or the body weight at harvest, a trait of great economic importance for the grow-out farmers that may easily be recorded in the breeding nucleus and that responds well to selection. However, the evolvement of Atlantic salmon breeding programs has been accompanied by a gradual expansion of the breeding goal to include traits such as those related to sexual maturation, product quality and disease resistance [8]. This has happened as a response to the needs of the grow-out industry and the increased competition between breeding companies on the smolt market. The recent inclusion of resistance to salmon lice in the breeding goal for several salmon breeding programs may be an example of a trait with severe environmental and ethical aspects considered most important by Norwegian public to select for [56]. Most likely, the future seed market for other aquaculture species will request genetic material that is selected for other traits in addition to growth rates. In many Asian countries (e.g., China), aquaculture systems are intensifying as producers seek higher returns on scarce land, water, and coastal zone resources [3].

Such intensification often implies higher stocking densities and growth that may stress the fish and increase the disease problems. Hence, disease problems and other emerging concerns related to feed efficiencies and resilience to environmental fluctuation, become of interest in genetic improvement. Fish disease is a major cause of economic losses and fish welfare problems in aquaculture. This intensification is expected to continue to a large extent. The most significant disease causing losses in tilapia culture is *streptococcosis*. Worldwide losses from the infection have been estimated at $150 million in 2000 and more than $250 million in 2008 [57]. Selection to improve the genetic resistance against the disease has the potential for significant economic impact for grow-out farmers. Breeding programs that want to contribute to the future genetic development of aquaculture stocks will consequently need to constantly evaluate the breeding goals of their selection program and to be involved in continuous upgrading of the selection procedures.

5. CONCLUSIONS AND RECOMMENDATIONS

The biological and technical constraints that have probably prevented fish farmers from developing domesticated and genetically improved aquaculture stocks during the history of fish farming have gradually been solved during recent decades. It is now possible to run commercial scale selection programs for most aquaculture species, and thus considerably improve the production efficiency (growth rate, survival), the product quality, and the animal welfare in the remaining 90% of the global aquaculture production that is still based on largely wild type genetic material.

However, the implementation of selective breeding in aquaculture is progressing slowly. The technology is too resource demanding for small scale, on-farm initiatives, and requires interaction between specialized operations. Based on interviews and literature review during this study, it was concluded that the adoption of selective breeding programs in commercial aquaculture in Eastern and Southeastern Asia will often depend on long-term public commitment in collaboration with cooperative and private interests to resolve several key issues:

- Financial support of the breeding nucleus operation will usually be needed, at least during the initial five to ten generations of selection, and, in particular, for species with long generation intervals. The early phase support should prioritize established aquaculture species with large production volumes, such as carps, that are not easily subject to investment from private breeders due to long generation interval and delayed profit.

- Training of more people in selective animal breeding technology should be prioritized at all levels, including government officers, university staff, breeding nucleus operators, hatchery operators, and grow-out farmers. Collaboration between different research groups in genetics (institutes and universities) and scholars within the fields of aquaculture and livestock breeding should be encouraged.

- If the private sector is expected to engage in selective breeding and genetic improvement of aquaculture stocks in less prosperous economies, development of appropriate business models are probably more important than legal systems to ensure fair benefit sharing between the breeding, multiplier and grow-out operators. Research is also needed to incorporate private breeders, cooperatives and small-scale farmers into traditional and 'guanxi' business management. Experience suggests opportunities for investing through cooperatives and other producer organizations.

ACKNOWLEDGEMENTS

Funding from Norglobal (CGIAR fellowship, project no 212952) and FUGE programs (ELSPRO, project no 208543) of the Research Council of Norway as well as Nofima is acknowledged. Thanks to the respondents who provided useful information and insight in the issues studied. The kind and crucial assistance by Azhar Hamzah, Fisheries Research Institute, Malaysia, and the staff at WorldFish, Penang, Malaysia as well as RIA1 and RIA2, Vietnam is highly appreciated. Comments and suggestions to the manuscript given by Kristin Rosendal and the two anonymous reviewers are also acknowledged.

APPENDIX

A. Questionnaires

A.1. Questionnaire for Fish Farmers

Personal information

1. What is your role/position?

Farm owner, Farm operator/manager

2. If employed by company/institution, is your organization (employer)profit, non-profit, public, private, cooperate, MNC, large national, SME, research
3. Name of organization, employer, location
4. What is your background (education, training and work experience)?

Farm characteristics

5. What fish species is (are) being cultured on your farm?

How important is the production of these species in your country or region (percent of total aquaculture production)? <20%, 20%–50% , 50%–80% or >80%

6. What is the total production on the farm?

Big (tonnes per year), medium (tonnes per year), small scale (tonnes per year), household farmers (kg per year or number of fish with average size)

7. Is the farming full time work?

Or part time work combined with other household income?

Is more than 50% of production on the farm sold?

8. What is the annual turnover (sales from production) on the farm? (Figures in currencies representing low, medium and large farm with options to check appropriate interval).

9. What are the annual running costs (USD), and what are the main cost factors (e.g., labour, feed, power)? (to reveal whether it is intensive (high input) or extensive (low-input))
10. Did you have disease outbreaks last years?
11. What type of feed is used?

Fish Breeding

12. How was/is the fish or seed of your breeders accessed?

Own breeding (integration), from private hatcheries, from public hatcheries, from middle men.

From hatchery every year, from hatchery every second year, from hatchery more than two years ago, from wild fish population last year, from wild fish population two years ago, from wild fish population more than two years ago

13. If you carry out fish breeding and reproduction on your farm, how is this carried out?

Mass spawning, pair matings, artificial mating

14. Do you select brood fish yourself? If so, for how long has your current fish population been selected (no. of generations and years of selection)?

How many broodfish do you select?

What trait(s) do you select for and how is the trait measured or recorded?

Body weight, other, please specify

Measured by scale or visually

Why were these chosen?

15. What are the main challenges and problems for breeding the fish today?

Fish seed

16. How is/was the fish seed disseminated to you?
As sexually mature brood fish, fingerlings, fertilized roe

17. What is the price of fish seed?
18. What is the price of harvested fish ex farm (price to farmer)?

What is the harvest weight?

19. What are the costs of the seed in percentage of all costs?

20. What quality or traits do you consider most important? Is price most important?

21. Are you interested in genetically improved fish? Is it easy or hard to provide genetically improved seed of your interest?

22. What characteristics of the seed would you prefer? Please assign which characteristic you prefer the most and least of the following (1–7):

All male, genetic quality (higher growth, survival, carcass yield, disease resistance), standardized size, price, service, education/training, DNA information or technology

23. Are there more than one competing hatchery or middle man, and how competitive are they with respect to price, quality and service? Are these

public, private or cooperative programs, owned by non-profit or profit organizations?

24. What price premium are you willing to pay for seed giving 20% faster growth?

<10%, 10%–20%, 20%–30%, >30%

25. What price premium are you willing to pay for seed requiring 20% less feed to reach harvest size?

<10%, 10%–20%, 20%–30%, >30%

26.What price premium are you willing to pay for seed with higher disease resistance such that the survival is increased with 20%?

<10%, 10%–20%, 20%–30%, >30%

Recommendations

27. What are your recommendations to governmental bodies, other public organizations and private actors and/or investors to facilitate new breeding programs and to increase dissemination and marketing of genetically improved fish to farmers?

28. Other comments?

A.2. Questionnaire for Hatchery Operators

Personal information

1. What is your position?

2. Name of Organization, employer, location

3. Is your organization (employer): Profit, non-profit, public, private, cooperate, MNC, large, national, SME, research, industry

4. What is your background (education, training and work experience)?

Breeding program

5. For what species is (are) the hatchery operating and disseminating in your jurisdiction or country?

How big is the grow out production of these species in your country or region? <20%, 20%–50%, 50%–80%, or >80%.

6. How was/is the fish or seed of for your breeders accessed?

Species?

From external breeding program every year, from external breeding program every second year, from external breeding program more than two years ago, from wild fish population last year, from wild fish population two years ago, from wild fish population more than two years ago

7. What breeding strategy do you apply?

Mass spawning, pair matings, artificial mating, crossing of different strains

8. Do you select brood fish? If so,

For how long has your fish population been selected (no. of generations and years of selection)? How many broodfish do you select? What trait(s) do you select for and how is it recorded/tested? Body weight, other, please specify, why were these chosen?

9. How many families are produced and selected among in your breeding population?

10. Are you applying any DNA technology or chromosome manipulation in your breeding program?

If not, would it be useful, and if so how/why?

11. How big is the genetic gain of your fish per generation and/or year (if available)?

12. How is the breeding program being funded or supported today?

Through income from sales only

Public support only

Through both sales and public support

13. How big are the investments made for the current breeding/selection operation (USD)?

14. What are the annual running costs (USD), and what are the main cost factors (e.g., labour, feed, power)?

15. How much are you willing to pay for seed/breeders giving 20% faster growing fish in the farms while the other traits are not changed?

<10%, 10%–20%, 20%–30%, >30%

16. How much are you willing to pay for seed/breeders requiring 20% less feed to reach harvest size in the farms while the other traits are not changed?

<10%, 10%–20%, 20%–30%, >30%

17. How much are you willing to pay for seed/breeders with higher disease resistance such that the fish survival at harvest on the farm is increased with 20% while the other traits are not changed?

<10%, 10%–20%, 20%–30%, >30%

18. What are the main limitations, challenges and problems for running the hatchery today?

Dissemination program

19. How are the fish disseminated to the farmers, through other hatcheries, middle men or direct to farmers? What infrastructure is needed for this?

20. How is the fish disseminated?

As sexually mature brood fish, Fingerlings, fertilized roe.

21. What are the costs of the dissemination?

22. What is the price of fish seed to grow out farmer?

23. How many fry are disseminated per year?

To how many farms or other hatcheries do you disseminate?

24. What type of grow-out farmers buy your seed?

What is their training/education, intensity (high or low input) and technology level?

	Percentage of customers	Education	Intensity/technology
Big	....	..	...
Medium,	....	...	...
Small scale	....	...	...
Household farmers	....	....	...

25. What quality or traits are the farmers requesting most? Is price more important to farmers?

26. How big is the demand from farmers? Is it easy or hard to market the genetically improved seed? Is all produced seed sold?

27. What would make it easier to market the seed? (e.g., all male, quality, standardization, price, service, education/training of middle men and farmers), DNA information or technology)

28. Are there competing hatcheries, and how competitive are these with respect to price, quality and service? Are these public, private or cooperative programs, owned by non-profit or profit organizations?

29. What price premium would you expect that farmers will pay for seed giving 20% faster growth?

<10%, 10%–20%, 20%–30%, >30%

30. What price premium would you expect that farmers will pay for seed requiring 20% less feed to reach harvest size?

<10%, 10%–20%, 20%–30%, >30%

31. What price premium would you expect that farmers will pay for seed with higher disease resistance such that the survival is increased with 20%?

<10%, 10%–20%, 20%–30%, >30%

Recommendations

32. What are your recommendations to governmental bodies, other public organizations and private actors and/or investors to facilitate new breeding programs and to increase dissemination and marketing of genetically improved fish to farmers?

33. Other comments?

A.3. Questionnaire for Academic Actors

Personal information

1. What is your position?

2. Name of Organization, employer, location

3. Is your organization (employer) profit, non-profit, public, private or cooperate (MNC, large national or SME), research or industry?

4. What is your background (education, training and work experience)?

5. How did you get your education and training? For example how was it funded?

6. Why did you choose this field of education and training?

7. Do you find your education useful for your current work?

Breeding program

8. For what species is (are) the fish breeding program(s) operating in your jurisdiction or country?

How big is the production of these species in your country?

9. How was the breeding program(s) established? Who took the initiatives and why? How was the upstart funded?

10. Who were the owner(s) and how was the program managed and organized in the first generations of breeding?

11. How big was the investment made to get it up and running?

12. For how long has the program been running (no. of generations and years of selection)?

13. Who are the owner(s) and how is the program managed and organized during the last generations (e.g., private, public, cooperative)?

14. What type of testing and selection strategy is applied and what causes the possible limitations?

15. How was the fish or seed of the base population accessed?

16. How many families are produced and tested in the nucleus breeding population?

17. What traits are recorded and selected for? Why were these chosen?

18. Are you applying any DNA technology or chromosome manipulation in your breeding program?

19. How big is the genetic gain per generation and/or year (if available)?

20. How is the program being funded or supported today?

21. How big are the investments made for the current operation (USD)?

22. What are the annual running costs (USD), and what are the main cost factors (e.g., labour, feed, power)?

23. What are the main challenges and problems for running the program today?

Dissemination program

24. How are the fish disseminated to the farmers, through hatcheries or direct to farmers? What infrastructure is needed for this?

25. What are the costs of the dissemination?

26. What is the price of fish seed to grow-out farmers?

27. How many fry are disseminated?

28. Who are the grow-out farmers buying the seed? Big, medium, small scale or household farmers? What is their training/education, intensity (high or low input) and technology level?

29. What quality or traits are the farmers requesting most? Is price more important to farmers?

30. How big is the demand from farmers? Is it easy or hard to market the genetically improved seed? Is all produced seed sold?

31. What would make it easier to market? (e.g., all male, quality, standardization, price, service, education/training of actors (hatchery operators and farmers), DNA information or technology?)

32. Are there competing breeding programs, and how competitive are these with respect to price, quality and service? Are these public, private or cooperative programs, owned by non-profit or profit organizations?

Recommendations

33. What are your recommendations to governmental bodies, other public organizations and private actors and/or investors to facilitate new breeding programs and to increase dissemination and marketing of genetically improved fish to farmers?

34. Other comments?

REFERENCES

1. FAO. Food and Agriculture Organization of the United Nations. Fisheries and aquaculture statistics, 2014. Available online: http://www.fao.org/ fishery/ statistics/en (accessed on 20 December 2014).
2. Mullon, C.; Fréon, P.; Cury, P. The dynamics of collapse in world fisheries. *Fish Fish.* **2005**, *6*, 111–120.
3. Fishery Bureau, Ministry of Agriculture, People's Republic of China. *China Fisheries Yearbook*; China Agriculture Press: Beijing, China, 2013. Cited by Cao,

L.; Naylor, R.; Henriksson, P.; Leadbitter, D.; Metian, M.; Troell, M.; Zhang, W. China's aquaculture and the world's wild fisheries. *Science* **2015**, *347*, 133–135.
4. Project Team for Research into the Sustainable Development Strategy of China's Cultivation Industry. In *Study on the Sustainable Development Strategy of China's Cultivation Industry—Aquaculture*; China Agriculture Press: Beijing, China, 2013. Cited by Cao, L.; Naylor, R.; Henriksson, P.; Leadbitter, D.; Metian, M.; Troell, M.; Zhang, W. China's aquaculture and the world's wild fisheries. *Science* **2015**, *347*, 133–135.
5. FAO. *National Aquaculture Sector Overview Viet Nam*, Available online: http: //www. fao.org/fishery/countrysector/naso_vietnam/en (accessed on 13 April 2015).
6. Nguyen, M.D. *Value Chain Analysis Vietnam*; Nong Lam University: Ho Chi Minh, Vietnam, 2011; p. 14. Available online: http://www.fao.org/fileadmin/ user_ upload /fisheries/docs/Vietnam_edited.doc (accessed on 13 April 2015).
7. Gjedrem, T.; Robinson, N.; Rye, M. The importance of selective breeding in aquaculture to meet future demands for animal protein: A review. *Aquaculture* **2012**, *350–353*, 117–129.
8. Gjedrem, T. The first family-based breeding program in aquaculture. *Rev. Aquac.* **2010**, *2*, 2–15.
9. Bentsen, H.B.; Gjerde, B.; Ponzoni, R.W. Genetic improvement of farmed tilapias: Response to five generations of selection for increased body weight at harvest in *Oreochromis niloticus* and the further impact of the project. Manuscript in preparation. 2015.
10. Gjedrem, T.; Olesen, I. Basic statistical parameters. In *Selection and Breeding Programs in Aquaculture*; Gjedrem, T., Ed.; Springer: Dordrecht, The Netherlands, 2005; pp. 45–72. ISBN: 978-1-4020-3341-4 (Print) 978-1-4020-3342-1 (Online).
11. Gjedrem, T. Improvement of productivity through breeding schemes. *GeoJournal* **1985**, *10*, 233–241.
12. Bentsen, H.B. Application of breeding and selection theory on farmed fish. In Proceedings of the 4th World Congress of Genetics Applied to Livestock Production, Edinburgh, Scotland, 23–27 July 1990; Volume XVI, pp. 149–158.
13. Gjedrem, T.; Thodesen, J. Selection. In *Selection and Breeding Programs in Aquaculture*; Gjedrem, T., Ed.; Springer: Dordrecht, The Netherlands, 2005; pp. 89–111. ISBN: 978-1-4020-3341-4 (Print) 978-1-4020-3342-1 (Online).
14. Gjedrem, T. Selective breeding to improve aquaculture production. *World Aquac.* **1997**, *28*, 33–45.
15. Ponzoni, R.W.; Nguyen, N.H.; Khaw, H.L. Investment appraisal of genetic improvement programs in Nile tilapia (*Oreochromis niloticus*). *Aquaculture* **2007**, *269*, 187–199.
16. Ponzoni, R.W.; Nguyen, N.H.; Khaw, H.L.; Ninh, N.H. Accounting for genotype by environment interaction in economic appraisal of genetic improvement programs in common carp *Cyprinus carpio*. *Aquaculture* **2008**, *285*, 47–55.
17. Olesen, I.; Rosendal, G.K.; Tvedt, M.W.; Bryde, M.; Bentsen, H.B. Access to and protection of aquaculture genetic resources – Structures and strategies in Norwegian aquaculture. *Aquaculture* **2007**, *272 S1*, S47–S61.
18. Rosendal, G.K.; Olesen, I.; Tvedt, M.W. Evolving legal regimes, market structures and biology affecting access to and protection of aquaculture genetic resources. *Aquaculture* **2013**, *402–403*, 97–105.
19. Refstie, T.; Gjedrem, T. Reproductive traits in aquatic animals. In *Selection and Breeding Programs in Aquaculture*; Gjedrem, T., Ed.; Springer: Dordrecht, The Netherlands, 2005; pp. 113–120. ISBN: 978-1-4020-3341-4 (Print) 978-1-4020-3342-1 (Online).
20. Eknath, A.E.; Doyle, R.W. Effective population size and rate of inbreeding in aquaculture of Indian major carps. *Aquaculture* **1990**, *85*, 293–305.
21. Ponzoni, R.W.; Khaw, H.L.; Yee, H.Y. *GIFT: The Story since Leaving ICLARM (Now Known as The WorldFish Center). Socioeconomic, Access and Benefit Sharing and*

Dissemination Aspects; Fridtjof Nansen Institute Report 14/2010; Fridtjof Nansen Institute: Lysaker, Norway, 2010; pp. 1–47. ISBN: 978-82-7613-602-9 (Print) 978-82-7613-603-6 (Online) ISSN 1504–9744.

22. Knibb, W.; Whatmore, P.; Lamont, R.; Quinn, J.; Powell, D.; Elizur, A.; Anderson, T.; Remilton, C.; Nguyen, N.H. Can genetic diversity be maintained in long term mass selected populations without pedigree information?—A case study using banana shrimp *Fenneropenaeus merguiensis*. *Aquaculture* **2014**, *428–429*, 71–78.
23. Bentsen, H.B.; Olesen, I. Designing aquaculture mass selection programs to avoid high inbreeding rates. *Aquaculture***2002**, *204*, 349–359.
24. Ponzoni, R.W.; Khaw, H.L.; Nguyen, N.H.; Hamzah, A. Inbreeding and effective population size in the Malaysian nucleus of the GIFT strain of *Nile tilapia* (*Oreochromis niloticus*). *Aquaculture* **2010**, *302*, 42–48.
25. Moav, R.; Wohlfarth, G.W. Two-way selection for growth rate in the common carp (*Cyprinus caprio* L). *Genetics* **1976**, *82*, 83–101.
26. Hulata, G.; Wohlfarth, G.W.; Halevy, A. Mass selection for growth rate in the *Nile tilapia* (*Oreochromis niloticus*). *Aquaculture* **1986**, *57*, 177–184.
27. Teichert-Coddington, D.R.; Smitherman, R.O. Lack of response by *T. nilotica* to mass selection for rapid early growth. *Trans. Am. Fish. Soc.* **1988**, *117*, 297–300.
28. Huang, C.M.; Liao, I.C. Response to mass selection for growth rate in *O. niloticus*. *Aquaculture* **1990**, *85*, 199–205.
29. Moav, R.; Hulata, G.; Wohlfarth, G. Genetic differences between the Chinese and European races of the common carp. I. Analysis of genotype-environment interactions for growth rate. *Heredity* **1975**, *34*, 323–340.
30. Gjedrem, T.; Gjøen, H.M.; Gjerde, B. Genetic origin of Norwegian farmed Atlantic salmon. *Aquaculture* **1991**, *98*, 41–50.
31. Eknath, A.E.; Bentsen, H.B.; Ponzoni, R.W.; Rye, M.; Nguyen, N.H.; Thodesen, J.; Gjerde, B. Genetic improvement of farmed tilapias: Composition and genetic parameters of a synthetic base population of *Oreochromis niloticus* for selective breeding. *Aquaculture* **2007**, *273*, 1–14.
32. Eknath, A.E.; Tayamen, M.M.; Palada-de Vera, M.S.; Danting, J.C.; Reyes, R.A.; Dionisio, E.E.; Capili, J.B.; Bolivar, H.L.; Abella, T.A.; Circa, A.V.; *et al.* Genetic improvement of farmed tilapias: The growth performance of eight strains of *Oreochromis niloticus* tested in different farm environments. *Aquaculture* **1993**, *111*, 171–188.
33. King, G.; Keohane, R.O.; Verba, S. *Designing Social Inquiry. Scientific Inference in Qualitative Research*; Princeton University Press: Princeton, NJ, USA, 1994; pp. 1–264.
34. Ministry of Agriculture. Seed subsidy: A successful case of China's agricultural and rural development policies with Chinese characteristics. **2008**. Available online: http://www.xinhua.org (accessed on 15 April 2015,Chinese text). Cited by Huang, J.; Wang, X.; Rozelle, S. The subsidization of farming households in China's agriculture. *Food Policy* **2013**, *41*, 124–132.
35. Acosta, B.O.; Gupta, M.V. The genetic improvement of farmed tilapias project: Impact and lessons learned. In *Success Stories in Asian Aquaculture*; De Silva, S.S., Davy, F.B., Eds.; Springer: London, UK, 2010; pp. 149–171.
36. Gjerde, B.; Su, X. Genetic Improvement of Freshwater Fish, Final Report of World Bank Aquaculture Improvement Project (CN-PE-69949). 2001, 1–51.
37. WorldFish Center. *Genetic Improvement of Carp Species in Asia (RETA 5711)*; Final report to the Asian development Bank; WorldFish Center: Penang, Malaysia, 2001; pp. 1–158. Available online: http://www.worldfishcenter.org/resource_centre/WF_322.pdf (accessed 21 January 2015).
38. Ninh, H.N.; Ponzoni, R.W.; Nguyen, H.N.; Woolliams, J.A.; Taggart, J.B.; McAndrew, B.J.; Penman, D.J. A comparison of communal and separate rearing of families in selective breeding of common carp (*Cyprinus carpio*): Responses to selection. *Aquaculture* **2013**, *408–409*, 152–159.

39. Phillips, M.; Rogers, W.; Downing, W.; Beveridge, M.C.M.; Padiyar, P.A.; Karim, M.; Subasinghe, R. Inclusive aquaculture - business at the bottom of the aquatic pyramid. *FAO Aquac. Newsl.* **2011**, *48*, 44–46. Available online: http://www.worldfishcenter.org/resource_centre/WF_3043.pdf (accessed on 21 January 2015).
40. Thodesen, J.; Ma, D.Y.; Rye, M.; Wang, Y.X.; Li, S.J.; Bentsen, H.B.; Gjedrem, T. Genetic improvement of tilapias in China: Genetic parameters and selection responses in growth, pond survival and cold-water tolerance of blue tilapia (*Oreochromis aureus*) after four generations of multi-trait selection. *Aquaculture* **2013**, *396–399*, 32–42.
41. Thodesen, J.; Ma, D.Y.; Rye, M.; Wang, Y.X.; Li, S.J.; Bentsen, H.B.; Yazdi, M.H.; Gjedrem, T. Genetic improvement of tilapias in China: Genetic parameters and selection responses in growth, survival and external color traits of red tilapia (*Oreochromis* spp) after four generations of multi-trait selection. *Aquaculture* **2013**, *416–417*, 354–366.
42. Olesen, I.; Groen, A.F.; Gjerde, B. Definition of animal breeding goals for sustainable production systems. *J. Anim. Sci.***2000**, *78*, 570–582.
43. Lerner, I.M. *Population Genetics and Animal Improvement as Illustrated by the Inheritance of Egg Production*; Cambridge University Press: Cambridge, UK, 1950; pp. 1–74.
44. Bentsen, H.B. Genetic effects of selection on polygenic traits with examples from Atlantic salmon, *Salmo salar* L. *Aquac. Res.* **1994**, *25*, 89–102.
45. Cock, J.; Gitterle, T.; Salazar, M.; Rye, M. Breeding for disease resistance of Penaeid shrimps. *Aquaculture* **2009**, *286*, 1–11.
46. Dey, M.M.; Eknath, A.E.; Sifa, L.; Hussain, M.G.; Tran, M.T.; Nguyen, V.H.; Aypa, S.; Pongthana, N. Performance and nature of genetically improved farmed tilapia: A bioeconomic analysis. *Aquac. Econ. Manag.* **2000**, *4*, 83–106.
47. Lovett, S.; Simmons, L.C.; Kali, R. Guanxi *versus* the Market: Ethics and Efficiency. *J. Int. Bus. Stud.* **1999**, *30*, 231–247.
48. Chen, M. *Asian Management Systems: Chinese, Japanese and Korean Styles of Business*; Cengage Learning: Boston, MA, USA, 2004; pp. 1–278. ISBN: 1861529414.
49. Turley, J. *Connecting with China: Business Success through Mutual Benefit and Respect*; John Wiley & Sons: Chichester, West Sussex, UK, 2010; pp. 1–254.
50. Kagawa, M.; Bailey, C. Trade Linkages in Shrimp Exports: Japan, Thailand and Vietnam. *Dev. Policy Rev.* **2006**, *24*, 303–319.
51. Tran, N.; Bailey, C.; Wilson, N.; Phillips, M. Governance of global value chains in response to food safety and certification standards: The case of shrimp from Vietnam. *World Dev.* **2013**, *45*, 325–336.
52. Eknath, A.E.; Hulata, G. Use and exchange of genetic resources of Nile tilapia (*Oreochromis niloticus*). *Rev. Aquac.***2009**, *1*, 197–213.
53. Rosendal, G.K.; Olesen, I.; Walløe Tvedt, M. *Access to, equity and protection of genetic resources in Ghana: The case of tilapia (O. niloticus)*; FNI Report 15/2012; Fridtjof Nansen Institute: Lysaker, Norway, 2012; pp. 1–28. ISBN: 978-82-7613-657-9 (Online) ISSN 1893-5486.
54. Hussain, M.G.; Sarwar, M.R.; Akhand, R.I.; Sarker, S.K.; Rahman, A.; Amin, A.S.; Rashid, M. *Assessment of the carp seeds value chain in southern Bangladesh*; WorldFish Center Report, WorldFish Center: Penang, Malaysia, 2012; pp. 1–56.
55. Rosendal, G.K.; Olesen, I.; Bentsen, H.B.; Walløe Tvedt, M.; Bryde, M. Access to and legal protection of aquaculture genetic resources—Norwegian perspectives. *J. World Intellect. Prop.* **2006**, *9*, 392–412.
56. Grimsrud, K.M.; Nielsen, H.M.; Navrud, S.; Olesen, I. Households' willingness-to-pay for improved fish welfare in breeding programs for farmed Atlantic salmon. *Aquaculture* **2013**, *372–375*, 19–27.

57. Klesius, P.; Shoemaker, C.A.; Evans, J.J. Streptococcus: A worldwide fish health problem. In Proceedings of the 8th International Symposium on Tilapia in Aquaculture, Cairo, Egypt, 12–14 October 2008; Volume 1, pp. 83–107.

CHAPTER 14

Identification of Multiple Genetically Distinct Populations of Chinook Salmon (*Oncorhynchus Tshawytscha*) in a Small Coastal Watershed

Chante D. Davis, John Carlos Garza, Michael A. Banks

Hatfield Marine Science Center, Oregon State University, 2030 SE Marine Science Drive, Newport, Corvallis, OR, USA

NOAA Southwest Fisheries Science Center and University of California, Santa Cruz, 110 McAllister Way, Santa Cruz, CA 95060, USA

ABSTRACT

Management and restoration planning for Pacific salmon is often characterized by efforts at broad multi-basin scales. However, finer-scale genetic and phenotypic variability may be present within individual basins and can be overlooked in such efforts, even though it may be a critical component for long-term viability. Here, we investigate Chinook salmon (*Oncorhynchus tshawytscha*) within the Siletz River, a small coastal watershed in Oregon, USA. Adult Chinook salmon were genotyped using neutral microsatellite markers, single nucleotide polymorphisms and "adaptive" loci, associated with temporal variation in migratory behavior in many salmon populations, to investigate genetic diversity based upon both spatial and temporal variation in migratory and reproductive behavior.

Results from all three marker types identified two genetically distinct populations in the basin, corresponding to early returning fish that spawn above a waterfall, a spring-run population, and later returning fish spawning below the waterfall, a fall-run population. This finding is an important consideration for management of the species, as spring-run populations generally only have been recognized in large watersheds, and highlights the need to evaluate population structure of salmon within smaller watersheds, and thereby increase the probability of successful conservation of salmon species.

Keywords: Population genetics, Conservation, Coastal watershed, Chinook salmon, Electronic supplementary material

INTRODUCTION

Genetic and life-history diversity contribute to the resilience of native species in dynamic environments. Chinook salmon (*Oncorhynchus tshawytscha*) in western North America historically exhibited at least four behavioral life histories that are associated with the season of adult upstream migration (fall, spring, winter, and summer runs) and maturation status (Yoshiyama et al. 1998). Contributing to evolution of these migratory patterns is their homing fidelity to natal spawning rivers that allow for reproductive isolation and development of unique evolutionary trajectories (Waples 2001; Quinn 2004). Anthropogenic activities including harvest, waterway development, hatchery production, and land use practices have altered salmon populations and their associated freshwater ecosystems; the result is often reduced life-history and genetic variability (Yoshiyama et al. 1998; Brenkman et al. 2012).

To counter the effects of declining populations, substantial focus has been directed toward increasing abundance (supplementation programs) while also maintaining genetic diversity within evolutionarily significant units (ESUs) or other management units (Waples 1998; Olsen et al. 2000; Lindley et al. 2004; Eldridge and Killebrew 2007). As genetic techniques have improved, they have been used increasingly to enable better management (Clemento et al. 2014; Shafer et al. 2015). Salmonid fishes were among the first groups of organisms to be studied with molecular techniques, when the available methods were limited to the evaluation of genetically determined variation in a handful of proteins that could be reliably stained for detection (Utter and Hodgins 1972). One important application of molecular techniques has been genetic stock identification (GSI), which uses a "baseline" reference dataset of genotypes from individuals of known origin to identify the most likely provenance of individuals of unknown origin on the basis of their genotype (Milner et al. 1985). A North America-wide baseline dataset has been developed for Chinook salmon that includes 42 major reporting units (Seeb et al. 2007) and, more recently, baseline datasets using single nucleotide polymorphisms (SNPs) have been developed for regional applications, including in the California Current ecosystem (Clemento et al. 2014).

Pacific salmon management and restoration planning is characterized by efforts at broad multi-basin scales; however, fine-scale genetic structure that may be present within small watersheds (< 120 river km) is often overlooked in recovery planning. Homing to natal stream of origin is an influential driver of isolation among spawning groups and contributes to the genetic structure that has been observed at the landscape scale within salmon species, as do other ecological factors (i.e., habitat fragmentation, water development, episodic landslides or fires etc.). However, proper management of this fine-scale life history variability is a critical component in the long-term population persistence in a dynamic landscape.

In addition to geographically structured genetic variation, salmon species demonstrate substantial life-history variation (for detailed descriptions of life history variation see (Groot and Margolis 1991; Waples 2001). In Chinook salmon, much of this variation manifests as run-timing "ecotypes" which are characterized by mean date of freshwater entry and reproductive status (Waples et al. 2004; Moran et al. 2013). The 'fall-run', which enter freshwater as reproductively mature adults in late summer and fall, is regionally dominant in the southern extreme of the species range in North America. The 'spring-run' ecotype, which enter freshwater as reproductively immature adults in spring and early summer and hold in deep pools before maturing in the fall, are much less abundant and are derived from proximate fall-run populations in coastal basins (Kinziger et al. 2013).

Here, we investigated if fine-scale population structure of Chinook salmon exists in the Siletz River, a small watershed in coastal Oregon. Previous studies that examined salmon populations from the Siletz River using data from neutral genetic variation included only samples from the fall-run, and were part of large-scale efforts to characterize genetic structure in the species and provide methods for identification of fish of unknown origin (Seeb et al. 2007; Moran et al. 2013; Clemento et al. 2014). We studied salmon of different life-history types and used data from three types of molecular marker to: i) identify patterns of neutral genetic variation within the watershed, ii) evaluate whether temporal patterns of life-history variation are correlated with variation at genetic markers known to be associated with run-timing, and iii) determine phylogeographic patterns within the river and compare them to other coastal basins in the

North Oregon Coast region. We then address the implications of our results for current conservation and management activities.

METHODS

Study area

The Siletz River system (~109 river km) has headwaters in the Oregon Coast Range and meets the ocean at Siletz Bay. The geology of the watershed is principally volcanic. A waterfall that may be a barrier to upstream migration by aquatic organisms is located at river kilometer 103.8 (Fig. 1). A fish ladder was constructed in 1953 to allow fish passage during the winter for migratory fishes and Oregon Department of Fish and Wildlife (ODFW) operates it year round to control fish passage. In 1994, management practices began limiting access of anadromous fish that compete for spawning habitat with wild summer steelhead (Siletz basin fish management plan; ODFW 1997). Substantial changes to water flow and suitable salmon spawning habitat also occurred as a result of historical logging and splash-damming and alteration to the floodplains.

As many as three Chinook salmon ecotypes may exist in the basin: spring-, summer-, and fall-run (Stan van de Wetering, Siletz Tribe, pers. comm). The largest of these is the fall-run, which enters the river in September. The spring-run enters the river in May and moves the farthest upstream; while the summer-run ecotype returns in July and spawns between mid-September and mid-October (Stan van de Wetering, Siletz Tribe, pers. comm.).

Sample collection

Tissue samples from Chinook salmon were collected by ODFW during their yearly carcass surveys and through monitoring at the trap on the fish ladder (Fig. 1). Samples that were collected during the fall carcass survey consisted of a small section of the least degraded flesh from each carcass. Body condition and geographic (reach ID) location was recorded for all samples. Collections occurred September through December in 2011 and 2012 and were considered to be from the fall-run ecotype (*SIFA*). These samples were collected downstream of the fish trap (Reach IDs 2–23; Fig. 1) and a total of 565 samples of sufficient quality for successful genotyping were collected.

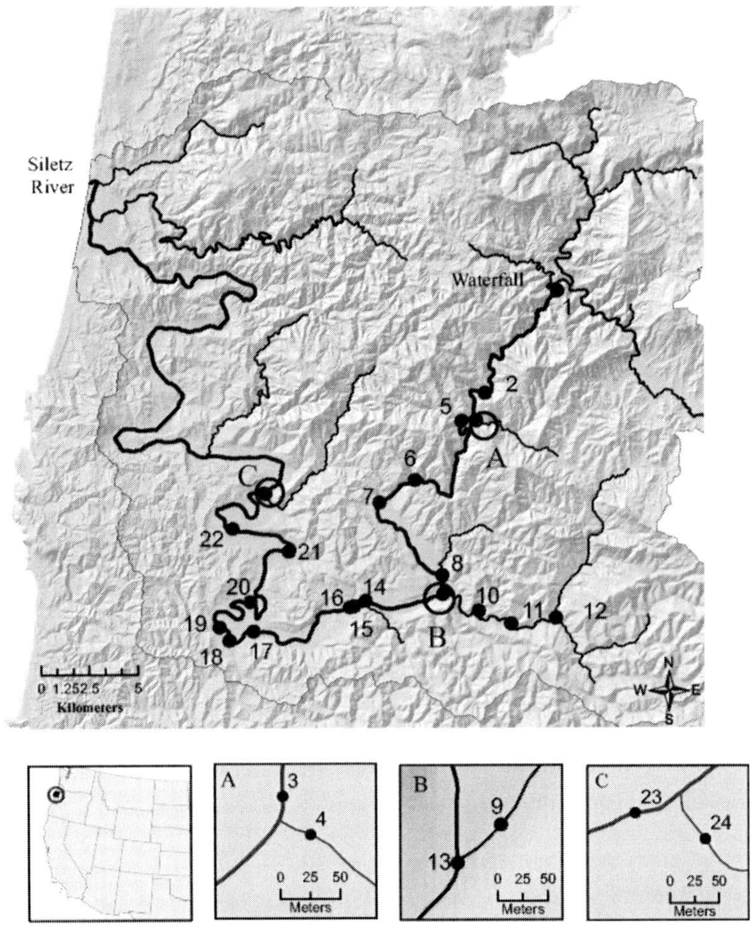

Figure. 1 Map of sample sites for Chinook salmon from the Siletz River basin. Sampling locations are identified numerically. Each dot represents the downstream reach identification for ODFW sampling locations within Siletz River. Numbered sites indicate locations where genetic samples were collected for analysis. Sites 3, 4, 9, 13, 23 and 24 are located around river confluences; each inlay expands the area at these sites

Samples from carcasses of fish that appeared in the main stem Siletz River prior to October 15th in 2012 and 2013 were hypothesized to be a unique group, based on adult return and spawn time and were designated lower river early (*SILE*) fish. These samples were collected downstream at sampling locations Reach IDs 17–21 (Fig. 1). There were 55 such samples of sufficient quality for successful genotyping

The fish ladder and trap are operated year round, and enable the upstream passage of all early returning Chinook salmon. Fish that entered the trap were considered to be the spring-run ecotype (*SISP*) and were passed over the waterfall and allowed access to spawning grounds located upstream (Fig. 1). Scales were removed for age analysis and genotyping from all passed fish for which it was possible (i.e. body condition). During the sampling years, 700 fish were passed upstream at the fish trap and, of these, 258 individuals were sampled and genotyped.

Genetic analysis

Total genomic DNA was isolated from each tissue sample following the extraction protocol of Ivanova et al. (2006). Samples were genotyped with up to three types of molecular marker: presumably neutral microsatellites, putatively adaptive microsatellites, associated with run-timing variation, and single nucleotide polymorphisms (SNPs). *Neutral microsatellite markers*are not known to be associated with phenotypic expression and are considered to be selectively neutral; therefore, we were able to infer demographic processes that shaped population structure. Polymerase chain reaction (PCR) was used to amplify 21 neutral microsatellites following published thermocycling protocols. These included 10 microsatellites that are standardized range-wide (Moran et al. 2013) and an additional 11 microsatellites (Nelson and Beacham 1998; Naish and Park 2002; Williamson et al. 2002). The standardized microsatellite panel was developed for range-wide genetic stock identification (GSI) of Chinook salmon and included genotypes from populations throughout North America (Seeb et al. 2007; Moran et al. 2013). Fluorescent PCR products were electrophoresed on a 96-capillary DNA sequencer (3730XL; Applied Biosystems Inc.) and genotypes called using GENEMAPPER v3.7 software (Applied Biosystems).

Putatively adaptive microsatellite markers are associated with phenotypic expression and may be useful for interpreting adaptation of individuals to their local environments. The circadian clock gene network has been identified in salmon as contributing to the genetic control of adult migration timing in salmon (O'Malley et al. 2007, 2010). Three circadian clock gene markers, *Ots515NWFSC (Ots515), Cryptochrome3 (Cry3)*, and *OtsClock1b (clock1b)*, were amplified via published PCR and thermal cycling protocols. *Ots515* is a QTL-linked marker that is associated with spawn time and body weight in rainbow trout (O'Malley et al. 2007). *Cry3* is linked to flavoproteins that mediate circadian rhythms in plants (O'Malley et al. 2010). *Clock1b* contains a polyglutamine repeat tract that has been shown to vary in Chinook salmon (O'Malley et al. 2010).

Single nucleotide polymorphisms (SNPs) are sites in the genome that have two nucleotides segregating and are ubiquitous in vertebrates. A panel of 96 SNPs specific for Chinook salmon (Clemento et al. 2014) were genotyped on 96.96 Dynamic Arrays with an EP1 System (Fluidigm Corp., South San Francisco, CA), and genotypes were called with Fluidigm Genotyping Analysis software v2.1.1. These SNPs have been used previously to construct a GSI "baseline" database that is comprised of genotypes from more than 69 populations, including fall-run from the Siletz River. A total of 188 samples, 94 fall-run and 94 spring-run, chosen representatively from the larger sets of samples was genotyped with these markers to determine phylogenetic relationships among proximate coastal basins.

Statistical analysis of genetic variation

Loci were assessed for genotyping problems including null alleles or allelic dropout using MICROCHECKER (Van Oosterhout et al. 2004). Observed allele frequencies were tested for evidence of deviations from Hardy-Weinberg expectations (HWE) and for significant linkage disequilibrium (LD) between loci with GENETIX (Belkhir et al. 1996–2004). Summary statistics of genetic diversity were calculated for each hypothesized population. Characterization of the genetic diversity among loci was evaluated using allelic richness (A_r) a measure of allelic number that corrects for unequal sample sizes using a rarefaction method with HP_RARE(Kalinowski 2005). The number of alleles (A), and observed (H_o) and expected (H_e) heterozygosity were calculated in GENALEX (Peakall and Smouse 2012).

Spatial structure was evaluated in three ways: pairwise estimates of the fixation index (F_{ST}), exact tests for genic and genotypic frequencies, and individual assignment tests. Pairwise F_{ST}values (θ; Weir and Cockerham 1984) were calculated in the program GENALEX and the data set was permuted 1000 times to determine if the values differed significantly from zero, an indication that populations may be genetically distinct. Exact tests for differences in genic and genotypic frequencies were performed with the program GENEPOP (Raymond and Rousset 1995), which applies a Markov Chain Monte Carlo algorithm to account for small sample sizes or low-frequency alleles. Significant values of genic and genotypic frequencies may occur even though sufficient power to detect genetic differentiation through other methods is not possible, as is the case with populations where high gene flow exists.

A model-based Bayesian clustering method was used as an additional method to identify the degree of differentiation between the hypothesized populations. The software package STRUCTURE v2.2 (Pritchard et al. 2000) estimates the likelihood for hypothesized values of *k*, the number of genetically distinct clusters or populations from which the sampled individuals were drawn. This method allows the data to define the clusters and assigns individuals to the *k*clusters without a priori information about their sampling locations. Five independent runs were performed for each value of *k* (2–6), using 50,000 burn-in and 150,000 retained iterations. An additional STRUCTURE run to assess the association of SISP and SIFA to the Oregon Coast reporting unit was conducted using published data provided in (Clemento et al. 2014). Five independent runs were also performed for values of $k = 4–10$, using the same burn-in and iterations as above.

Phylogeography

Phylogeographic patterns of fish within the Siletz River were inferred with a dendrogram based on SNP genotypes, constructed using chord distances (Cavalli-Sforza and Edwards 1967) and with the topology determined using the neighbor-joining algorithm, in the PHYLIP package (Felsenstein 1993). Majority-rule consensus values were calculated from 1000 bootstrap replicates of the data by the PHYLIP components SEQBOOT and CONSENSE. Only bootstrap values above 80% were reported.

RESULTS

Neutral genetic variation

Null alleles or other problems were identified in three microsatellite loci (*Ots209, Ots211,* and *Omm1080*), due to the presence of more homozygous individuals than would be expected. Significant departures from HWE existed for seven markers (*Ots9, Ots104, OtsG409, Ogo4, Ots208, Ots249,* and *OtsG83*) after adjusting for multiple comparisons, the remaining 11 loci were used for analysis. Allelic richness and heterozygosity of SISP was greater than that of either SIFA or SILE (Table 1). Linkage disequilibrium was found in all populations; the largest fraction of locus pairs in LD was within SISP (Table 1). Overall accuracy of individual assignment to population of origin using the microsatellite data was greater than 83% (Table 2). Pairwise F_{ST} values across years (2011 and 2012) were low (SIFA, $F_{ST} = 0.001, p = 0.006$; SISP, $F_{ST} = 0.003, p = 0.001$) and marginally significantly different from zero, but did not likely represent biologically meaningful differentiation (Hedrick 1999). Data were therefore pooled across years within groups for subsequent analyses.

Table 1 Summary of genetic markers for 11 microsatellites, 3 adaptive loci and 91 SNPs by population (SIFA = fall-run, SILE = lower river early, SISP = spring-run) across all marker types. A_r = allelic richness, A = number of alleles, H_o and H_e are observed and expected heterozygosity, respectively, LD = percentage of locus pairs with significant genotypic linkage disequilibrium within populations

	Microsatellites					Pooled Adaptive Loci					SNPs				
	A_r	A	H_o	H_e	LD	A_r	A	H_o	H_e	LD	A_r	A	H_o	H_e	LD
SIFA	18.5	28.3	0.87	0.87	7.0%	19.5	36.0	0.70	0.70	-	1.9	1.9	0.33	0.32	5.8%
SILE	18.0	18.8	0.86	0.87	10%	20.4	21.3	0.71	0.70	-	-		-	-	-
SISP	19.6	25.6	0.90	0.90	25%	18.2	28.0	0.71	0.69	-	2	1.9	0.35	0.35	10.0%

Table 2 Individual assignment to population of origin for each molecular marker. Accuracy reflects the percent of correct assignment to population of origin. The percent correct assignment of populations below the waterfall to either of the populations sampled below the waterfall is in parentheses

	Microsatellites				Pooled Adaptive Loci				SNPs		
	SIFA	SILE	SISP	Accuracy	SIFA	SILE	SISP	Accuracy	SIFA	SISP	Accuracy
SIFA	490	53	22	87(96)	381	121	63	67 (89)	76	6	92
SILE	27	19	9	35(83)	24	19	12	34(78)			
SISP	20	11	227	88	31	28	199	77	13	76	85

Of the 96 SNP loci, four loci (*Ots_108735–302, Ots_118175–479, Ots_128302–57,* and *Ots_Pr12*) did not yield genotypes within the SISP population, and *OkiOts_120255* functions to discriminate Chinook salmon from closely related coho salmon (*O. kisutch*); the remaining 91 loci were used for subsequent analyses (suppl. 1). There were no departures from HWE among loci following correction for multiple comparisons. Allelic richness and heterozygosity of SISP was again greater than that of either SIFA or SILE (Table 1). There was no evidence of linkage disequilibrium within SIFA; however, a small amount of LD was present in the SISP group (Table 1). The SNP dataset had similar ability to accurately assign individuals to population of origin as the microsatellite dataset (accuracy >85%; Table 2).

Pairwise F_{ST} differed significantly from zero between SILE and SISP, but not between SIFA and SILE (Table 3). Model-based clustering analysis with STRUCTURE provided evidence of two major genetic groups in the Siletz River that corresponded to the fall-run and spring-run Chinook salmon ecotypes (Fig. 2a and b) and was consistent with the pattern identified by other analyses with both datasets.

Table 3 Matrix of pairwise F_{ST} values calculated from the microsatellite dataset (below diagonal) and the SNP dataset (above diagonal). F_{ST} values significantly different from zero are denoted by an asterisk

	SIFA	SILE	SISP
SIFA	-	-	0.009*
SILE	0.00253	-	-
SISP	0.02075*	0.01709*	-

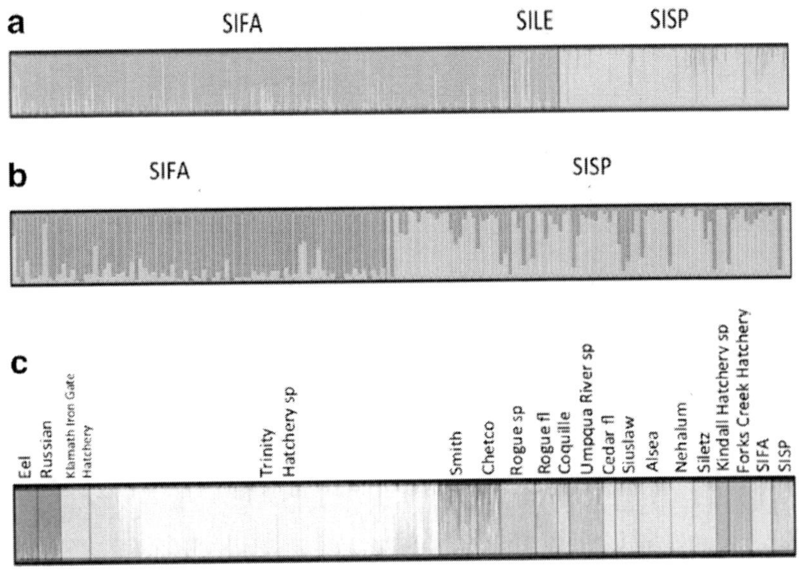

Figure. 2 Fractional ancestry plots obtained from STRUCTURE for a) microsatellite markers, b) SNPs, and c) SNPs using Clemento et al. (2014) genotypes for the North Oregon Coast Reporting Unit. Each vertical bar represents an individual's genotype and the probability of being assigned to one of k (k = 2 or 6) genetically distinct clusters

Temporal adaptive genetic variation

Variation at the three markers associated with circadian clock genes provided further evidence for differentiation between migratory ecotypes of Chinook salmon in the Siletz River. All pairwise F_{ST} values and exact tests of genic and genotypic divergence between SISP and both SIFA and

SILE, except those for locus *Clock1b*, were significant, indicating differentiation of SISP and both SIFA and SILE (Table 4). In contrast, these loci provided minimal evidence for differentiation between the earlier returning (SILE) and later returning lower river fish (SIFA), with non-significant F_{ST} values and significant tests of genic and genotypic differentiation at loci *OTS515* only (Table 4).

Table 4 Genic and genotypic exact test results and pairwise F_{ST} values for all populations (SIFA = fall-run, SILE = lower river early, SISP = spring-run)

	SIFA vs. SILE			SIFA vs. SISP			SILE vs. SISP		
	Genotypic	Genic	F_{ST}	Genotypic	Genic	F_{ST}	Genotypic	Genic	F_{ST}
Clock1b	0.719	0.339	0.000	0.042*	0.040*	0.003*	0.104	0.073	0.004
Ots515	0.003*	0.010*	0.003	0.000*	0.000*	0.009*	0.001*	0.000*	0.008*
Cry3	0.094	0.068	0.001	0.000*	0.000*	0.008*	0.001*	0.001*	0.008*

* Significant values

Phylogeography

A neighbor-joining dendrogram was created with the SNP dataset and was compared to a larger published study of Chinook salmon by Clemento et al. (2014), that used the same loci and in which "reporting groups" of populations from the same geographical regions were identified. Siletz River fall-run Chinook salmon were grouped with the North Oregon Coast reporting unit in that study, as they were in several other genetic studies (Narum et al. 2008; Moran et al. 2013). Our analysis also placed SIFA in the North Oregon Coast reporting unit (bootstrap 83%; Fig. 3), which is consistent with the fact that the SIFA samples were collected from the same general location as the fall-run Siletz River fish that were analyzed in these previous studies. The Siletz River spring-run ecotype (SISP) population also branched with the North Oregon Coast reporting unit. Model-based clustering analysis with STRUCTURE also supports this finding (Fig. 2c).

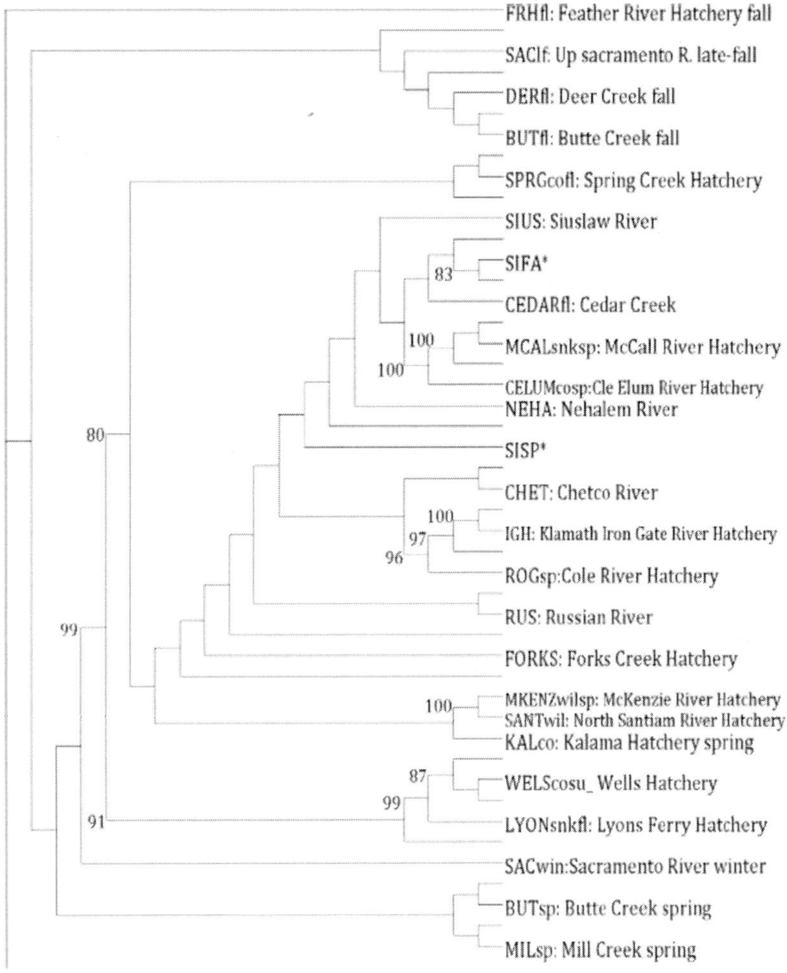

Figure. 3 Mid-point rooted neighbor-joining dendrogram constructed with chord distances calculated from SNP data. All SNP data other than those from SIFA and SISP are from Clemento et al. (2014). Bootstrap values greater than 80% (out of 1000 bootstrap resamplings) are reported

DISCUSSION

Here, we used genetic data to identify a previously unrecognized population of early returning (i.e., spring-run) Chinook salmon within the Siletz River, a basin that is <120 km from source to ocean exit. The use of three different types of molecular genetic data to investigate population genetic structure of Chinook salmon in this watershed allowed us to resolve fine-scale structure not previously recognized. Patterns of genetic variation within the watershed indicated that individuals spawning downstream of the waterfall (i.e., SILE and SIFA) are a fall-run population, and that a genetically distinct population of the spring-run ecotype (i.e., SISP) spawns upstream of the waterfall. Concordance among results from analyses of the multiple marker types lent strength to the resulting conclusion of significant structure in the Siletz River that corresponds to fish with different freshwater entry timing and to fish spawning above and below a waterfall. Pairwise F_{ST} values for populations spawning upstream and downstream of the waterfall were significant for all three molecular markers. The presumably neutral microsatellite loci that were used in this study were highly polymorphic and provided substantial power for resolution of genetic structure. However, persistent LD was found between some of these markers. The SNP markers were biallelic, and therefore less polymorphic per locus, but more numerous and did not have significant LD. The adaptive genetic markers correlate with behavioral variation in some salmon populations that have a temporal component and could therefore potentially discriminate such populations in the absence of other genetic differentiation. Previous work has shown that the combination of data from multiple marker types improved resolution of population structure in salmon, especially among populations with potentially high gene flow, as can be expected in smaller watersheds (Narum et al. 2008; Hess et al. 2011; DeFaveri et al. 2013; Garvin et al. 2013). Below, we discuss the patterns of genetic differentiation as inferred from the three sets of molecular markers and summarize the implications of our findings for continued conservation efforts.

Temporal variation

In contrast to the finding of significant differentiation between fall- and spring-run ecotypes returning above and below the waterfall with all three types of molecular marker, none of the datasets found differentiation between the lower river, early fall returning fish (SILE) and the lower river, fall-run ecotype (SIFA). There is no physical barrier between these two groups, but the timing of their return to freshwater and spawning dates suggested that heritable behavioral differences might exist. None of the pairwise F_{ST} values between SIFA and SILE were significant, indicating that these two groups of fish are likely experiencing high levels of gene flow. However, significant exact tests for genic and genotypic frequencies may indicate slight differentiation associated with this early spawning phenotype at a locus (*Ots 515*) that has been found to be associated with spawn time (O'Malley et al. 2007). We did not identify fine-scale genic and genotypic diversity using *Clock1b* and *Cry 3*, but these markers may only be informative for adaptive variation across basins, as suggested by O'Malley and Banks (2008). Although the existence of an early fall-run is informally acknowledged in some coastal rivers, only two non-fall-run populations of Chinook salmon (two spring-run populations in the upper Umpqua River) are formally delineated in the Oregon Coastal Multi-Species Conservation and Management Plan for the purposes of management and recovery (ODFW 2014). The finding of a genetically unique, wild population of spring-run Chinook salmon in the Siletz River indicates that greater life history diversity exists within this species in these smaller coastal rivers.

The adaptive genetic markers provided signals of structure that were similar to the other types of genetic markers, although variation in them has been found in other studies to be strongly associated with temporal variation in migration by salmon (O'Malley et al. 2007). Clock genes are part of the molecular mechanism of long-term timekeeping for tracking season-specific activities in response to photoperiod in many animals (Leder et al. 2006; Bradshaw and Holzapfel 2007; Liedvogel et al. 2009). Timing of freshwater entry and reproductive maturity in salmonid fishes is a complex array of interrelated behavioral and physiological traits that has heritable components and exhibits phenotypic plasticity in response to environmental variability (Carlson and Seamons 2008; Abadia-Cardoso et al. 2013). Chinook salmon populations have been able to exploit a wide range of habitats because of evolution at this trait, often in the face of ongoing gene flow (Waples 2001; O'Malley et al. 2013). Using markers associated with the circadian clock gene network might have provided additional insight in the differentiation among these closely related groups and they did discriminate the spring-run population above the waterfall from all downstream fish, although with genetic differentiation similar to the other markers.

Conservation implications

It is important for species to maintain genetic variability in order to respond to dynamic environments. The extent and scale of intraspecific genetic diversity is therefore a crucial consideration from both conservation and management perspectives (Manel et al. 2010; Funk et al. 2012). Chinook salmon populations are characterized by hierarchical genetic structure (Seeb et al. 2007; Moran et al. 2013) with evidence of isolation across their range in the north Pacific (> 10,000 km), within larger river systems (> 1000 km), and regionally among watersheds (< 1000 km). Finer-scale structure in Chinook salmon within smaller coastal watersheds (e.g., the Siletz River <120 km) has been relatively unstudied.

Matala et al. (2012) identified spatial structuring of Chinook salmon in the South Fork Salmon River (SFSR, ~90 km), which is part of the Columbia

River system. Chinook salmon that spawned within the main stem SFSR were significantly different from individuals returning to two other tributaries within the subbasin. To reach these isolated tributaries, salmon must travel hundreds of kilometers up the main stem Columbia River. The F_{ST} values that we report between fish above and below the waterfall in the Siletz River are of similar magnitude to those reported among these Columbia River tributaries, but on a much smaller geographic scale. This demonstrates that such life history variation and genetic differentiation is not limited to large river systems and can be found in smaller watersheds.

Population structure described solely on the basis of divergence at one type of molecular marker, particularly presumably neutral ones, may fail to identify distinct populations that warrant separate management. Life history diversity of salmon is most often associated with spatial diversity and larger river systems (e.g., Columbia River or Sacramento-San Joaquin river systems) that typically have both numerous genetically distinct salmon populations and a greater number of life history strategies associated with them (Groot and Margolis 1991; Taylor 1991; Waples 2001).

Identifying and preserving genetic and phenotypic diversity is an important component of formulating strategies to maintain resiliency and fitness of salmon, particularly in smaller watersheds (McElhany et al. 2000). It has become evident that genetic data can inform many decisions relating to management strategies, especially those that are aimed at maintaining abundance and genetic diversity in natural salmon populations (Olsen et al. 2000; Eldridge and Killebrew 2007; Grandjean et al. 2009; Brenkman et al. 2012; Matala et al. 2012). Without such information there is great risk of losing important life history variation that enables resilience of anadromous species to changing environments.

CONCLUSIONS

We have demonstrated how the application of molecular genetic data from multiple types of markers provides strong support for existence of two genetically and phenotypically distinct salmon populations in a small coastal watershed where only one is currently recognized. Much of fishery management and conservation is based upon status of larger, regional management units (e.g., the North Oregon Coast Chinook Salmon ESU). These management units are combinations of unique spawning groups from multiple, smaller river basins (e.g., the Siletz, Alsea, Coquille, and Siuslaw Rivers). Management solely at such larger scales may not take into account fine-scale genetic and phenotypic variability that is present within smaller watersheds, such as has been demonstrated here. This fine-scale variability is a necessary component of long-term resilience and its maintenance should be explicitly considered to ensure successful conservation and management.

REFERENCES

1. Abadia-Cardoso A, Anderson EC, Pearse DE, Garza JC (2013) Large-scale parentage analysis reveals reproductive patterns and heritability of spawn timing in a hatchery population of steelhead (*Oncorhynchus mykiss*). Mol Ecol 22:4733–4746. doi:10.1111/mec.12426
2. Belkhir K, Borsa P, Chikhi L, Raufaste N, Bonhomme F (1996-2004) Genetix 4.05, 4.05 edn. Laboratoire Génome, Populations, Interactions, CNRS UMR 5000, Université de Montpellier II, Montpellier
3. Bradshaw WE, Holzapfel CM (2007) Evolution of animal photoperiodism. Annu Rev Ecol Evol Syst 38:1–25. doi:10.1146/annurev.ecolsys.37.091305.110115
4. Brenkman SJ, Duda JJ, Torgersen CE, Welty E, Pess GR, Peters R, McHenry ML (2012) A riverscape perspective of Pacific salmonids and aquatic habitats prior to large-scale dam removal in the Elwha River, Washington, USA. Fish Manag Ecol 19:36–53. doi:10.1111/j.1365-2400.2011.00815.x
5. Carlson SM, Seamons TR (2008) A review of quantitative genetic components of fitness in salmonids: implications for adaptation to future change. Evol Appl 1:222–238. doi:10.1111/j.1752-4571.2008.00025.x
6. Cavalli-Sforza LL, Edwards AWF (1967) Phylogenetic analysis: models and estimation procedures. Am J Hum Genet 19:233–257
7. Clemento AJ, Crandall ED, Garza JC, Anderson EC (2014) Evaluation of a single nucleotide polymorphism baseline for genetic stock identification of Chinook Salmon (*Oncorhynchus tshawytscha*) in the California current large marine ecosystem. Fish Bull 112:112–130. doi:10.7755/fb.112.2-3.2
8. DeFaveri J, Viitaniemi H, Leder E, Merila J (2013) Characterizing genic and nongenic molecular markers: comparison of microsatellites and SNPs. Mol Ecol Resour 13:377–392. doi:10.1111/1755-0998.12071
9. Eldridge WH, Killebrew K (2007) Genetic diversity over multiple generations of supplementation: an example from Chinook salmon using microsatellite and demographic data. Conserv Genet 9:13–28. doi:10.1007/s10592-007-9298-y
10. Felsenstein J (1993) Phylip (phylogeny inference package), version 3.69 edn. Depatment of genetics, University of Washington, Seattle
11. Funk WC, McKay JK, Hohenlohe PA, Allendorf FW (2012) Harnessing genomics for delineating conservation units. Trends Ecol Evol 27:489–496. doi: 10.1016/j.tree.2012.05.012
12. Garvin MR et al (2013) Recent physical connections may explain weak genetic structure in western Alaskan chum salmon (*Oncorhynchus keta*) populations. Ecology and Evolution 3:2362–2377. doi:10.1002/ece3.628
13. Grandjean F, Verne S, Cherbonnel C, Richard A (2009) Fine-scale genetic structure of Atlantic salmon (*Salmo salar*) using microsatellite markers: effects of restocking and natural recolonization. Freshw Biol 54:417–433. doi:10.1111/j.1365-2427.2008.02116.x
14. Groot C, Margolis L (1991) Pacific Salmon life histories. University of British Columbia press, University of British Columbia, Vancouver, British Columbia
15. Hedrick PW (1999) Perspective: highly variable loci and their interpretation in evolution and conservation. Evolution 53:313–318
16. Hess JE, Matala AP, Narum SR (2011) Comparison of SNPs and microsatellites for fine-scale spplication of genetic stock identification of Chinook salmon in the

Columbia river basin. Mol Ecol Resour 11:137–149. doi:10.1111/j.1755-0998.2010.02958.x10.5061/dryad.7986
17. Ivanova NV, deWaard JR, Hebert PDN (2006) An inexpensive, automation-friendly protocol for recovering high-quality DNA. Mol Ecol Notes 6:1–5. doi:10.1111/j.1471-8286.2006.01428.x
18. Kalinowski ST (2005) HP-RARE 1.0: a computer program for performing rarefaction on measures of allelic richness. Mol Ecol Notes 5:187–189. doi:10.1111/j.1471-8286.2004.00845.x
19. Kinziger AP, Hellmair M, Hankin DG, Garza JC (2013) Contemporary population structure in Klamath River basin Chinook salmon revealed by analysis of microsatellite genetic data. Trans Am Fish Soc 142:1347–1357. doi:10.1080/00028487.2013.806351
20. Leder EH, Danzmann RG, Ferguson MM (2006) The candidate gene, clock, localizes to a strong spawning time quantitative trait locus region in rainbow trout. J Hered 97:74–80. doi:10.1093/jhered/esj004
21. Liedvogel M, Szulkin M, Knowles SC, Wood MJ, Sheldon BC (2009) Phenotypic correlates of clock gene variation in a wild blue tit population: evidence for a role in seasonal timing of reproduction. Mol Ecol 18:2444–2456. doi:10.1111/j.1365-294X.2009.04204.x
22. Lindley ST et al. (2004) Population structure of threatened and endangered Chinook salmon ESUs in California's Valley basin NOAA-NMFS-SWFSC-360. National Marine Fisheries Service, Southwest Fisheries Science Center
23. Manel S et al (2010) Perspectives on the use of landscape genetics to detect genetic adaptive variation in the field. Mol Ecol 19:3760–3772. doi:10.1111/j.1365-294X.2010.04717.x
24. Matala AP, Narum SR, Young W, Vogel JL (2012) Influences of hatchery supplementation, spawner distribution, and habitat on genetic structure of Chinook salmon in the South fork Salmon River, Idaho. N Am J Fish Manag 32:346–359. doi:10.1080/02755947.2012.678961
25. McElhany P, Ruckelshaus MH, Ford MJ, Wainwright TC, Bjorkstedt EP (2000) Viable salmonid populations and the recovery of evolutionarily significant units. NOAA-NMFS-NWFSC-42.
26. Milner GB, Teel DJ, Utter F, Winans GA (1985) A genetic method of stock identification in mixed populations of Pacific salmon, *Oncorhynchus* spp. Mar Fish Rev 47:1–8
27. Moran P et al (2013) Divergent life-history races do not represent Chinook salmon coast-wide: the importance of scale in quaternary biogeography. Can J Fish Aquat Sci 70:415–435. doi:10.1139/cjfas-2012-0135
28. Naish KA, Park LK (2002) Linkage relationships for 35 new microsatellite loci in Chinook salmon *Oncorhynchus tshawytscha*. Anim Genet 33:312–318. doi:10.1046/j.1365-2052.2002.t01-4-00886.x
29. Narum SR et al (2008) Differentiating salmon populations at broad and fine geographical scales with microsatellites and single nucleotide polymorphisms. Mol Ecol 17:3464–3477. doi:10.1111/j.1365-294X.2008.03851.x
30. Nelson RJ, Beacham TD (1998) Isolation and cross species amplification of microsatellite loci useful for study of Pacific salmon. Anim Genet 30:225–244. doi:10.1046/j.1365-2052.1999.00404-4.x
31. Olsen JB, Bentzen P, Banks MA, Shaklee JB, Young S (2000) Microsatellites reveal population identity of individual pink salmon to allow supportive breeding of a population at risk of extinction. Trans Am Fish Soc 129:232–242. doi:10.1577/1548-8659(2000)129<0232:mrpioi>2.0.co;2
32. O'Malley KG, Banks MA (2008) A latitudinal cline in the Chinook salmon (*Oncorhynchus tshawytscha*) clock gene: evidence for selection on PolyQ length variants. Proceedings of the Royal Society B-Biological Sciences 275:2813–2821. doi:10.1098/rspb.2008.0524

33. O'Malley KG, Camara MD, Banks MA (2007) Candidate loci reveal genetic differentiation between temporally divergent migratory runs of Chinook salmon (*Oncorhynchus tshawytscha*). Mol Ecol 16:4930–4941. doi:10.1111/j.1365-294X.2007.03565.x
34. O'Malley KG, McClelland EK, Naish KA (2010) Clock genes localize to quantitative trait loci for stage-specific growth in juvenile coho salmon, *Oncorhynchus kisutch*. J Hered 101:628–632. doi:10.1093/jhered/esq063
35. O'Malley KG, Jacobson DP, Kurth R, Dill AJ, Banks MA (2013) Adaptive genetic markers discriminate migratory runs of Chinook salmon (*Oncorhynchus tshawytscha*) amid continued gene flow. Evol Appl 6:1184–1194. doi:10.1111/eva.12095
36. Oregon Department of Fish and Wildlife (1997) Siletz river basin fish management plan. Oregon Department of fish and Wildlife. Salem, Oregon
37. Oregon Department of Fish and Wildlife (2014) Coastal multi-species conservation and management plan. Oregon Department of Fish and Wildlife. Salem, Oregon
38. Peakall R, Smouse P (2012) GenAlEx 6.5: genetic analysis in excel. Population genetic software for teaching and research - an update. Bioinformatics:1-3 doi:10.1093/bioinformatics/bts460
39. Pritchard JK, Stephens M, Donnelly P (2000) Inference of population structure using multilocus genotype data. Genetics 155:945–959
40. Quinn TP (2004) The behavior and ecology of Pacific Salmon and Trout. University of Washington Press, Seattle
41. Raymond M, Rousset F (1995) GENEPOP (version 1.2): population genetics software for exact tests and ecumenicism. J Hered 86:248–249
42. Seeb LW et al (2007) Develoment of a standardized DNA database for Chinook salmon. Fisheries 31:540–552
43. Shafer ABA et al (2015) Genomics and the challenging translation into conservation practice. Trends Ecol Evol 30:78–87. doi:10.1016/j.tree.2014.11.009
44. Taylor EB (1991) A review of local adaptation in Salmonidae, with particular reference to Pacific and Atlantic salmon. Aquaculture 98:185–207
45. Utter FM, Hodgins HO (1972) Biochemical genetic variation at six loci in four stocks of rainbow trout. Trans Am Fish Soc 101:494–502. doi:10.1577/1548-8659(1972)101<494:bgvasl>2.0.co;2
46. Van Oosterhout C, Hutchinson WF, Wills DPM, Shipley P (2004) MICRO-CHECKER: software for identifying and correcting genotyping errors in microsatellite data. Mol Ecol Notes 4:535–538. doi:10.1111/j.1471-8286.2004.00684.x
47. Waples RS (1998) Evolutionarily significant units, distinct population segments, and the endangered species act: reply to Pennock and Dimmick. Conserv Biol 12:718–721. doi:10.1111/j.1523-1739.1998.97524.x
48. Waples R (2001) Characterizing diversity in salmon from the Pacific northwest. J Fish Biol 59:1–41. doi:10.1006/jfbi.2001.1764
49. Waples RS, Teel DJ, Myers JM, Marshall AR (2004) Life-history divergence in Chinook salmon: historic contingency and parallel evolution. Evolution 58:386–403. doi:10.1554/03-323
50. Weir BS, Cockerham CC (1984) Estimating F-statistics for the analysis of population structure. Evolution 38:1358–1370. doi:10.2307/2408641
51. Williamson KS, Cordes JF, May B (2002) Characterization of microsatellite loci in Chinook salmon (*Oncorhynchus tshawytscha*) and cross-species amplification in other salmonids. Mol Ecol Notes 2:17–19. doi:10.1046/j.1471-8286.2002.00129.x
52. Yoshiyama RM, Fisher FW, Moyle PB (1998) Historical abundance and decline of Chinook salmon in the Valley region of California. N Am J Fish Manag 18:487–521. doi:10.1577/1548-8675(1998)018<0487:HAADOC>2.0.CO;2

CHAPTER 15

The Chromosomal Constitution of Fish Hybrid Lineage Revealed by 5S rDNA FISH

Chun Zhang†, Lihai Ye†, Yiyi Chen, Jun Xiao, Yanhong Wu, Min Tao, Yamei Xiao and Shaojun Liu*

Key Laboratory of Protein Chemistry and Fish Developmental Biology of the Ministry of Education of China, College of Life Sciences, Hunan Normal University, Changsha 410081, China

ABSTRACT

Background: The establishment of the bisexual fertile fish hybrid lineage including the allodiploid and allotetraploid hybrids, from interspecific hybridization of red crucian carp (*Carassius auratus* red var. 2n = 100, 2n = AA) (♀) × common carp (*Cyprinus carpio* L. 2n = 100, 2n = BB) (♂), provided a good platform to investigate genetic relationship between the parents and their hybrid progenies.

Results: The chromosomal inheritance of diploid and allotetraploid hybrid progenies in successive generations, was studied by applying 5S rDNA fluorescence in situ hybridization. Signals of 5S rDNA distinguished the chromosomal constitution of common carp (B-genome) from red crucian carp (A-genome), in which two strong signals were observed on the first submetacentric chromosome, while no major signal was found in common carp. After fish hybridization, one strong signal of 5S rDNA was detected in the same locus on the chromosome of diploid hybrids. As expected, two strong signals were observed in $4nF_3$ tetraploid hybrids offspring and it is worth mentioning that two strong signals were detected in a separating bivalent of a primary spermatocyte in $4nF_3$. Furthermore, the mitosis of heterozygous chromosomes was shown normal and stable with blastular tissue histological studies.

Conclusions: We revealed that 5S rDNA signal can be applied to discern A-genome from B-genome, and that 5S rDNA bearing chromosomes can be stably passed down in successive generations. Our work provided a significant method in fish breeding and this is important for studies in fish evolutionary biology.

Keywords: Interspecific hybridization Heterozygous chromosomes FISH 5S rDNA

BACKGROUND

In general, interspecific hybridization and polyploidy in plants were potent evolutionary mechanisms [1, 2]. Recently, new genetic evidence suggested that hybrid speciation was more common in plants and animals than we thought, and it has played a very constructive part in animal evolution [3, 4]. As fertile hybrids, the integrity and inheritance of heterozygous genome was a focused issue, which concerned to mitotic and meiotic stability of hybrid offspring and formation of fertile progenies. Now the inheritance rule of heterozygous genome had been widely studied in hybrid plants, while as there was a limit to the material, related research only focused on such a few hybrid vertebrates as fish and frog [5, 6].

Through selecting and breeding for more than 20 years, the bisexual fertile allotetraploid hybrid fish (abbreviated as AT) ($4n = 200$) has been acquired, which resulted from fertilization of unreduced eggs and sperm produced by hybrids of red crucian carp (*Carassius auratus* red var. $2n = 100$) (♀) (abbreviated as RCC) × common carp (*Cyprinus carpio* L. $2n = 100$) (♂) (abbreviated as CC). It provided a unique material for exploring the inheritance rule of heterozygous chromosomes in hybrid lineage [7, 8, 9, 10]. The production procedure of AT was as follows: RCC females were mated with CC males to produce F_1 fish (abbreviated as $2nF_1$) which were then mated with each other to produce F_2 fish (abbreviated as $2nF_2$). The males and females of diploid F_2 hybrids could generate unreduced diploid eggs and diploid sperm, respectively, which were fertilized to form the allotetraploid hybrids in F_3 fish (abbreviated as $4nF_3$). Fertile tetraploid female and male F_3 fish were found to generate diploid eggs and sperm, respectively. By self-breeding of F_3, the tetraploid F_4fish were produced. Until now, the F_3–F_{24} AT has been formed through successive breeding. Our previous study has indicated that the AT has inherited large amount of genetic material from their original parents with some variations

[11, 12, 13, 14, 15]. However, the chromosomes constitution and inheritance pattern of hybrid lineage were still unknown. Identification of all genome chromosomes was crucial to understanding the genome constitution and inheritance rule of the hybrid genome. It has been widely studied in hybrid plant, using by fluorescence in situ hybridization (abbreviated as FISH), and genomic in situ hybridization (abbreviated as GISH) [16, 17, 18]. However, due to the lack of chromosome-specific molecular probes, the current reported studies were few in hybrid fish related to chromosomal localization.

The 5S rDNA in higher eukaryotes was organized in tandem repeat units that consisted of highly conserved 120 bp transcribing sequence and variable non-transcribed spacers (abbreviated as NTS) [19]. Variations in NTS, related to insertions-deletions, minirepeats and pseudo genes, were often species specific and have successfully been served as markers in evolutionary studies [20, 21, 22]. In addition, due to the numerous copies of repeated sequences, their chromosomal localization was easily detected by FISH. Analysis of 5S rDNA sequences and chromosomal localization of them were an effective method of genetic diversity monitoring, which were widely used to explore the phylogeny relationship among closer species and polyploidy origin [23, 24, 25]. Masaru and Hideo [26] also found that the 5S rDNA could be candidates for phylogenetic molecular markers for the crucian carp.

In the current study, based on our establishment of $2nF_1$, $2nF_2$, $4nF_3$ and $4nF_{22}$ hybrid lineage of RCC × CC, a comparative analysis of 5S rDNA fragments among all samples were carried by PCR and related sequences were screened out as probes to determine the chromosomal localization of the 5S rDNA for $2nF_1$, $2nF_2$, and $4nF_3$ fish by FISH, which elucidated the inheritance rules of heterozygous chromosomes in course of polyploid hybrid fish origin and propagation. Furthermore, cytological observation of early embryos in blastula stagewas used to determine the mitotic stability in hybrid offspring.

METHODS

Ethics statement

All experiments were approved by Animal Care Committee of Hunan Normal University and followed guidelines statement of the Administration of Affairs Concerning Animal Experimentation of China. All samples are raised in natural ponds and all dissections are performed under sodium pentobarbital anesthesia, and all efforts are made to minimize suffering.

Fishes and genomic DNA samples

Specimens of RCC ($2n = 100$), CC ($2n = 100$), $2nF_1$ ($2n = 100$), $2nF_2$ ($2n = 100$), $4nF_3$ ($4n = 200$) and $4nF_{22}$ ($4n = 200$) were obtained from the Engineering Center of Polyploid Fish Breeding of National Education Ministry located at Hunan Normal University. Total genomic DNA was isolated from the peripheral blood cells according to the standard phenol-chloroform extraction procedures described by Sambrook et al. [27] with minor modifications.

PCR amplification, cloning and sequencing

One pair of primers (5'-TATGCCCGATCTCGTCTGATC-3' and 5'-CAGGTTGGTATGGCCGTAAGC-3') [26] was synthesized to amplify the 5S rDNA repeats directly from genomic DNA by PCR. The genomic DNA of RCC, CC, $2nF_1$, $2nF_2$, $4nF_3$ and $4nF_{22}$ were used as a template for subsequent PCR (2 to 10 individuals for each of the 6 samples). The PCR cycling conditions were: 5 min at 94 C, 30 cycles of denaturation at 94 C for 30s, annealing at 56 C for 30s, and extension 72 C for 1 min, ending with 10 min of extension at 72 C. The PCR products were analyzed in 1.2 % agarose gels stained by ethidium bromide, purified by Gel Ex- traction Kit (Sangon), cloned into the pMD18-T vector (Takara), and transferred into *E. coli* DH5α. Then the positive clones were sequenced by Sangon. The sequences were analyzed by ClustalW2.

Chromosome preparation

The kidney cells in RCC, CC, $2nF_1$, $2nF_2$ and $4nF_3$ were used for the chromosome observation at the metaphase of mitosis and the testis cells in $4nF_3$ were used to observe the process of meiosis. The concanvalin A was injected into the abdominal cavity of the samples for one to three times at the dosage of 2-8 μg/g. The interval time was 12–24 h. Two to six hours prior to harvest, colchicines (2-4 μg/g) were used to arrest the chromosome at the metaphase. All the kidney tissue in the above samples was ground in 0.8 % NaCl. The hypotonic treatment was accomplished with 0.075 mol/L KCl for 40–60 min, followed by fixation in 3:1 methanol-acetic acid (three changes). The cells were spread on clean slides.

Probe preparation of the 5S rDNA sequence and fluorescence in situ hybridization

FISH is used to assess chromosomal location of 5S rDNA of RCC, CC, $2nF_1$, $2nF_2$, and $4nF_3$ fish. The hybridization was performed according to the method described by Masaru and Hideo with minor modifications [26]. In brief, the purified PCR product of related 5S rDNA fragments of RCC labeled with Dig-11-dUTP by PCR DIG Probe Synthesis Kit (Roche, Germany) were used as probes. After pretreatment with 2 × SSC for 30 min, 70 %, and 100 % ethanol for 5 min each, the slides with chromosome metaphase spreads of all samples were denatured in 70 % deionized formamide /2 × SSC for 2 min at 75 C, dehydrate in a 70 % (−20 C, to avoid DNA renaturation) and 100 % ethanol series for 5 min each, and then air-dry. The probe was prepared by adding labeled DNA with 20 × SSC, deionized formamide and 50 % dextran sulphate and denaturing in boiling water for 5 min. Hybridizations were allowed to proceed under a sealed cover slip in a moist chamber at 37 C overnight. The next day the slides were washed; twice for 15 min in 2 × SSC with 50 % formamide and then in 2 × SSC and 1 × SSC for 5 min each. After a series of post-hybridization washing were performed, the spectrum signals were achieved with 10ul of 1ug ml^{-1} FITC-conjugated anti-digoxigenin antibody from sheep (Roche, Germany). Then the slides were washed three times for 5 min each in TNT (formulas of eluate: 0.1 M Tris–HCl, 0.15 M NaCL,0.05 % Tween 20). After slides were placed in 10ul anti-fade solution containing 0.5 μg/ml of 4, 6-diamidino-2-phenylindole (abbreviated as DAPI), they were viewed using a Leica inverted DMIRE2 microscope image system (Leica, Germany). Images were captured with CW4000 FISH software (Leica, Germany). Good-quality metaphase spreads were photographed and used for analysis of karyotypes.

Cytological observations of early embryos

After 6 h after fertilization, the F_3 embryos in blastula stage were fixed in Smith solution for 4 to12 h, and then are washed by alcohol for 2 to 3 times. The paraffin-embedded sections were cut at 6 μm and stained with Harris hematoxylin and eosin. The structure of zygote was observed under a light microscope and photographed with Pixera Pro 600 ES.

RESULTS

Analysis of 5S rDNA fragments

The electrophoretic band pattern of DNA fragments amplified with the primers of 5 s rDNA was distinctive between RCC and CC. RCC exhibited two major DNA bands (approximately 200 bp and 340 bp) and some repeated ladder-like bands, CC exhibited other two major DNA bands (approximately 200 bp and 400 bp) and some repeated ladder-like bands as well. As their hybrid offspring, $2nF_1$ and 4n F_{22} exhibited three major DNA bands (approximately 200 bp, 340 bp and 400 bp) and some repeated ladder-like bands (Fig. 1). After ligation of sized DNA fragments and transformation, a total number of 100 clones were sequenced to examine the different patterns of 5SrDNA, including 30 clones from 5 RCC individuals, 20 clones from 5 CC individuals, 20 clones from 5 2 nF_1 individuals and 30 clones from 5 4 nF_{22} individuals. consensus sequences of the 5 s rDNA repeat units were obtained. All of the typical sequences were deposited at GenBank under the accession numbers (KM359661- KM359679) (Table 1).

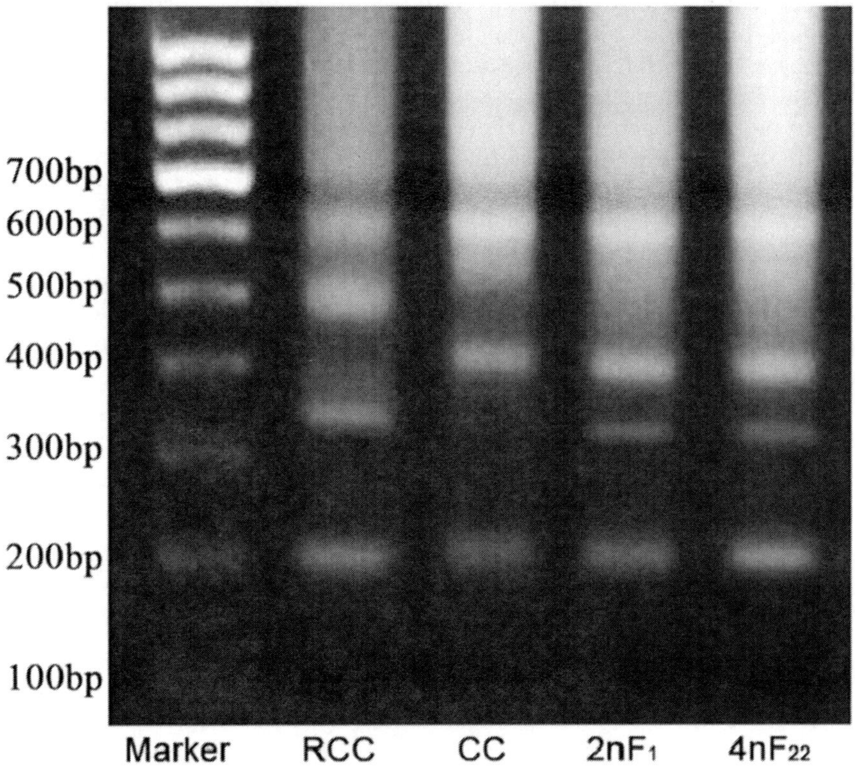

Figure. 1 DNA bands amplify from RCC, CC and their hybrid offspring. Marker: DNA ladder markers (100 bp increments); RCC: two DNA bands (200, 340 bp) and some repeated ladder-like bands; CC: two DNA bands (200, 400 bp) and some repeated ladder-like bands; $2nF_1$ and $4nF_{22}$: three DNA bands (200, 340 and 400 bp) and some repeated ladder-like bands

Table 1 Size (bp), type, and GenBank accession number for each of the 5SrDNA repeat units (coding sequence plus NTS) found in RCC, CC, $2nF_1$, and $4nF_{22}$

Species	Length(bp)			Accession no.
	5 s repeat	Coding sequence	NTS	
RCC	201	119	82 = 0*138 + 82 [RCC-a]	KM359661
	202	120	82 = 0*138 + 82 [RCC-b]	KM359662
	340	121	219 = 1*137 + 82	KM359663
	339	121	218 = 1*136 + 82	KM359664
	477	121	356 = 2*137 + 82	KM359665
	613	120	493 = 3*137 + 82	KM359666
CC	202	120	82 = 0*138 + 82 [CC-a]	KM359667
	204	121	83 = 0*138 + 83 [CC-b]	KM359668
$2nF_1$	203	121	82 = 0*138 + 82	KM359671
	201	120	82 = 0*138 + 82	KM359672
	341	120	219 = 1*137 + 82	KM359673
	339	120	219 = 1*137 + 82	KM359674
$4nF_{22}$	201	119	82 = 0*138 + 82	KM359675
	202	120	82 = 0*138 + 82	KM359676
	338	119	219 = 1*137 + 82	KM359677
	477	120	357 = 2*138 + 82	KM359678
	639	122	504 = 3*137 + 93	KM359679

[RCC-a, RCC-b, CC-a, CC-b]: indicate the different approximately 82 bp-units. The alignment data of them was showed in Table 2

Nucleotide sequencing and BlastN sequences corresponded to 5 s rDNA repeat units. All units consist of a 120 bp coding region and a variable NTS. In RCC, four kinds of fragments of 5 s rDNA (202 bp, 340 bp, 477 bp and 613 bp) were characterized by different lengths of NTS. In CC, there was only one kind of fragment of 5 s rDNA (202 bp) with 120 bp coding region and one 82 bp NTS unit. The 400 bp fragment from amplified DNA bands was just two repeats of 202 bp unit. In $2nF_1$ and $4nF_{22}$, like their original parents, two kinds of fragments of 5 s rDNA (approximately 200 bp and 340 bp) both contained similar coding region and NTS (Table 1).

The NTS sequence of 5S rDNA also showed an extensive variation between RCC and CC. In RCC, different-sized 5S rDNA units contained one 120 bp coding region and one 82 bp repeat unit, which were spaced at regular interval with zero, one, two, or three 138 bp repeat units (Interposed Region: IPR), respectively showing 202 bp, 340 bp, 477 bp and 613 bp 5S rDNA repeat units. While in CC, only around 200 bp-sized 5S rDNA units consisting of one 120 bp coding region and one monomeric 82 bp repeat unit were identified. The similarity of 82 bp-unit among different clones of RCC was 95.1 % and the similarity of 82 bp-unit among different clones of CC was 91.5 %, while the 82 bp-unit of RCC was different from 82 bp-unit of CC with lower similarity of 52.4 %-68.3 % (Table 2). Sequence alignments showed high similarity of 340 bp units between RCC, $2nF_1$ and $4nF_{22}$, even in gibel carp, *Carassius auratus*(abbreviated as Cag) [25] (Fig. 2 and Table 3), so the 340 bp-fragments were screened out as a probe to assess chromosomal location of 5S rDNA of RCC, CC, $2nF_1$, $2nF_2$, and $4nF_3$ fish.

Table 2 Nucleotide similarities of 82 bp-units of 5S rDNA sequence between RCC and CC

Sequence name	RCC-a	RCC-b	CC-a	CC-b
RCC-a	-	95.1 %	68.3 %	53.7 %
RCC-b	-	-	68.3 %	52.4 %
CC-a	-	-	-	91.5 %
CC-b	-	-	-	-

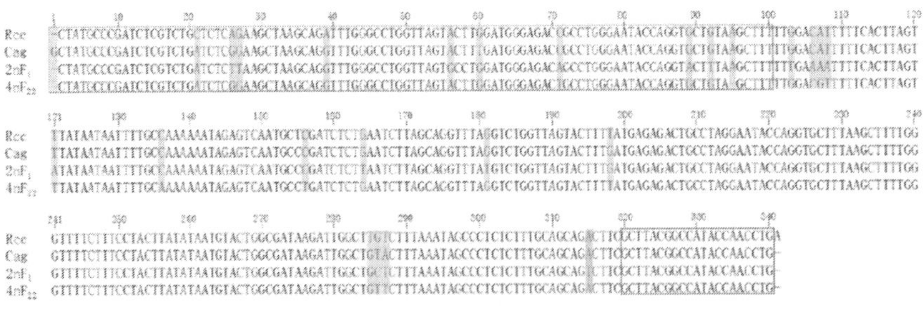

Figure. 2 Sequence alignments of 340 bp-fragments of RCC, Cag, $2nF_1$ and $4nF_{22}$. The 120 bp coding regions were included in the boxes. The sequence data of RCC, Cag, $2nF_1$ and $4nF_{22}$ were available from GenBank under accession number KM359663, DQ659260, KM359673, KM359677

Table 3 Nucleotide similarities of 340 bp- fragments among RCC, Cag, $2nF_1$ and $4nF_{22}$

sequence name	Cag-340	$2nF_1$-340	$4nF_{22}$-340
RCC-340	97.1 %	94.4 %	97.3 %
Cag-340	-	95.0 %	97.9 %
$2nF_1$-340	-	-	95.6 %

5S rDNA localization

The diploid RCC, CC, $2nF_1$, and $2nF_2$ samples showed 100 chromosomes which were classified into 22 metacentric (abbreviated as M), 34 submetacentric (abbreviated as SM), 22 subtelelocentric (abbreviated as ST) and 22 telocentric (abbreviated as T) chromosomes. The tetraploid $4nF_3$ samples revealed 200 chromosomes, which were classified into 44 metacentric (M), 68 submetacentric (SM), 44 subtelelocentric (ST) and 44 telocentric(T) chromosomes [7, 28]. No difference was observed between the females and the males.

Two signals of 340 bp-fragment were clearly detected in the first SM chromosomes pair of RCC, with a few weak signals being detected in other chromosomes (Fig. 3a, b), while the CC fish has no major signal. As to their hybrid lineage offspring, one strong signal was found in the first SM chromosome of $2nF_1$ and $2nF_2$ fish (Fig. 3c, d) and two strong signals were detected in the same chromosomes of AT (Fig. 3e, f), also with a few weak signals being detected in other chromosomes.

Chromosomal localization of 340 bp-fragment in metaphase of meiosis spreads in primary spermatocytes of $4nF_3$ indicated that two strong signals being located on a separating bivalent (Fig. 3g, h).

Cytological observation of early embryos in blastula stage

Cytological observation of mitosis in F_3 early embryo cells showed that F_3 performed normal and stable mitosis, with normal distribution of chromosomes and structure of spindle body during metaphase (Fig. 4a), anaphase (Fig. 4b), and telophase (Fig. 4c) of mitosis, not being affected by the genetic attributes of heterozygous genome.

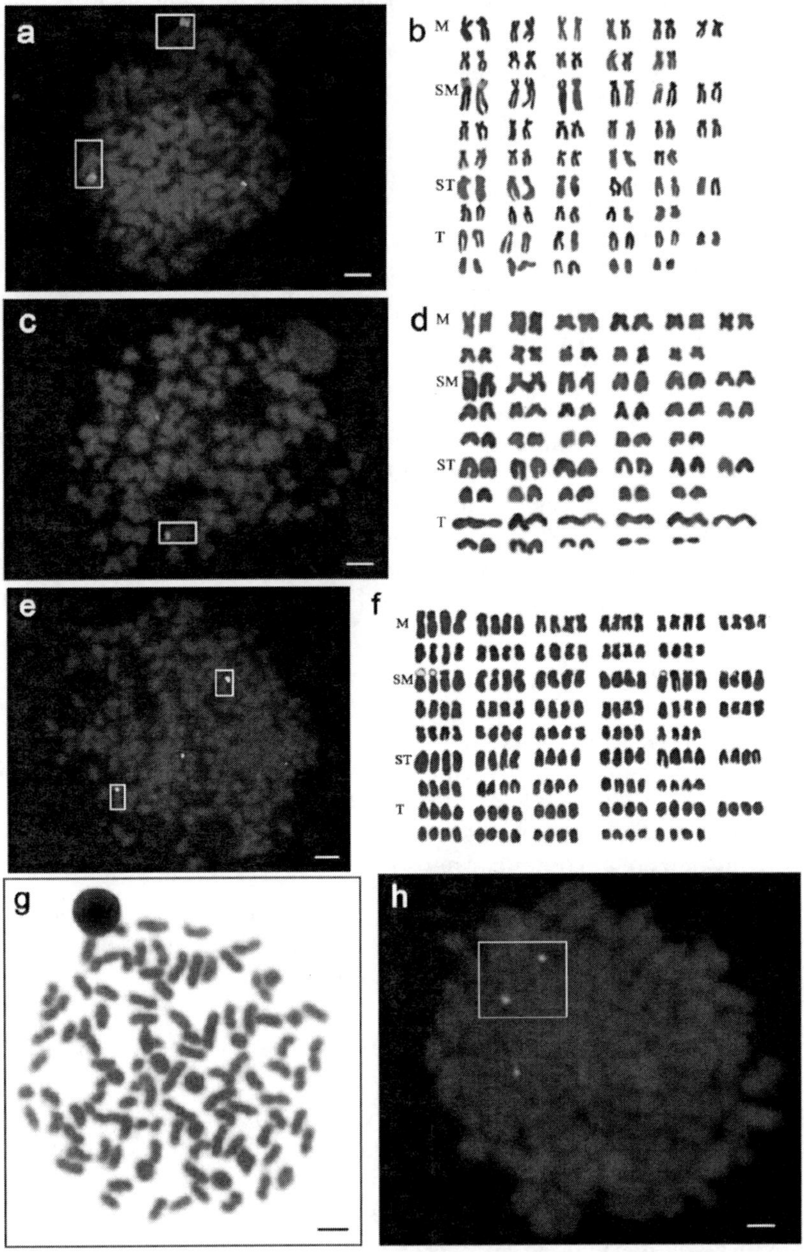

Figure. 3 Mitotic metaphase spreads and karyotypes of RCC, 2 nF_2 and 4 nF_3 fish with FISH signals of 5S rDNA. (**a**) Two strong and a few weak signals were detected in RCC fish; the boxes indicated two major signals of 5S rDNA (green). (**b**) The karyotype of RCC fish revealed two strong signals

located on the first pair of larger-sized SM chromosomes. (**c**) One strong signal and a few weak signals in $2nF_2$ fish (green),the box indicated the one major signal of 5S rDNA (green). (**d**) The karyotype of $2nF_2$ fish show one strong signal on the first group of larger-sized SM chromosomes. (**e**) The FISH analysis of $4nF_3$ fish exhibited the double signal feature of two strong and a few weak signals (green), the boxes indicated the two major signals (green). (**f**) The karyotype of $4nF_3$ fish showed two strong signals on the first group of larger-sized SM chromosomes. All metaphase chromosomes were counterstained with DAPI and appear blue. (**g**) The chromosome pairing in Meiosis I of spermatocytes in $4nF_3$ showed 100 bivalents under light microscope; (**h**) Two signals of 5S rDNA located on a separating bivalent. M: metacentric; SM: submetacentric; ST subtelelocentric; T: telocentric. (**a-f**) Bar = 3 μm; (**g,h**) Bar = 5 μm

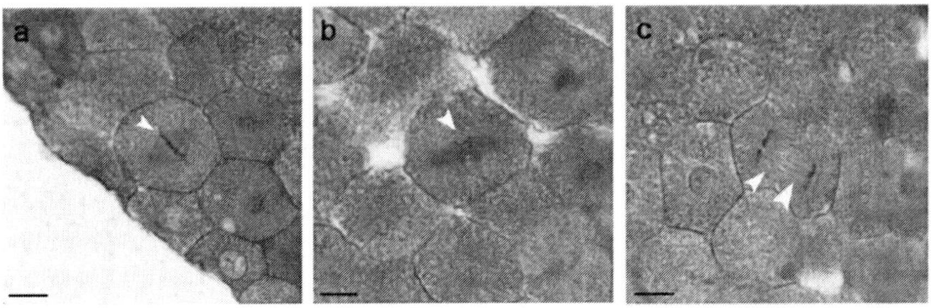

Figure. 4 Cytological observations of early embryos in blastula stage. (**a**) Early embryo cells in metaphase of mitosis; (**b**) Early embryo cells in anaphase of mitosis; (**c**) Early embryo cells in telophase of mitosis; the arrowheads showed the normal distribution of chromosomes and structure of spindle body. Bar = 10 μm

DISCUSSION

Different 5 s rDNA classes were reported in several mammals and fish species, and generally two different type of 5 s rDNA units have been characterized by distinct NTS types and base substitution [29, 30]. In our work, the construction of different-sized 5S rDNA units were further elucidated as containing one 120 bp coding region and one 82 bp repeat unit, which were spaced at regular interval with zero, one, two, or three 138 bp-IPR repeat units (Table 1), this result revealed the assembly rule of NTS in RCC, which enriched the current data of molecular organization analysis of 5 s rDNA.

Modern cytogenetic analyses, such as FISH and GISH have been widely applied for studies in many processes of chromosome evolution, including structural rearrangements, as well as extensive studies on phylogenetic and genomic relationships [31]. In this work, chromosome inheritance of hybrid lineage progenies of RCC × CC were investigated by detecting the signal of 340 bp fragments of 5 s rDNA on chromosomes in successive generations, i.e. RCC, CC, $2nF_1$, $2nF_2$, and $4nF_3$ fish. The 340 bp fragments was highly conserved between RCC and their hybrid lineage progenies, but not in CC genome, which made the 340 bp fragments in RCC a suitable indecator to monitor chromosomal inheritance in RCC, CC and their hybrid progeny, as it was recently showed in other hybrid fish [32, 33]. Besides that, our results revealed that the chromosomal localization of the 340 bp fragments of RCC could serve as a suitable genetic marker to distinguish the chromosomal constitution of CC from RCC, as two strong detected signals on the first pair of larger-sized SM chromosomes were observed, but not in CC chromosomes. The other weak FISH signals were also 5S rDNA locus with less copy of repeat units.

Since there have been very few studies on chromosomal inherence behavior in fish hybridization, we looked into the 5S rDNA bearing chromosome in successive hybrid generation. The strong signal with same locus as observed in RCC were also found in $2nF_1$ and $2nF_2$ fish and the number of signal doubled in $4nF_3$ fish. Furthermore, two strong signals were also detected in a separating bivalent of a primary spermatocyte in $4nF_3$, which promised that two 5S rDNA sites would equally distributed to each gamete and made every gamete acquire one 5S rDNA bearing chromosome. Those results may suggest that $2nF_1$ and $2nF_2$ fish contained one 5S rDNA bearing chromosome from A-genome and one 5S rDNA bearing chromosome from B-genome, and $4nF_3$ fish possessed two 5S rDNA bearing chromosomes from A-genome and two 5S rDNA bearing chromosomes from B-genome. Additional studies was made in order to investigate the mitosis, results demonstrated that in hybridization the mitosis process maintained normal and stable, which demonstrated the compatibility of genome between RCC and CC, and promised the genetic stability of hybrid progeny. Based on the fish analysis and cytological observation, we speculate that chromosomes of RCC and CC may equally pass down to hybrid generation (Fig. 5).

The chromosomal rearrangements having occurred less frequently was a promising result supporting the development of stable allotetraploid hybrid lines, which were also proved in allohexaploid *Brassica* line [34] and the lineage of *X.laevis* [35]. However, the allotetraploid *Festuca pratensis* × *Lolium perenne* hybrid of three generations shared various rDNA loci profiles with chromosomal rearrangements [31], indicating a tendency of *F. pratensis* genome-like chromosomes to be less stable in hybrid of three generations. In our work, the strong FISH signals of 5S rDNA in RCC passed down stably in successive hybrid generation, which implied that chromosomal rearrangements have occurred less frequently in area of highly repeated sequence of 5S rDNA. Thus it can be seen that genomic contribution was different in different type of hybridized combinations, which will affect the genetic stability of hybrid progenies. Further studies were necessary to prove or reject this hypothesis as well as for deeper understanding of the mechanisms responsible for the fate of heterozygous chromosomes in hybrid lineage progenies.

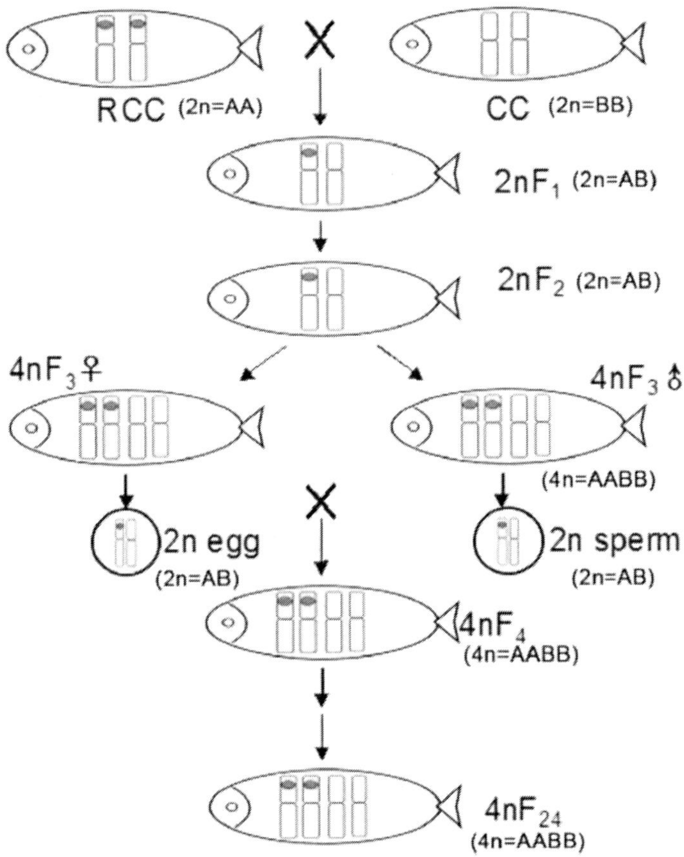

Figure. 5 Schematic presentation of chromosome transmission in hybrid fish from RCC, CC to $2nF_1$, and to $2nF_2$, as well as to $4nF_3$ and $4nF_{24}$

CONCLUSIONS

Few studies focused on chromosomal inheritance in successive generations of hybrid fish. We executed the 5S rDNA FISH analysis in successive generations of diploid and allotetraploid hybrid progenies and revealed 5S rDNA signal can be applied to discern A-genome from B-genome, and that 5S rDNA bearing chromosomes respectively coming from A-genome and B-genome can be stably passed down in successive hybrid generations.

ABBREVIATIONS

AT: Allotetraploid hybrid fish*:* **RCC;** Red crucian carp;**CC:** Common carp*;***$2nF_1$:** The first generation of diploid hybrid fish of RCC × CC*;***$2nF_2$:** The second generation of diploid hybrid fish of RCC × CC*;***$4nF_3$:**The third generation of tetraploid hybrid fish of RCC × CC*;* **Cag:** Gibel carp, *Carassius auratus;* **FISH:** Fluorescence in situ h ybridization*;* **GISH:** Genomic in situ hybridization*;***NTS:** Non-transcribed spacers*;* **IPR:** Interposed region*;* **DAPI:** 4, 6-diamidino-2-phenylindole*;* **M:** Metacentric*;* **SM:** Submetacentric*;* **ST:** Subtelelocentric*;* **T:** Telocentric

ACKNOWLEDGMENTS

This work is supported by the National Natural Science Foundation of China (30901100/31430088), the Cooperative Innovation Center of Engineering and New Products for Developmental Biology of Hunan Province (20134486), major international cooperation projects of the National Natural Science Foundation of China (31210103918), training Program of the Major Research Plan of the National Natural Science Foundation of China (91331105), the National High Technology Research and Development Program of China (2011AA100403), the National Key Basic Research Program of China (2012CB722305), The fund of education department of Hunan province (15A116), The fund of sci-tec innovation and entrepreneur for Hunan young talents, the Project Supported by Science and Technology Program of Hunan Province (No:2015JC3065; No:2014FJ3084) and the construct program of the key discipline in Hunan province and China.

REFERENCES

1. Otto SP, Whitton J. Polyploid incidence and evolution. Annu Rev Genet. 2000;34:401–37.
2. Leitch AR, Leitch IJ. Genomic plasticity and the diversity of polyploidy plants. Science. 2008;320:481–3.
3. Mallet J. Hybrid speciation. Nature. 2007;446:279–83.
4. Van de PY, Maere S, Meyer A. The evolutionary significance of ancient genome duplications. Nat Rev Genet. 2009;10:725–32.
5. Choleva L, Janko K, De Gelas K, Bohlen J, Šlechtová V, Rábová M. Synthesis of clonality and polyploidy in vertebrate animals by hybridization between two sexual species. Evolution. 2012;66:2191–203.
6. Pruvost NBM, Hoffmann A, Reyer HU. Gamete production patterns, ploidy, and population genetics reveal evolutionary significant units in hybrid water frogs (Pelophylax esculentus). Ecol Evol. 2013;3:2933–46.
7. Liu SJ, Liu Y, Zhou GJ, Zhang XJ, Luo C, Feng H, et al. The formation of tetraploid stocks of red crucian carp × common carp hybrids as an effect of interspecific hybridization. Aquaculture. 2001;192:171–86.
8. Liu SJ. Distant Hybridization Leads to Different Ploidy Fishes. Sci China Life Sci. 2010;53:416–25.
9. Sun YD, Liu SJ, Zhang C, Li JZ, Huang WR, Zhang J, et al. The chromosome number and gonadal structure of F_9-F_{11}allotetraploid crucian carp. Acta Genetica Sinica. 2003;30:414–8 (In Chinese).
10. Liu Y, Zhou GJ. Cytologicalstudy on the gonadal development of F_1 hybrids produced by crossing carassius auratus with *cyprinus carpio*. Acta Hydrobiological Sinica. 1986;10:101–11 (In Chinese).
11. Li JZ, Lu SQ, Liu SJ, Liu ZH, Zhang XJ, Liu Y. RAPD analysis of genetic variation between the allotetraploid hybrid of red crucian carp × common carp and their original parents. J Fish China. 2003;27:403–8.
12. Li JZ, Liu SJ, Zhang XJ, Lu SQ, Liu Y. Microsatellite Marker analysis of genetic variation between the allotetraploid crucian carp and the original parents. Acta Genetica Sinica. 2005;32:378–83.
13. Liu LG, Yan JP, Liu SJ, Liu D, You CP, Zhong H, et al. Evolutionary analysis of allotetrapoid hybrids of red crucian carp common carp, based on ISSR, AFLP molecular markers and cloning of cyclins genes. Chinese Sci Bull. 2009;54:2849–61.
14. Chen L, Li W, Liu SJ, Tao M, Long Y, Duan W, et al. Novel genetic markers derived from the DNA fragments of Sox genes. Mol Cell Probe. 2009;23:157–65.
15. Wang J, Ye LH, Liu QZ, Peng LY, Liu W, Yi XG, et al. Rapid genomic DNA changes in allotetraploid fish hybrids. Heredity. 2015;114:601–9.
16. Leflon M, Eber F, Letanneur JC, Chelysheva L, Coriton O, Huteau V, et al. Pairing and recombination at meiosis of *Brassica rapa*(AA) × *Brassica napus* (AACC) hybrids. Theor Appl Genet. 2006;113:1467–80.
17. Kosmala A, Zwierzykowski Z, Gasior D, Rapacz M, Zwierzykowska E, Humphreys MW. GISH/FISH mapping of genes for freezing tolerance transferred from *Festuca pratensis* to *Lolium multiflorum*. Heredity. 2006;96:243–51.
18. Xiong ZY, Pires CJ. Karyotype and identification of all homoeologous chromosomes of allopolyploid brassica napus and its diploid progenitors. Genetics. 2011;187:37–49.
19. Long EB, David IB. Repeated genes in eukaryotes. Annu Rev Biochem. 1980;49:727–64.
20. Martins C, Wasko AP. Orgnization and evolution of 5S ribosomal DNA in the fish genome. In: Williams CR, editor. Focus on Genome research. New York: Nova; 2004. p. 289–318.

21. Qin QB, He WG, Liu SJ, Wang J, Xiao J, Liu Y. Analysis of 5S rDNA organization and variation in polyploid hybrids from crosses of different fish subfamilies. J Exp Zool (Mol Dev Evol). 2010;314:403–11.
22. He W, Qin QB, Liu SJ, Li TL, Wang J, Xiao J, et al. Organization and variation analysis of 5S rDNA in different ploidy-level hybrids of Red Crucian Carp × Topmouth Culter. PLoS One. 2012;7(6), e38976.
23. Ruth BP, Kent MR. Location of repetitive DNAs to zebrafish (Danio rerio) chromosomes by fluorescence in situ hybridization (FISH). Chromosome Res. 2000;8:27–35.
24. Singh M, Kumar R, Nagpure NS, Kushwaha B, Gond I, Lakra WS. Chromosomal localization of 18S and 5S rDNA using FISH in the genus Tor(Pisces, Cyprinidae). Genetica. 2009;137:245–52.
25. Zhu HP, Ma DM, Gui JF. Triploid origin of the gibel carp as revealed by 5S rDNA localization and chromosome painting. Chromosome Res. 2006;14:767–76.
26. Masaru M, Hideo F. Characterization of repetitive DNA sequences carrying 5S rDNA of the triploid ginbuna. Genes Genet Syst. 1998;73:9–20.
27. Sambrook J, Fritsch EF, Maniatis T. Molecular cloning: a laboratory manual. 2nd ed. New York: Cold Spring Harbor Laboratory Press; 1989.
28. Levan A, Fredga K, Sandberg AA. Nomenclature for centromeric position on chromosomes. Hereditas. 1964;52:197–201.
29. Martins C, Galetti PMJ. Two 5S Rdna arrays in Neotropical fish species: is it a general rule for fishes? Genetica. 2001;111:439–46.
30. Campo D, Machado-Schiaffino G, Horreo JL, Garcia-Vazquez E. Molecular Organization and Evolution of 5S rDNA in the Genus Merluccius and Their Phylogenetic Implications. J Mol Evol. 2009;68:208–16.
31. Ksiazczyk T, Zwierzykowska E, Molik K, Taciak M, Kraiewski P, Zvierzykowski Z. Genome-dependent chromosome dynamics in three successive generations of the allotetraploid *Festuca pratensis* × *Lolium perenne* hybrid. Protoplasma. 2015;252:985–96.
32. He WG, Xie LH, Li TL, Liu SJ, Xiao J, Hu J, et al. The formation of diploid and triploid hybrids of female grass carp × male blunt snout bream and their 5S rDNA analysis. BMC Genetics. 2013;14:110.
33. Qin QB, Wang YD, Wang J, Dai J, Xiao J, Hu FZ, et al. The autotetraploid fish derived from hybridization of Carassius auratus red var. (Female) × Megalobrama amblycephala (Male). Biology of reproduction. 2014;91(4):93. 1–11.
34. Mason AS, Nelson MN, Takahira J, Cowling WA, Moreira Alves G, Chaudhuri A, et al. The fate of chromosomes and Alleles in an allohexaploid *Brassica* Population. Genetics. 2014;197:273–83.
35. Uno Y, Nishida C, Takagi C, Ueno N, Matsudal Y. Homoeologous chromosomes of *Xenopus laevis* are highly conserved after whole-genome duplication. Heredity. 2013;111(5):430–6.

CHAPTER 16

Fishing for Adaptive Epistasis using Mitonuclear Interactions

David M. Rand

Department of Ecology and Evolutionary Biology, Brown University, Providence, Rhode Island, United States of America

ABSTRACT

Mitochondria provide us with the energy to get up and do what needs to be done. They do this with 37 genes encoded in their own genome and more than 1,200 genes from the nuclear genome, whose gene products are targeted to mitochondria after being translated on cytosolic ribosomes. This complex intergenomic communication is a hallmark of eukaryotic life and has been under continuous refinement for more than 10^9 years, following the fusion of eubacterial and archaebacterial cells [1]. How these two genomes cooperate in the face of conflicting modes of genetic transmission, distinct evolutionary pressures of mutation and recombination, and oxidative damage brought on by the very activity of mitochondria, has puzzled and motivated biologists for decades. A central challenge is to understand the coordinated modes of communication that maintain biochemical, physiological, and evolutionary coadaptation of the nuclear and mitochondrial compartments. The traditional view is that central administration (the nucleus) maintains the activities of these hardworking laborers (mitochondria) by sending regulatory agents (nuclear gene products) to organelles spread across the cellular landscape. A growing body of literature points to diverse ways that these energy factories voice their concerns about the government of mitonuclear interactions and send retrograde signals back to the nucleus, altering its control over mitochondria [2]. Studies of these patterns and processes hope to resolve some pretty fundamental questions about life on earth: the causes of aging and disease [3–5], the metabolic bases of adaptation and Darwinian fitness [6], the genetics of speciation [7], the origin of sex and recombination [8, 9], the battle of the sexes, and genomic conflict [10].

Most questions about mitonuclear interactions focus on the oxidative phosphorylation pathway (OXPHOS), the ATP producing process that includes four multisubunit protein complexes of the electron transport chain (ETC) and a fifth complex that phosphorylates ADP to produce ATP. These OXPHOS complexes, embedded in the mitochondrial inner membrane, include all of the 13 proteins encoded in animal mitochondrial DNA (mtDNA) plus ~78 proteins encoded on nuclear chromosomes. Two other classes of mtDNA genes, transfer- and ribosomal- RNAs, are needed to express mtDNA encoded proteins, and their functions require nuclear encoded proteins for transfer RNA (tRNA) charging, ribosome function, and RNA processing. Because these sets of genes involve physical interactions between mtDNA- and nuclear-encoded gene products, they are logical targets for studies of mitonuclear communication, coordination, and coevolution. But mitochondria do much more than make ATP; for example, they maintain redox and calcium balance, mediate apoptosis, regulate amino acid and lipid metabolism via the Krebs cycle and beta-oxidation, and influence TOR signaling, among many other functions [2]. So, there are additional nuclear genes that might influence mitonuclear coadaptation by indirect interactions with mtDNA-encoded functions.

Testing hypotheses of molecular coadaptation in the age of genomics is potentially rather easy. One can sequence the genomes or transcriptomes of different organisms and ask if the interacting genes or proteins show parallel patterns of molecular evolution. Because mtDNA genes usually evolve faster than nuclear genes, those that interact with mtDNA-encoded partners should show elevated rates of divergence, implying coevolution. For functional tests, genetic replacement of a foreign mtDNA in to the nuclear background of a different species should disrupt mitonuclear coadaptation and cause fitness defects. There are some nice cases in support of these molecular [11] and genetic predictions [12], but there are also some robust examples that have not uncovered the predicted patterns from coevolution [13–15]. So, why has putatively ongoing mitonuclear coevolution not left consistent footprints of selection that should be easy to discover? Are there systems that are more likely to reveal the expected patterns?

In this issue of *PLOS Genetics*, Baris et al. [16] use the estuarine fish *Fundulus heteroclitis* to search for a signature of adaptive mitonuclear interactions, taking advantage of a natural experiment that is a geographic legacy of glacial retreat along the east coast of North America. They sampled a population of fish in Mantoloking, New Jersey, harboring two distinct mtDNA haplotypes that diverged during a period of geographic isolation imposed by the Hudson River basin that drained meltwater from the Wisconsin ice sheet. Following glacial retreat, "northern" and "southern" populations came in contact, and today, the New Jersey population has a ~50:50 ratio of two mtDNA haplotypes and a ~80:20 ratio of nuclear alleles typically found in Maine and Georgia. Baris et al. used genotyping by sequencing (GBS) to sample the nuclear genomes of fish from the New Jersey population and stratified the analysis of nuclear allelic variation by the mtDNA haplotype that each fish carried. The question they ask is whether there is evidence for population subdivision of nuclear alleles within a single population, associated with the divergent mtDNA haplotypes segregating in that population (Fig 1). Assuming neutral alleles and random mating, there should be no such mitonuclear association, especially because there is no physical linkage between nuclear chromosomes and cytoplasmic mtDNA.

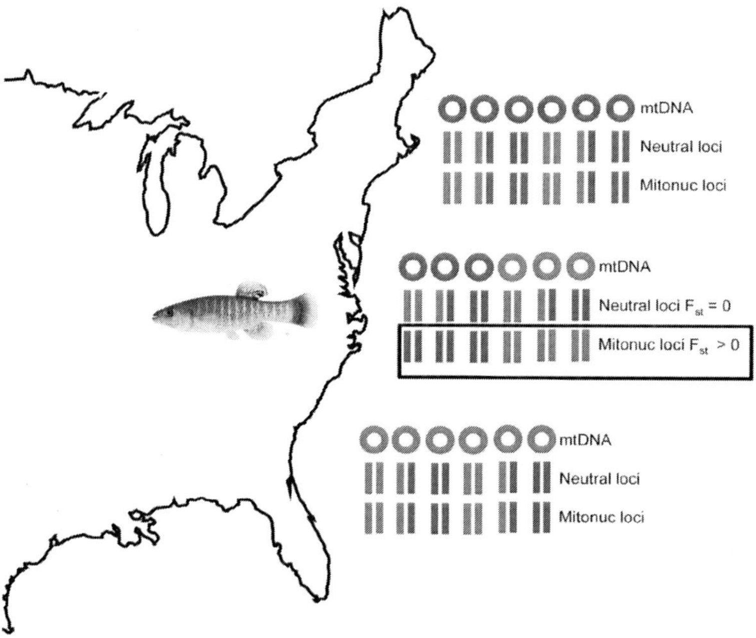

Figure 1. Nonrandom mitonuclear associations in a mixed population. In this simplified cartoon, the northern and southern populations of *F. heteroclitis* each have one mitochondrial DNA (mtDNA) haplotype (blue in the north, red in the south), and alleles at neutral nuclear loci and mitonuclear loci (Mitonuc loci) associated with mitochondrial function are in Hardy-Weinberg equilibrium ([HWE], for red and blue alleles). The New

Jersey population has both mtDNAs in equal frequency, and the neutral loci show the same HWE frequencies in each mtDNA background, but the mitonuclear loci show significant differences in genotype frequencies in the two mtDNA haplotype backgrounds ($F_{st} > 0$, see boxed "Mitonuc loci"). The latter suggests a selective maintenance of mitonuclear interactions for a subset of nuclear loci that have epistatic fitness interactions with mtDNA haplotypes. Map image is from http://d-maps.com/carte.php?num_car=11867&lang=en. Fish image is from http://www2.dnr.cornell.edu/ cek7/nyfish/Cyprinodontidae/mummichog.html.

Baris et al. found a few hundred single nucleotide polymorphisms (SNPs) spread across the nuclear chromosomes of *F. heteroclitis* that have significantly different allele frequencies in the two mtDNA "populations" within this single collecting locality (*cf.* 349 SNPs using $p < 0.01$ and 236 or 72 SNPs using a false discovery rate of 10% and 1%). Importantly, these SNPs show higher levels of allele frequency difference between mtDNA partitions within this population (measured as F_{st} "outliers") than they do between the focal New Jersey population and geographically isolated populations in Maine and Georgia, a clear rejection of neutral expectations. Baris et al. quantified OXPHOS activity (ADP-dependent state-3 respiration) of mitochondria extracted from heart tissue from four mitonuclear genotypes that define pure parental and admixed hybrid individuals (e.g., mtDNA-Nuclear genotypes as: North-North, North-Mixed, South-Mixed, South-South) and show that these mitonuclear genotypes explain a significant portion of the variation in OXPHOS function.

It is tempting to invoke adaptive mitonuclear epistasis to explain these data, in which mtDNA haplotypes offer a sort of alternative metabolic niche for nuclear alleles within a population, akin to the classical Levene model of balancing selection via opposing selection in alternative habitats [17]. Baris et al. make a concerted effort to test alternative explanations for F_{st} outlier status of SNPs partitioned by mtDNA, by testing 1,000 alternative data partitions of >9,000 nuclear SNPs. None of these "nuclear-nuclear" F_{st} outlier tests uncovered as many outlier SNPs as the original mitonuclear outlier test, suggesting that the main result is not littered with many false positives.

One unexpected result from these analyses is that that none of the 349 F_{st} outlier loci map to known nuclear genes encoding subunits of the OXPHOS complexes. If this study really has identified loci important in mitonuclear coadaptation, it suggests that long-range *cis-* or *trans-*regulatory factors, or downstream pathway effects, are mediating the epistatic interactions between mtDNA haplotypes and nuclear loci other than those encoding subunits of OXPHOS complexes. Understanding the genetic basis of fitness variation in the wild is a central goal of ecological and evolutionary genetics, and the genes of central metabolism have long fascinated biologists as a logical place to search for the source of this variation [6]. Given the critical role that mitochondrial function plays in organismal performance, and the increasing knowledge of the diverse roles of mitonuclear communication in regulating homeostasis, there is a real hope that we can track down the biochemical bases of these kinds of nonneutral patterns in nature. The Baris et al. study offers a nice example of how admixed populations with divergent mtDNAs might serve as a natural genetic screen for the footprints of mitonuclear epistasis and coevolution that could point to unanticipated targets of selection on metabolic function.

REFERENCES

1. Martin WF, Garg S, Zimorski V. Endosymbiotic theories for eukaryote origin. Philos Trans R Soc Lond B Biol Sci. 2015;370(1678):20140330. Epub 2015/09/02. pmid:26323761
2. Quiros PM, Mottis A, Auwerx J. Mitonuclear communication in homeostasis and stress. Nature reviews Molecular cell biology. 2016;17(4):213–26. Epub 2016/03/10. pmid:26956194
3. Kauppila TE, Kauppila JH, Larsson NG. Mammalian Mitochondria and Aging: An Update. Cell Metab. 2017;25(1):57–71. Epub 2017/01/18. pmid:28094012
4. Latorre-Pellicer A, Moreno-Loshuertos R, Lechuga-Vieco AV, Sanchez-Cabo F, Torroja C, Acin-Perez R, et al. Mitochondrial and nuclear DNA matching shapes metabolism and healthy ageing. Nature. 2016;535(7613):561–5. Epub 2016/07/08. pmid:27383793
5. Wallace DC. Genetics: Mitochondrial DNA in evolution and disease. Nature. 2016;535(7613):498–500. Epub 2016/07/08. pmid:27383787
6. Marden JH. Nature's inordinate fondness for metabolic enzymes: why metabolic enzyme loci are so frequently targets of selection. Mol Ecol. 2013;22(23):5743–64. Epub 2013/10/11. pmid:24106889
7. Burton RS, Barreto FS. A disproportionate role for mtDNA in Dobzhansky-Muller incompatibilities? Mol Ecol. 2012;21(20):4942–57. Epub 2012/09/22. pmid:22994153
8. Radzvilavicius AL, Blackstone NW. Conflict and cooperation in eukaryogenesis: implications for the timing of endosymbiosis and the evolution of sex. Journal of the Royal Society, Interface. 2015;12(111):20150584. Epub 2015/10/16. pmid: 26468067
9. Garg SG, Martin WF. Mitochondria, the Cell Cycle, and the Origin of Sex via a Syncytial Eukaryote Common Ancestor. Genome biology and evolution. 2016;8(6):1950–70. Epub 2016/06/28. pmid:27345956
10. Beekman M, Dowling DK, Aanen DK. The costs of being male: are there sex-specific effects of uniparental mitochondrial inheritance? Philos Trans R Soc Lond B Biol Sci. 2014;369(1646):20130440. Epub 2014/05/28. pmid:24864311
11. Barreto FS, Burton RS. Evidence for compensatory evolution of ribosomal proteins in response to rapid divergence of mitochondrial rRNA. Mol Biol Evol. 2013;30(2):310–4. Epub 2012/09/21. pmid:22993236
12. Ellison CK, Burton RS. Interpopulation hybrid breakdown maps to the mitochondrial genome. Evolution. 2008;62(3):631–8. Epub 2007/12/18. pmid:18081717

13. Montooth KL, Meiklejohn CD, Abt DN, Rand DM. Mitochondrial-nuclear epistasis affects fitness within species but does not contribute to fixed incompatibilities between species of Drosophila. Evolution. 2010;64(12):3364–79. Epub 2010/07/14. pmid:20624176
14. Adrion JR, White PS, Montooth KL. The Roles of Compensatory Evolution and Constraint in Aminoacyl tRNA Synthetase Evolution. Mol Biol Evol. 2016;33(1):152–61. Epub 2015/09/30. pmid:26416980
15. Mossman JA, Biancani LM, Zhu CT, Rand DM. Mitonuclear Epistasis for Development Time and Its Modification by Diet in Drosophila. Genetics. 2016;203(1):463–84. Epub 2016/03/12. pmid:26966258
16. Baris TZ, Wagner DN, Dayan DI, Du X, Blier P, Pichaud N, et al. Evolved Genetic and Phenotypic Differences due to Mitochondrial-Nuclear Interactions. PLoS Genet. 2017;13: e1006517.
17. Levene H. Genetic equilibrium when more than one ecological niche is available. Am Nat. 1953;87:331–3.

CHAPTER 17

Epigenetics—Potential for Programming Fish for Aquaculture?

Hooman Moghadam[1], Turid Mørkøre[1] and Nick Robinson[1,2,*]

[1]*Nofima, P.O. Box 5010, 1432 Ås, Norway*

[2]*Flinders University, Sturt Road, Bedford Park, SA 5042, Australia*

ABSTRACT

Epigenetic marks affecting the expression of genes are triggered by environmental stimuli, can persist throughout life or across multiple generations and can affect an individuals phenotype. In recent years there has been a revival of interest about the possible role of epigenetics in affecting complex or quantitative traits. This growing interest is partly driven by the increasing affordability of ultra-high throughput sequencing methods for studying the epigenome. In this review we focus on some of the possible applications of epigenetic knowledge to the improvement of aquaculture. DNA methylation, in which a methyl group is added to the C5 carbon residue of a cytosine by DNA methyltransferase, has been the most widely studied epigenetic mechanism to date, and methods used to obtain and analyse genome-wide DNA methylation data are outlined. The influence of epigenetic processes on the estimation of breeding values and accuracy of genomic selection for genetic improvement of aquatic species is explored. The possibility of tightly controlling nutritional stimuli found to affect epigenetic processes in order to tailor the development of fish for aquaculture is also discussed. Complex experiments will be required in order to gain a better understanding of the role of epigenetics in affecting quantitative traits in fish.

Keywords: epigenetics; aquaculture; DNA methylation; high-throughput sequencing; quantitative genetics; nutritional programming

1. INTRODUCTION

Traditionally we have viewed the influence of a given gene on an animal's phenotype as a functional consequence of the order in which its nucleotide bases are arranged. However, over the past few years, data mainly from model organisms and a few non-model species have demonstrated the importance of additional genomic modifications that can affect an individual's phenotype. These modifications, generally referred to as "epigenetics" (Greek for "upon" genetics), might persist during mitosis and meiosis and can impact transcriptional regulation of a gene across various somatic cell types as well as germ cells. Epigenetic marks such as DNA methylation, histone modifications and chromatin conformational changes can be triggered by environmental stimuli and can persist throughout life or across multiple generations. However, since DNA methylation is the most widely investigated epigenetic mechanism and since the majority of the data and literature are associated with this particular epigenetic type, in this short review, we will mainly focus on the methylation impacts on gene expression and phenotype.

Methylation of DNA is a process in which a methyl group ($-CH_3$) is added to the carbon 5 position of a cytosine residue, mostly within a CpG dinucleotide, although the mechanism has also been reported outside of the CpG context [1]. Studies in mammals suggest that approximately 70%–80% of such dinucleotide sites remain methylated in non-embryonic cells [2]. Interestingly, the distribution of CpG dinucleotides seem to be underrepresented throughout the length of the genome, possibly due to high mutability of the methylated cytosine bases and much higher frequency at which the deamination of a methylated cytosine can create thymine [3,4,5]. This process has therefore been suggested as a major evolutionary force in shaping the vertebrate genome. On the other hand, dense repeats of CpG dinucleotides also exist within the genome, referred to as CpG islands. These islands are an important feature of the vertebrate genome, are usually found at the 5′ ends of almost all housekeeping genes as well as a number of tissue specific genes [6], are mainly associated with the promoter region [7], generally remain unmethylated throughout an individual's life, and as a result are less prone to spontaneous deamination [4,6].

The enzymatic activities of DNA methyltransferases (DNMTs), a large enzyme family which are evolutionary conserved in animals and plants (e.g., [8,9,10]), regulate methylation processes, either through *de novo* marking of a cytosine nucleotide or by the maintenance of existing methylation patterns. Through conformational changes to DNA and histone structures, the level of methylation at the promoter region is usually inversely correlated with the transcriptional activity of a gene [11]. Although CpG islands, particularly the ones that are in close proximity to promoters, are rarely methylated, methylation of these sites can significantly reduce or completely silence the expression of a gene by blocking the binding of transcription factors to DNA and recruiting methyl binding proteins [12,13]. However, the exact mechanisms of such processes are still not very well understood [12,14]. We also lack a sufficient understanding of the spatio-temporal alterations of a gene's methylation status, even in well-studied model species, although such mechanisms have been suggested as one of the key components affecting development, genomic imprinting, sex chromosome inactivation and transposable element silencing [15]. Although DNA methylation is generally considered as a stable genomic modification, particularly for imprinted genes, a number of investigations have revealed that genome-wide DNA demethylation, either through active or passive processes, is also a common feature of some cells during certain developmental stages [16].

Recent findings from different studies have also started to converge on the view that the environment in which an individual experiences during early life stages, even prior to its birth, is an important determinant for its future health and metabolism. Through epigenetic marks, environmental cues can be transmitted to the genomic machinery of a cell and regulate patterns of gene transcription. Such exposures have been shown to mark their effects in a locus-specific manner throughout the length of the genome. For instance it was shown that the increased stress experienced by rat pups with reduced grooming and licking by their mother can affect the expression of the glucocorticoid receptor gene (GR), and that this is associated with differences in DNA methylation at GR promoter, which is thought to affect behavioural and hypothalamic–pituitary axis responses to stress later in life [17]. However, in addition to environmental factors, an individual's genotype also influences the methylation landscape and the transcriptional patterns of associated genes. Recent findings suggest that the genetic background of an organism may affect how it responds to particular environmental stimuli. For example it was shown that the methylome profile of neonatal humans is a function of the health of the uterine environment and its interaction with genetic variation (G × E) [18]. Teh *et al.* (2014) [18] showed that the variability of 75% of differentially methylated regions could best be explained by G × E interactions. Such findings reveal the importance of analysing DNA methylation along with genetic variation and gene expression through integrative approaches. Considering the heritable nature of many epigenetic marks and their impacts on gene expression, in future breeding plans we expect that it will not only become important to understand how an epigenetic mark might influence a phenotype through regulation of a target gene, but also how genetic variation within a gene might interact with a specific methylation site [19].

2. OBTAINING AND ANALYZING DNA METHYLATION DATA

Recent advances in genomic technologies and bioinformatics have made it possible to obtain genome-wide signatures of DNA methylation data, with a low cost per sequenced base, high yield and in a short time span. These types of studies can be done on a large number of samples, and depending on the method, can generate data with resolution in base pairs [20]. The methylation profile of a genome can be studied with microarray or high-throughput sequencing using methods such as DNA methylation enrichment and capture [14,21,22], bisulfite treatment [23] or methylation sensitive digestion [24,25]. Recent comparative studies suggest that all of these methods generate accurate methylation data [26,27], albeit with various resolution and sensitivity.

Below we will briefly introduce some of the most popular of these methods and some of their associated challenges. We also focus on high-throughput sequencing approaches, as for some important aquaculture species a draft sequence genome assembly is already available, and for many others we expect that such resources will become available in the near future. Using a sequencing approach for methylation studies is advantageous because such data can also be used for identifying variations including single-nucleotide polymorphisms (SNPs) and insertion/deletion at the genome level. High Performance Capillary Electrophoresis (e.g., [28]), 5-Methylcytosine Immunolocation (e.g., [29]) and Methylation-sensitive Amplified Polymorphism (e.g., [30]) are other potentially useful methods that could be considered for non-model species where fully sequenced genomes are unavailable.

2.1. Methylated DNA Immunoprecipitation Sequencing (MeDIP-Seq)

MeDIP-Seq relies on the purification and enrichment of fragmented DNA using specific antibodies with high affinity against 5-methylcytosine [14].

DNA fragments are usually about 300–1000 bp in size and one of the main advantages of this method is that the constructed libraries, unlike restriction enzyme based methods, are not biased towards any specific sequence pattern, as the DNA fragmentation process is completely random. The fragments are treated with proteinase K and the purified fractions are then sequenced and aligned to the reference genome. The depth and the pattern of the read coverage can be used as a proxy to infer methylated regions of the genome. To obtain absolute methylation levels, Down *et al.* [21] developed an algorithm known as the Bayesian tool for methylation analysis (BATMAN), allowing more accurate assessment of the sequenced fragments and methylation profile. In human embryonic stem cells, this method was able to identify about 80% of CpGs with maximum resolution of 150 bp [27]. For a more comprehensive coverage, MeDIP-Seq can also be complemented by other approaches, e.g., by treating DNA with methyl-sensitive restriction enzymes such as HpaII, Hin6I and AciI (to obtain unmethylated CpG sites), followed by high-throughput sequencing of DNA libraries [31].

2.2. Methylated DNA Binding Domain Sequencing (MBD-Seq)

The MBD-Seq enrichment method is similar to the immunoprecipitation approach, as sequence fragments are enriched based on their methylation content. Unlike the immunoprecipitation methods, however, the fragment precipitation and enrichment is achieved using methyl binding proteins such as MBD2 [32], which have high affinity against methylcytosine. Compared to MeDIP-Seq, MBD-Seq results in a greater enrichment of DNA fragments with higher CpG density [27]. This discrepancy might in part be a function of salt concentration during elution of methyl-CpG containing DNA fragments [33].

2.3. Whole Genome Shotgun Bisulfite Sequencing (BS-Seq)

This method is regarded as the current gold standard for detecting methylated cytosine bases throughout the genome. It involves shotgun sequencing of DNA following treatment of DNA fragments with sodium bisulfite. Bisulfite treatment of the DNA sequences selectively converts the unmethylated cytosyine residues to uracils while leaving the vast majority of the methylated bases intact. Since DNA polymerase reads uracils as thymines, the unmethylated Cs appear as Ts in the sequence data. The methylation profile of the sequenced loci can then be inferred by aligning these reads to a reference genome and assessing the relative percentage of methylation at individual CpG sites. This is a powerful approach for investigating the methylome as it provides resolution at a single nucleotide level and has long been used for targeted analysis of candidate loci through either direct Sanger sequencing [34] or more recently using pyrosequencing [35]. However, sequencing the entire genome using a BS-Seq approach is still expensive, particularly for large sample sizes. Other limitations of this method include the conversion efficiency which can be incomplete and unstable [36] resulting in false methylation assignments, potentially high levels of DNA degradation [37,38,39] and an inability to distinguish between 5-methylcytosine and 5-hydroxymethylcytosine, a putatively important intermediate in the DNA demethylation pathway [40], in the absence of a complementary Tet-assisted bisulfite sequencing [41].

2.4. Reduced Representation Bisulfite Sequencing (RRBS)

Reduced representation bisulfite sequencing (RRBS) provides an alternative, cost-effective approach to BS-Seq. Still providing bp resolution, this method targets and enriches regions of the genome such as promoters, which have moderate to high levels of CpG [42]. RRBS relies on DNA digestion using a methylation-insensitive restriction enzyme such as MspI (C|CGG) followed by repair to the sticky ends of the double stranded digested DNA and addition of an A base to the ends of both strands. Sequencing adapters are then ligated to DNA fragments and desired sizes (40–250 bp) are separated using gel-electrophoresis, excised and purified. Purified DNA is subjected to bisulfite treatment, PCR amplification, purification and sequencing [42]. Compared to BS-Seq, RRBS's coverage of the genome-wide CpG sites is much lower (about 12%) except for the CpG islands [27]. However, both of these methods allow accurate estimation of genome-wide methylation at the nucleotide level.

2.5. Sequence Assessment and Analysis

Prior to sequence alignment, it is important to assess the quality of the reads and to identify and correct for any irregular pattern that might exist. For instance, there might be contamination from PCR primers, sequencing adapters and barcodes, which need to be either trimmed or filtered out from the sequence data. Low quality bases, which can mostly be found at the ends of the reads, should also be trimmed, or sometimes the entire read should be removed. FastQC [43] can be used to detect contamination, low quality scores or other types of biases. Trimming or filtering can be done using tools such as TRIMMOMATIC [44] or PRINSEQ [45].

In immunoprecipitation-based methods, the assessment of the methylation status of a site is generally deduced based on the mapping patterns of sequenced reads for any particular locus across different replicates and conditions. The mapping can be done using any short-read aligner such as BOWTIE [46] or BWA [47]. Immunoprecipitation-based methods are expected to yield higher mapping efficiency compared to sequence reads generated following bisulfite treatment, as nucleotides are not modified by the library preparation process. The number of mapped reads for any pre-defined regions of the genome such as gene, promoter, genomic bin, etc. is obtained and scored on a binary scale [27]. Although locus-specific coverage can be interpreted as methylation level, the pattern is heavily biased and dependent on the CpG content of that genomic region [21]. This is perhaps one of the main limitations in the enrichment-based methods, as methylation levels cannot be easily determined. To mitigate this bias and to normalize the data, algorithms such as BATMAN [21], MEDME [48] and MEDIPS [49,50] have been developed. The normalization algorithms can also be applied to other enrichment-based methods including MBD-Seq [20]. Regions with different coverage between the treatment groups can then be identified using tools like MEDIPS [50].

On the other hand through BS based methods, the frequencies of Cs and Ts at each position of bisulfite treated sequences can be used to obtain the level of DNA methylation for a particular site. However, nucleotide conversion poses a unique challenge for efficient mapping of such reads to the genome. Overall, there are two common approaches, implemented in different softwares, for handling bisulfite sequence alignment. Aligners such as BSMAP [51] and GSNAP [52] use the concept of a wildcard where they either change the alignment scoring matrix or convert all the genomic Cs to Ys, so they can match all the Cs and Ts in the sequenced data. Alternatively, aligners like Bismark [53] and BS-Seeker [54] operate by replacing all the Cs in the sequence reads as well as both strands of the genomic DNA into Ts. In this way, these tools reduce the complexity of the sequence data and facilitate the mapping process. Due to incomplete bisulfite conversion, sequencing errors or sample heterogeneity, the methylation status of any nucleotide is generally reported as a percentage. Tools such as methylKit [55] can then use these percentages as well as the read coverage to normalized data across samples and to identify differentially methylated bases between different conditions.

By scanning the alignment sequence data, in addition to the methylated regions, one can also identify genomic variation such as single nucleotide changes (SNPs) and insertion/deletions (INDELS). There are a number of established variant calling tools including GATK [56] and FreeBayes [57], which are ideal for processing non-bisulfite reads. For bisulfite treated sequences on the other hand, the correct identification and distinction of C to T SNPs from those of the bisulfite conversion can be made using tools such as Bis-SNP [58].

3. INFLUENCE ON PHENOTYPIC VARIANCE, ESTIMATION OF BREEDING VALUE AND ACCURACY OF GENOMIC SELECTION

This topic has been recently reviewed in relation to sheep and cattle genetic improvement [59], but because of the different biology and breeding systems of aquatic species, epigenetics could be of greater influence for aquaculture species, and could have greater ramifications for the estimation of breeding values and for the implementation of genomic selection. Therefore, we will elaborate on the topics discussed by [59] and add our thoughts in relation to genetic improvement for aquaculture.

The phenotypic variance of any trait is composed of heritable genetic effects and environmental effects. Heritable effects can be influenced by additive and non-additive (dominant, interactive and epigenetic) genetic components [60]. The causative variation affecting phenotype could be due to changes that result in structural modification of particular proteins (amino acid sequence, alternative splicing etc.) or due to the variations in the relative expression of genes affecting the level of production of transcripts or proteins in response to particular external or internal stimuli or at different stages of development and in different tissues, cell types, spatial locations in the body etc. Most traits of interest in aquaculture are quantitative by nature, that is the phenotypes of animals in a population show a typical Gaussian distribution which is thought to arise due to the additive effects of many genes and environmental effects. Non-heritable effects on phenotype are influenced by the environmental factors that can modify cellular processes such as gene expression, transcript splicing, half-life of enzymes and other proteins, etc.

To make genetic improvement we need accurate estimates of the additive genetic effect. This is the main component of the genetic variance causing the similarity observed between close relatives, and is the main component determining the response of the population to selection. The additive genetic effect is estimated from the degree of resemblance between close relatives, particularly the resemblance between sibs, or between parents and their offspring. Because epigenetic effects might also be vertically transmitted from parent to offspring and affect the resemblance between sibs (as is the case for the control of expression of the mouse agouti locus [61] which affects fur colour, weight and likelihood of developing cancer), epigenetics could in some cases affect the estimation of the additive effect component. Because epigenetic programming can be inherited, and because epigenetic programming can be reversed, or wiped, in subsequent generations, the influence of epigenetics could therefore cause inaccuracies when estimating breeding values.

The level of epigenetic influence will depend on population structure, on the size of the epigenetic component relevant to additive genetic component and on the persistence of the epigenetic markings. If the epigenetic effects are stable over many generations then we could consider them as part of the true additive genetic effect.

For the typical situation in aquaculture where there are simple pedigree structures consisting of either large full- or half-sibling families, the potential for epigenetics to bias the estimation of the additive genetic component is greater than for more complex pedigree structures. It is conceivable that for some traits, all offspring will inherit epigenetic marks from their parents, so that some offspring groups that appear to have inherited superior additive genetic variation (additive genetic variation causing permanent and accumulative effects on the phenotype of subsequent generations), might actually have inherited temporary epigenetic marks that have a strong influence on the trait. The Gaussian type distribution of phenotypes for a trait in a population might also be caused or influenced by epigenetic effects. This is because the epigenetic effects (e.g., degree of DNA methylation and consequent effect on gene expression) are quantitative in nature [19]. Unlike a single gene mutation effect on a trait, the degree of methylation at a single locus could potentially have a quantitative effect on a trait and lead to a Gaussian distribution of trait phenotypes.

Epigenetics could play an important role in influencing disease resistance traits. A consideration of epigenetic effects might help explain the age-dependant development of some diseases (those not well explained by accumulated mutation) and the mechanisms by which the environment might modulate genetic predisposition to disease [19]. When the mother encounters disease, she builds immunity and transmits immunity to offspring so that they are better able to resist the disease. The environment encountered during early development is also likely to affect latter predisposition of adult fish to disease. These influences may be less important for some aquatic species where the adaptive immune system is less well developed. The diet of the mother could also play an important role, or nutrition of the developing fish could be important. In the case of aquatic species, the window for a maternal dietary influence is short (*i.e.*, during gamete development in the ovary or testis), but epigenetic marking could be occurring as the gamete matures or might be inherited by the gamete at the time it is formed, and this might affect predisposition to the disease. Bacteria are known to directly alter epigenetic marks and to alter the machinery for epigenetic regulation, and this could also negatively or positively affect the ability of the host, or its offspring, to resist disease [62].

It has been suggested that epigenetic modification could directly influence expression and genetic disease predisposition or that the epigenetic marking may interact with specific DNA-sequence variants so that it affects the penetrance of disease-causing variants [19]. This type of interaction has been demonstrated in the case of the "masking" or "unmasking" of a Hsp90 gene mutation in *Drosophila melanogaster* [63], and is also the way that imprinting affects the expression of genes originating from one parent in mammals [64]. If such mechanisms were involved in controlling disease resistance then the association between the disease phenotype and QTL controlling the trait would be "buffered" and dependent on the state of the epigenetic marking. Therefore, for some complex traits, measurement of the epigenome (e.g., using DNA methylation sequencing) could be especially informative.

Estimation of the Epigenomic Variance Component and Calculation of GenEomic EBVs

In cases where there is a significant component explained by epigenetic effects and the pedigree structure affects the estimation of true additive genetic effects (as is possible for fish pedigrees containing many full-sibling offspring), correction of genetic solutions accounting for epigenetic effects would result in a more accurate prediction of breeding value and therefore result in more rapid genetic gain. In these cases it will be necessary to estimate the "true" additive genetic component (V_{Atrue}), where,

$$V_{Atrue} = V_{Auncorrected} - (V_{EpiMeth} + V_{EpisRNA} \;.....)$$

where $V_{Auncorrected}$ is the uncorrected additive genetic component estimate, $V_{EpiMeth}$ is the epigenetic component caused by methylation markings and $V_{EpisRNA}$ is the epigenetic component caused by small RNA molecules etc.

Future genomic selection could estimate and account for the effects of epigenetics by using the solutions for V_{Atrue} at every marker position across the genome (we will refer to such selection as "GenEomic Selection" here). The most efficient option for finding genome-wide epigenetic markings associated with phenotype (which ultimately would allow estimation of the epigenetic variance component) would be to utilise information from the same training families that are used for estimating genome-wide marker associations for the purpose of within-family genomic selection. Technologies for measuring epigenetic markings throughout the genome were discussed above. As well as following the inheritance of allelic variation (used for the estimation of the additive genetic component), experiments should be designed so that it is possible to search for differential methylation or other epigenetic marks between families and within families at loci spread over the whole genome. This could allow discrimination between the two main random effect components on the phenotype, one due to the additive genetic effects, and one due to the epigenetic effect components (*i.e.*, $y = \mu + V_A + V_{EpiMeth} + V_{EpiSrna}$ etc. $+V_R$). To discriminate between these effects it will be important to use genotype information from across the genome (genomic relationships) to estimate the actual genetic relatedness between full-sibs (*i.e.*, normally the relationship between full-sibs is assumed to be 0.5, which is the theoretical average relationship). Genes contributing to the additive genetic variance component should be inherited in a Mendelian fashion, and therefore, those individuals with a higher genomic relationship should have a more similar additive genetic component, whereas this will not hold for the epigenetic component.

The epigenetic component itself might consist of paternal and/or maternally inherited effects, sex specific effects on offspring and/or interactive effects (where the epigenome interacts with specific gene mutations). In some cases, the allele inherited from one parent might be silenced (imprinted) so that only the allele from the other parent is expressed. In order to be able to distinguish between these types of epigenetic effects the sex of all animals would need to be recorded and included as a fixed effect in the model for analysis and the changes to the epigenome would need to be mapped relative to QTL found to affect the trait.

It may be that the influence of epigenetics is small and that V_{Atrue} is almost the same as $V_{Auncorrected}$, or there may be a significant epigenetic component for some traits. The potential for epigenetics to influence quantitative traits, has been recently demonstrated in lines of Arabidopsis where several experimentally-induced DNA methylation changes have been shown to effectively act as epigenetic quantitative trait loci affecting flowering time and primary root length, accounting for 60%–90% of the heritability for the two traits [65]. Many of these inducible DNA-methylated regions were also found to be variable in natural populations of Arabidopsis. For this review we were unable to find any attempt to measure the epigenetic variance component for traits in an aquaculture species, probably because techniques such as whole genome methylation sequencing are new and expensive to apply across many animals. We simply do not have good estimates of the extent of the epigenetic variance component for traits in aquaculture species and the importance of this component is therefore uncertain.

On-the-other-hand it would be much easier to identify epigenetic effects and to quantify this variance component, if we could reduce the noise caused by additive genetic effects [59]. Reduction of the additive genetic effects might be possible in fish by creating inbred experimental lines [66], or completely homozygous double haploid fish [67], so long as these fish can be bred with other inbred or completely homozygous double haploid fish to produce a new generation for tracing the inherited epigenetic effects. Because it is often easier and faster to create these inbred or double haploid lines for fish than it is for livestock or other experimental model organisms, fish could be useful model species for studying epigenetics and estimating the epigenetic variance components affecting certain traits. In addition, it could be possible to create epigenetic recombinant inbred lines of fish, and to use these animals to try to map epigenetic-based quantitative trait loci affecting particular traits, in much the same way as has been done to for Arabidopsis [65].

4. NUTRIGENOMICS—PROGRAMMING THE FISH THROUGH NUTRITION?

The more we understand about epigenetics and how it influences phenotype, the more accurately we may be able to program the development of the fish by tightly controlling various components of the environment (including the diet of the fish) at different stages of development. Nutrients and bioactive food components can influence epigenetic phenomena either by directly inhibiting enzymes that catalyze DNA methylation or histone modifications, or by altering the availability of substrates necessary for those enzymatic reactions [68], and this is known as nutritional programming. There have been few studies demonstrating nutritional programming in fish [69,70], but if this phenomena does occur in fish it might be possible to manipulate programming to improve performance in the aquaculture environment. As development occurs in the water column this would allow very simple and controlled manipulation of groups of larvae or fingerlings for programming. Nutrients in the diet or other environmental influences could have a direct effect fuelling development, or they could be changing the epigenome in such a way that particular genes are more highly or lowly expressed after exposure, with some changes to the epigenome being inherited by subsequent sons and/or daughters through paternal and/or maternal lines.

Production efficiency and fish health have large effects on the profitability of aquaculture, and it would therefore be highly desirable to be able to control the development of these traits using nutrigenomic programming. Variation in fillet quality is also important, in particular soft texture and deviating appearance (colour and deformities) are quality problems that cause negative market responses. Fish with shortening of the body due to fusion of the vertebrae cause problems during machine filleting and are considered as second grade fish by the processing industry. Additionally, fish deformities have a detrimental effect on the consumers' image of aquaculture and, therefore, also on the market value of the juvenile fish [71]. Research into the factors influencing performance, health and flesh quality have identified important contributing factors, but may have overlooked major root causes of the problems, such as the possibility of epigenetic effects controlling cellular differentiation and development.

DNA methylation and other epigenetic controls of gene expression are known to be important in affecting cellular differentiation during development [72]. There are a number of nutritional stimuli that are known to affect the differentiation of fish cells during development and possibly do so by changing the epigenetic markings in particular cells. A lack of vitamins (A, D, C) in sea bass has been shown to disrupt the temporal sequence and coordination of growth factor gene expression which affects the differentiation of osteoblasts so that some are converted into adipocytes resulting in deformities [73]. The conversion of osteoblasts into adipocytes can also occur when dietary highly unsaturated fatty acid (HUFA) levels are too high during the early stages of the development of marine fish [71]. Salmon pre-adipocytes seem to possess the ability to differentiate into cells of the immune lineage (most probably macrophage-like cells) depending on environmental conditions [74].However, epigenetic cues affecting cellular differentiation during the development of fish have not yet been investigated.

Growth of skeletal muscle in fish is a plastic process that involves a combination of enlargement of already existing muscle fibres (hypertrophy) and the recruitment of new fibres (hyperplasia). The balance between these mechanisms determines the fiber density, which in turn affects muscle firmness and robustness to handling and processing [75,76]. The hypertrophy–hyperplasia balance depends on strain genetic differences and extrinsic factors such as early nutritional history [77]. The cellular development of fish muscle in general appears to be strongly influenced by environmental conditions faced by the egg, larvae and fry during development [78]. For instance, temperature during embryonic stages is known to affect muscle fibre recruitment in zebra fish (*Danio rerio*) and Atlantic cod (*Gadus morhua*) [79,80] and is also known to affect the development of deformities, deviant heart development and the expression of stress related genes in Atlantic salmon [81,82,83]. Although fry nutrition is assumed to be important, there are relatively few studies following the effect of fry nutrition on meat quality or health at harvest [84]. On the other hand, most research on optimising fillet quality has focused on the impact of dietary composition and feeding practices for larger fish. The expression of a number of genes in the adult salmon fillet has been correlated with fillet softness [85], but we know little about the expression of these genes during early development stages, what factors affect their expression and the precise role, if any, that these genes may play in determining fillet quality.

There are emerging findings that suggest that fatty acids, in particular polyunsaturated fatty acids (*n*-3 HUFA's), can modify the epigenome [86]. The *n*-3 HUFA's are important as precursors of eicosanoids, which are particularly essential for the development and growth of fish [87], for providing metabolic energy to the fish in the form of ATP [88] and for the formation of membrane structure [87]. Changes to the dietary lipid source of female salmon during gonadal development affect egg fatty acid composition, and as a consequence, can result in abnormal embryonic development [89]. Supplementation of broodstock diets with higher proportions of fish oil has been shown to influence survival rate of 3 day post-hatching larvae, egg diameter, length of fry, oil globule diameter of hatchlings, egg protein content, lipid content and the morphology and metabolism of adipocytes in sea bream [90,91]. Research in humans has shown that perinatal manipulation of alpha-linolenic acid uptake induces epigenetic changes in the liver of offspring [92] and that some fatty acids act as hormones and control the activity of transcription factors [93]. With the production of salmon for harvesting at 3–5 kg, fish have been shown to develop typical life-style related symptoms, such as arteriosclerosis, abnormal fat deposition in liver and heart which results in impaired organ function and various circulation problems [94,95]. Health problems such as these have been recently documented to affect flesh softness and ability to cope with stress of handling [96,97]. Early life exposures can, through epigenetic programming, determine the response to nutritional signals in later life [98], and high-fat exposure during early life phases could therefore program the fish and lead to a higher incidence of obesity, and related symptoms, in response to adult diets and decreased resilience to pathogen exposure and other stresses.

Therefore, the nutrition and health of the parents, egg, larvae and/or fry, and consequent programming prior to conception in the parents or during early development of the individual itself, could influence the type, number and distribution of osteoblast, adipocyte and myocyte precursor cells, gene expression and subsequent development of muscle and meat quality [77], but this needs to be further explored.

5. KEY CHALLENGES AND KNOWLEDGE GAPS

Understanding the epigenome and the changes affecting particular phenotypes will involve complex experiments using family material, like the experiments needed for finding quantitative trait loci. The cost of measuring epigenetic changes across the genome is becoming cheaper, but there are additional experimental costs and considerations to be made when studying the epigenome. For instance, at what stage(s) of development should the epigenome be assessed? To see if the effects are inherited it would be necessary to study the epigenome of mature parents and offspring at an early stage of development (egg or larvae). In some cases it may be necessary to follow epigenetic markings as the fish develops which would also necessitate repeated testing of fish.

There is currently very little knowledge about the genes that are programmed by nutritional treatments in early life stages (e.g., fry) and of epigenetic processes in early life stages affecting robustness and meat quality. Knowledge of these processes could drive changes in the formulation of new diets, could result in greater selection accuracy for the genetic improvement of important traits and could result in improved robustness and in greater consistency and quality at harvest. If the programming persists until maturation, it may result in heritable changes, and could affect the early stage development of subsequent generations. Histone methylation may be an epigenetic regulator in some molluscs [99]. We therefore need to determine whether programming persists until harvest and how programming is regulated in fish and shellfish.

We need to be able to discriminate between true additive effects and epigenetic effects to give greater accuracy for selective breeding to make genetic improvement. We also need to develop biomarkers for early life stages that can enable us to predict negative consequences (phenotypes) that only become apparent later in life. Epigenetic marks in the genome could be useful biomarkers for this purpose. Regular biomarker testing could provide an alert about variation in nutritional or environmental factors affecting fry and subsequent meat quality in the adult fish. Active surveillance of this type is commonly used to warn about the spread of agricultural diseases. If nutritional driven epigenetic effects are important, this type of surveillance could also help ensure consistent quality of aquaculture products from year to year, which in turn would have large effects on the industry in terms of reduced wastage and reduced production of low quality grade product, improved confidence of consumers and export markets for salmon products and higher profitability for producers, processors and exporters.

ACKNOWLEDGEMENTS

This review was undertaken using internal project funding from Nofima (project 10878).

REFERENCES

1. Yan, J.; Zierath, J.R.; Barres, R. Evidence for non-CpG methylation in mammals. *Exp. Cell Res.* **2011**, *317*, 2555–2561.
2. Tucker, K.L. Methylated cytosine and the brain: A new base for neuroscience. *Neuron* **2001**, *30*, 649–652.
3. Weber, M.; Hellmann, I.; Stadler, M.B.; Ramos, L.; Paabo, S.; Rebhan, M.; Schubeler, D. Distribution, silencing potential and evolutionary impact of promoter DNA methylation in the human genome. *Nat. Genet.* **2007**, *39*, 457–466.
4. Bird, A.P. CpG-rich islands and the function of DNA methylation. *Nature* **1986**, *321*, 209–213.
5. Krueger, F.; Kreck, B.; Franke, A.; Andrews, S.R. DNA methylome analysis using short bisulfite sequencing data. *Nat. Methods* **2012**, *9*, 145–151.
6. Deaton, A.M.; Bird, A. CpG islands and the regulation of transcription. *Genes Dev.* **2011**, *25*, 1010–1022.
7. Antequera, F.; Bird, A. Number of CpG islands and genes in human and mouse. *Proc. Natl. Acad. Sci. U.S.A.* **1993**, *90*, 11995–11999.
8. Monk, M. Changes in DNA methylation during mouse embryonic development in relation to X-chromosome activity and imprinting. *Philos. Trans. R. Soc. Lond. Biol. Sci.* **1990**, *326*, 299–312.
9. Bestor, T.; Laudano, A.; Mattaliano, R.; Ingram, V. Cloning and sequencing of a cDNA encoding DNA methyltransferase of mouse cells. The carboxyl-terminal domain of the mammalian enzymes is related to bacterial restriction methyltransferases. *J. Mol. Biol.* **1988**, *203*, 971–983.
10. Law, J.A.; Jacobsen, S.E. Establishing, maintaining and modifying DNA methylation patterns in plants and animals. *Nat. Rev. Genet.* **2010**, *11*, 204–220.
11. Robertson, K.D.; Wolffe, A.P. DNA methylation in health and disease. *Nat. Rev. Genet.* **2000**, *1*, 11–19.
12. Moore, L.D.; Le, T.; Fan, G. DNA methylation and its basic function. *Neuropsychopharmocology* **2013**, *38*, 23–38.
13. Jones, P.A. Functions of DNA methylation: islands, start sites, gene bodies and beyond. *Nat. Rev. Genet.* **2012**, *13*, 484–492.

14. Weber, M.; Davies, J.J.; Wittig, D.; Oakeley, E.J.; Haase, M.; Lam, W.L.; Schubeler, D. Chromosome-wide and promoter-specific analyses identify sites of differential DNA methylation in normal and transformed human cells. *Nat. Genet.* **2005**, *37*, 853–862.
15. Zentner, G.E.; Henikoff, S. High-resolution digital profiling of the epigenome. *Nat. Rev. Genet.* **2014**, *15*, 814–827.
16. Wu, S.C.; Zhang, Y. Active DNA demethylation: many roads lead to Rome. *Nat. Rev. Mol. Cell Biol.* **2010**, *11*, 607–620.
17. Weaver, I.C.; Cervoni, N.; Champagne, F.A.; D'Alessio, A.C.; Sharma, S.; Seckl, J.R.; Dymov, S.; Szyf, M.; Meaney, M.J. Epigenetic programming by maternal behavior. *Nat. Neurosci.* **2004**, *7*, 847–854.
18. Teh, A.L.; Pan, H.; Chen, L.; Ong, M.L.; Dogra, S.; Wong, J.; MacIsaac, J.L.; Mah, S.M.; McEwen, L.M.; Saw, S.M.; *et al.* The effect of genotype and in utero environment on interindividual variation in neonate DNA methylomes. *Genome Res.* **2014**, *24*, 1064–1074.
19. Bjornsson, H.T.; Fallin, D.M.; Feinberg, A.P. An integrated epigenetic and genetic approach to common human disease. *Trends Genet.* **2004**, *20*, 350–358.
20. Bock, C. Analysing and interpreting DNA methylation data. *Nat. Rev. Genet.* **2012**, *13*, 705–719.
21. Down, T.A.; Rakyan, V.K.; Turner, D.J.; Flicek, P.; Li, H.; Kulesha, E.; Graf, S.; Johnson, N.; Herrero, J.; Tomazou, E.M.; *et al.* A Bayesian deconvolution strategy for immunoprecipitation-based DNA methylome analysis. *Nat. Biotechnol.* **2008**, *26*, 779–785.
22. Aberg, K.A.; McClay, J.L.; Nerella, S.; Xie, L.Y.; Clark, S.L.; Hudson, A.D.; Bukszar, J.; Adkins, D.; Hultman, C.M.; Sullivan, P.F.; *et al.* MBD-seq as a cost-effective approach for methylome-wide association studies: demonstration in 1500 case-control samples. *Epigenomics* **2012**, *4*, 605–621.
23. Meissner, A.; Mikkelsen, T.S.; Gu, H.; Wernig, M.; Hanna, J.; Sivachenko, A.; Zhang, X.; Bernstein, B.E.; Nusbaum, C.; Jaffe, D.B.; *et al.* Genome-scale DNA methylation maps of pluripotent and differentiated cells. *Nature* **2008**, *454*, 766–770.
24. Irizarry, R.A.; Ladd-Acosta, C.; Carvalho, B.; Wu, H.; Brandenburg, S.A.; Jeddeloh, J.A.; Wen, B.; Feinberg, A.P. Comprehensive high-throughput arrays for relative methylation (CHARM). *Genome Res.* **2008**, *18*, 780–790.
25. Brunner, A.L.; Johnson, D.S.; Kim, S.W.; Valouev, A.; Reddy, T.E.; Neff, N.F.; Anton, E.; Medina, C.; Nguyen, L.; Chiao, E.; *et al.* Distinct DNA methylation patterns characterize differentiated human embryonic stem cells and developing human fetal liver. *Genome Res.* **2009**, *19*, 1044–1056.
26. Bock, C.; Tomazou, E.M.; Brinkman, A.B.; Muller, F.; Simmer, F.; Gu, H.; Jager, N.; Gnirke, A.; Stunnenberg, H.G.; Meissner, A. Quantitative comparison of genome-wide DNA methylation mapping technologies. *Nat. Biotechnol.***2010**, *28*, 1106–1114.
27. Harris, R.A.; Wang, T.; Coarfa, C.; Nagarajan, R.P.; Hong, C.; Downey, S.L.; Johnson, B.E.; Fouse, S.D.; Delaney, A.; Zhao, Y.; *et al.* Comparison of sequencing-based methods to profile DNA methylation and identification of monoallelic epigenetic modifications. *Nat. Biotechnol.* **2010**, *28*, 1097–1105.
28. Hasbun, R.; Valledor, L.; Rodriguez, J.L.; Santamaria, E.; Rios, D.; Sanchez, M.; Canal, M.J.; Rodriguez, R. HPCE quantification of 5-methyl-2′-deoxycytidine in genomic DNA: Methodological optimization for chestnut and other woody species. *Plant Physiol. Biochem.* **2008**, *46*, 815–822.
29. Covelo-Soto, L.; Moran, P.; Pasantes, J.J.; Perez-Garcia, C. Cytogenetic evidences of genome rearrangement and differential epigenetic chromatin modification in the sea lamprey (*Petromyzon marinus*). *Genetica* **2014**, *142*, 545–554.
30. Zhao, Y.; Chen, M.Y.; Storey, K.B.; Sun, L.N.; Yang, H.S. DNA methylation levels analysis in four tissues of sea cucumber *Apostichopus japonicus* based on fluorescence-labeled methylation-sensitive amplified polymorphism (F-MSAP)

during aestivation. *Comp. Biochem. Physiol B-Biochem. Mol. Biol.* **2015**, *181*, 26–32.

31. Maunakea, A.K.; Nagarajan, R.P.; Bilenky, M.; Ballinger, T.J.; D'Souza, C.; Fouse, S.D.; Johnson, B.E.; Hong, C.; Nielsen, C.; Zhao, Y.; *et al.* Conserved role of intragenic DNA methylation in regulating alternative promoters. *Nature* **2010**, *466*, 253–257.
32. Hendrich, B.; Bird, A. Identification and characterization of a family of mammalian methyl-CpG binding proteins. *Mol. Cell. Biol.* **1998**, *18*, 6538–6547.
33. Li, N.; Ye, M.; Li, Y.; Yan, Z.; Butcher, L.M.; Sun, J.; Han, X.; Chen, Q.; Zhang, X.; Wang, J. Whole genome DNA methylation analysis based on high throughput sequencing technology. *Methods* **2010**, *52*, 203–212.
34. Frommer, M.; McDonald, L.E.; Millar, D.S.; Collis, C.M.; Watt, F.; Grigg, G.W.; Molloy, P.L.; Paul, C.L. A genomic sequencing protocol that yields a positive display of 5-methylcytosine residues in individual DNA strands. *Proc. Natl. Acad. Sci. U.S.A.* **1992**, *89*, 1827–1831.
35. Colella, S.; Shen, L.; Baggerly, K.A.; Issa, J.P.; Krahe, R. Sensitive and quantitative universal Pyrosequencing methylation analysis of CpG sites. *Bio Techniques* **2003**, *35*, 146–150.
36. Pomraning, K.R.; Smith, K.M.; Freitag, M. Genome-wide high throughput analysis of DNA methylation in eukaryotes. *Methods* **2009**, *47*, 142–150.
37. Olek, A.; Oswald, J.; Walter, J. A modified and improved method for bisulphite based cytosine methylation analysis. *Nucleic Acids Res.* **1996**, *24*, 5064–5066.
38. Ji, L.; Sasaki, T.; Sun, X.; Ma, P.; Lewis, Z.A.; Schmitz, R.J. Methylated DNA is over-represented in whole-genome bisulpite sequencing data. *Front. Genet.* **2014**, *5*, 341.
39. Ehrich, M.; Zoll, S.; Sur, S.; van den Boom, D. A new method for accurate assessment of DNA quality after bisulfite treatment. *Nucleic Acids Res.* **2014**, *42*, 12331–12331.
40. He, Y.F.; Li, B.Z.; Li, Z.; Liu, P.; Wang, Y.; Tang, Q.; Ding, J.; Jia, Y.; Chen, Z.; Li, L.; *et al.* Tet-mediated formation of 5-carboxylcytosine and its excision by TDG in mammalian DNA. *Science* **2011**, *333*, 1303–1307.
41. Wen, L.; Li, X.; Yan, L.; Tan, Y.; Li, R.; Zhao, Y.; Wang, Y.; Xie, J.; Zhang, Y.; Song, C.; *et al.* Whole-genome analysis of 5-hydroxymethylcytosine and 5-methylcytosine at base resolution in the human brain. *Genome Biol.* **2014**, *15*, R49.
42. Gu, H.; Bock, C.; Mikkelsen, T.S.; Jager, N.; Smith, Z.D.; Tomazou, E.; Gnirke, A.; Lander, E.S.; Meissner, A. Genome-scale DNA methylation mapping of clinical samples at single-nucleotide resolution. *Nat. Methods* **2010**, *7*, 133–136.
43. Babraham Bioinformatics. Available online: http://www. bioinformatics. babraham.ac.uk/projects/fastqc/ (accessed on 22 April 2015).
44. Bolger, A.M.; Lohse, M.; Usadel, B. Trimmomatic: A flexible trimmer for Illumina sequence data. *Bioinformatics* **2014**, *30*, 2114–2120.
45. Schmieder, R.; Edwards, R. Quality control and preprocessing of metagenomic datasets. *Bioinformatics* **2011**, *27*, 863–864.
46. Langmead, B.; Trapnell, C.; Pop, M.; Salzberg, S.L. Ultrafast and memory-efficient alignment of short DNA sequences to the human genome. *Genome Biol.* **2009**, *10*, R25.
47. Li, H.; Durbin, R. Fast and accurate long-read alignment with Burrows-Wheeler transform. *Bioinformatics* **2010**, *26*, 589–595.
48. Pelizzola, M.; Koga, Y.; Urban, A.E.; Krauthammer, M.; Weissman, S.; Halaban, R.; Molinaro, A.M. MEDME: An experimental and analytical methodology for the estimation of DNA methylation levels based on microarray derived MeDIP-enrichment. *Genome Res.* **2008**, *18*, 1652–1659.
49. Chavez, L.; Jozefczuk, J.; Grimm, C.; Dietrich, J.; Timmermann, B.; Lehrach, H.; Herwig, R.; Adjaye, J. Computational analysis of genome-wide DNA methylation

during the differentiation of human embryonic stem cells along the endodermal lineage. *Genome Res.* **2010**, *20*, 1441–1450.
50. Lienhard, M.; Grimm, C.; Morkel, M.; Herwig, R.; Chavez, L. MEDIPS: genome-wide differential coverage analysis of sequencing data derived from DNA enrichment experiments. *Bioinformatics* **2014**, *30*, 284–286.
51. Xi, Y.; Li, W. BSMAP: Whole genome bisulfite sequence MAPping program. *BMC Bioinform.* **2009**, *10*, 232.
52. Wu, T.D.; Nacu, S. Fast and SNP-tolerant detection of complex variants and splicing in short reads. *Bioinformatics***2010**, *26*, 873–881.
53. Krueger, F.; Andrews, S.R. Bismark: a flexible aligner and methylation caller for Bisulfite-Seq applications. *Bioinformatics* **2011**, *27*, 1571–1572.
54. Chen, P.Y.; Cokus, S.J.; Pellegrini, M. BS Seeker: precise mapping for bisulfite sequencing. *BMC Bioinform.* **2010**, *11*, 203.
55. Akalin, A.; Kormaksson, M.; Li, S.; Garrett-Bakelman, F.E.; Figueroa, M.E.; Melnick, A.; Mason, C.E. methylKit: A comprehensive R package for the analysis of genome-wide DNA methylation profiles. *Genome Biol.* **2012**, *13*, R87.
56. McKenna, A.; Hanna, M.; Banks, E.; Sivachenko, A.; Cibulskis, K.; Kernytsky, A.; Garimella, K.; Altshuler, D.; Gabriel, S.; Daly, M.; *et al.* The Genome Analysis Toolkit: a MapReduce framework for analyzing next-generation DNA sequencing data. *Genome Res.* **2010**, *20*, 1297–1303.
57. Garrison, E.; Marth, G. Haplotype-based variant detection from short-read sequencing. *arXiv* **2012**, *1207.3907*, 1–9.
58. Liu, Y.; Siegmund, K.D.; Laird, P.W.; Berman, B.P. Bis-SNP: combined DNA methylation and SNP calling for Bisulfite-seq data. *Genome Biol.* **2012**, *13*, R61.
59. Goddard, M.E.; Whitelaw, E. The use of epigenetic phenomena for the improvement of sheep and cattle. *Front. Genet.***2014**, *5*.
60. Falconer, D.S.; Mackay, T.F.C. *An Introduction to Quantitative Genetics*, 4th ed.; Addison Wesley Longman Limited: Edinburgh Gate, UK, 1996; p. 464.
61. Morgan, H.D.; Sutherland, H.G.; Martin, D.I.; Whitelaw, E. Epigenetic inheritance at the agouti locus in the mouse. *Nat. Genet.* **1999**, *23*, 314–318.
62. Bierne, H.; Hamon, M.; Cossart, P. Epigenetics and Bacterial Infections. *Cold Spring Harb. Perspect. Med.* **2012**, *2*, a010272.
63. Rutherford, S.L.; Lindquist, S. Hsp90 as a capacitor for morphological evolution. *Nature* **1998**, *396*, 336–342.
64. Barlow, D.P.; Bartolomei, M.S. Genomic Imprinting in Mammals. *Cold Spring Harb. Perspect. Biol.* **2014**, *6*, a018382.
65. Cortijo, S.; Wardenaar, R.; Colome-Tatche, M.; Gilly, A.; Etcheverry, M.; Labadie, K.; Caillieux, E.; Hospital, F.; Aury, J.M.; Wincker, P.; *et al.* Mapping the Epigenetic Basis of Complex Traits. *Science* **2014**, *343*, 1145–1148.
66. Whitlock, M.C.; Fowler, K. The Changes in Genetic and Environmental Variance With Inbreeding in *Drosophila melanogaster*. *Genetics* **1999**, *152*, 345–353.
67. Streisinger, G.; Walker, C.; Dower, N.; Knauber, D.; Singer, F. Production of clones of homozygous diploid zebra fish (*Brachydanio rerio*). *Nature* **1981**, *291*, 293–296.
68. Choi, S.-W.; Friso, S. Epigenetics: A New Bridge between Nutrition and Health. *Adv. Nutr.* **2010**, *1*, 8–16.
69. Geurden, I.; Mennigen, J.; Plagnes-Juan, E.; Veron, V.; Cerezo, T.; Mazurais, D.; Zambonino-Infante, J.; Gatesoupe, J.; Skiba-Cassy, S.; Panserat, S. High or low dietary carbohydrate: protein ratios during first-feeding affect glucose metabolism and intestinal microbiota in juvenile rainbow trout. *J. Exp. Biol.* **2014**, *217*, 3396–3406.
70. Fang, L.; Liang, X.F.; Zhou, Y.; Guo, X.Z.; He, Y.; Yi, T.L.; Liu, L.W.; Yuan, X.C.; Tao, Y.X. Programming effects of high-carbohydrate feeding of larvae on adult glucose metabolism in zebrafish, *Danio rerio*. *Br. J. Nutr.* **2014**, *111*, 808–818.
71. Zambonino-Infante, J.; Koumoundouros, G.; Tandler, A. The influence of nutrition at the later larval stage in marine fish. In *Control of Malformations in Fish*

Aquaculture: Science and Practice, Finefish; Bæverfjord, G., Helland, S., Hough, C., Eds.; Federation of European Aquaculture Producers (FEAP): Liege, Belgium, 2009; pp. 85–92.
72. Reik, W. Stability and flexibility of epigenetic gene regulation in mammalian development. *Nature* **2007**, *447*, 425–432.
73. Mazurais, D.; Darias, M.J.; Gouillou, M.F.; Le Gall, M.M.; Huelvan, C.; Desbruyeres, E.; Quazuguel, P.; Cahu, C.; Zambonino-Infante, J.-L. Dietary vitamin mix levels influence the ossification process oin European sea bass (*Dicentrarchus labrax*) larvae. *Am. J. Physiol. Regul. Comp. Physiol.* **2008**, *294*, R520–R527.
74. Škugor, S.; Škugor, A.; Todorčević, M.; Torgersen, J.; Ruyter, B.; Krasnov, A. Exposure to lipopolysaccharide induces immune genes in cultured preadipocytes of Atlantic salmon. *Fish Shellfish Immunol.* **2010**, *29*, 817–824.
75. Johnston, I.A.; Alderson, R.; Sandham, C.; Dingwall, A.; Mitchell, D.; Selkirk, C.; Nickell, D.; Baker, R.; Robertson, B.; Whyte, D.; *et al.* Muscle fibre density in relation to the colour and texture of smoked Atlantic salmon (*Salmo salar* L.). *Aquaculture* **2000**, *189*, 335–349.
76. Mørkøre, T.; Ruhonen, K.; Kiessling, A. Variation in texture of farmed Atlantic salmon (*Salmo salar* L.). Relevance of muscle fibre cross-sectional area. *J. Texture Stud.* **2009**, *40*, 1–15.
77. Bjørnevik, M.; Beattie, C.; Hansen, T.; Kiessling, A. Muscle growth in juvenile Atlantic salmon as influenced by temperature in the egg and yolk sac stages and diet protein level. *J. Fish Biol.* **2003**, *62*, 1–17.
78. Brooks, S.; Tyler, C.R.; Sumpter, J.P. Quality in fish: what makes a good egg? *Rev. Fish Biol. Fish.* **1997**, *7*, 387–416.
79. Johnston, I.A.; Andersen, O. Number of muscle fibres in adult Atlantic cod varies with temperature during embryonic development and pantophysin (PanI) genotype. *Aquat. Biol.* **2008**, *4*, 167–173.
80. Johnston, I.A.; Lee, H.T.; Macqueen, D.J.; Paranthaman, K.; Kawashima, C.; Anwar, A.; Kinghorn, J.R.; Dalmay, T. Embryonic temperature affects muscle fibre recruitment in adult zebrafish: Genome-wide changes in gene and microRNA expression associated with the transition from hyperplastic to hypertrophic growth phenotypes. *J. Exp. Biol.* **2009**, *212*, 1781–1793.
81. Takle, H.; Bæverfjord, G.; Helland, S.; Kjorsvik, E.; Andersen, Ø. Hyperthermia induced atrial natriuretic peptide expression and deviant heart development in Atlantic salmon *Salmo salar* embryos. *Gen. Comp. Endocrinol.* **2006**, *147*, 118–125.
82. Takle, H.; Bæverfjord, G.; Andersen, Ø. Identification of stress sensitive genes in hyperthermic Atlantic salmon (*Salmo salar*) embryos by RAP-PCR. *Fish Physiol. Biochem.* **2004**, *30*, 275–281.
83. Takle, H.; Bæverfjord, G.; Lunde, M.; Kolstad, K.; Andersen, Ø. The effect of heat and cold exposure on HSP70 expression and development of deformities during embryogenesis of Atlantic salmon (*Salmo salar*). *Aquaculture* **2005**, *249*, 515–524.
84. Bæverfjord, G.; Hjelde, K.; Helland, S.; Refstie, S. The effects of minerals on juvenile Atlantic salmon. In *Control of Malformations in Fish Aquaculture: Science and Practice, Finefish*; Bæverfjord, G., Helland, S., Hough, C., Eds.; Federation of European Aquaculture Producers (FEAP): Liege, Belgium, 2009; pp. 67–71.
85. Larsson, T.; Mørkøre, T.; Kolstad, K.; Østbye, T.-K.; Afanasyev, S.; Krasnov, A. Gene Expression Profiling of Soft and Firm Atlantic Salmon Fillet. *PLoS ONE* **2012**, *7*, e39219.
86. Burdge, G.C.; Lillycrop, K.A. Fatty acids and epigenetics. *Curr. Opin. Clin. Nutr. Metab. Care* **2014**, *17*, 156–161.
87. Sargent, J.R.; Bell, M.V.; Bell, J.G.; Henderson, R.J.; Tocher, D.R. Evolution and roles of (*n*-3) polyunsaturated fatty acids in marine organisms. In *Phospholipids: Characterisation, Metabolism and Novel Biological Applications*; Cevic, G., Paltauf,

F., Eds.; American Oil Chemistry Society Press: Champaign, IL, USA, 1995; pp. 248–259.

88. Sargent, J.R.; Henderson, R.J.; Tocher, D.R. The lipids. In *Fish Nutrition*, 2nd ed.; Halver, J.E., Ed.; Academic Press: New York, NY, USA, 1989; pp. 153–218.
89. Pickova, J.; Kiessling, A.; Petterson, A.; Dutta, P.C. Fatty acid and carotenoid composition of eggs from two nonanadromous Atlantic salmon stocks of cultured and wild origin. *Fish Phys. Biochem.* **1999**, *21*, 147–156.
90. Cruz-Garcia, L.; Sanchez-Gurmaches, J.; Bouraoui, L.; Saera-Vila, A.; Perez-Sanchez, J.; Gutierrez, J.; Navarro, I. Changes in adipocyte cell size, gene expression of lipid metabolism markers, and lipolytic responses induced by dietary fish oil replacement in gilthead sea bream (*Sparus aurata* L.). *Comp. Biochem. Physiol. Mol. Integr. Physiol.* **2011**, *158*, 391–399.
91. Zakeri, M.; Kochanian, P.; Marammazi, J.G.; Yavari, V.; Savari, A.; Haghi, M. Effects of dietary *n*-3 HUFA concentrations on spawning performance and fatty acids composition of broodstock, eggs and larvae in yellowfin sea bream, *Acanthopagrus latus*. *Aquaculture* **2011**, *310*, 388–394.
92. Niculescu, M.D.; Lupu, D.S.; Craciunescu, C.N. Perinatal manipulation of alpha-linolenic acid intake induces epigenetic changes in maternal and offspring livers. *FASEB J.* **2013**, *27*, 350–358.
93. Benatti, P.; Peluso, G.; Nicolai, R.; Calvani, M. Polyunsaturated fatty acids: Biochemical, nutritional and epigenetic properties. *J. Am. Coll. Nutr.* **2004**, *23*, 281–302.
94. Seierstad, S.L.; Svindland, A.; Larsen, S.; Rosenlund, G.; Torstensen, B.E.; Evensen, O. Development of intimal thickening of coronary arteries over the lifetime of Atlantic salmon, *Salmo salar* L., fed different lipid sources. *J. Fish Dis.* **2008**, *31*, 401–413.
95. Kjaer, M.A.; Vegusdal, A.; Gjoen, T.; Rustan, A.C.; Todorevic, M.; Ruyter, B. Effect of rapeseed oil and dietary *n*-3 fatty acids on triacylglycerol synthesis and secretion in Atlantic salmon hepatocytes. *Biochim. Et Biophys. Acta MoL.Cell Biol. Lipids* **2008**, *1781*, 112–122.
96. Mørkøre, T.; Prytz, K. Fôrtilskudd gir friskere og fastere laks. *Norsk Fiskeoppdrett* **2011**, *2*, 46–48.
97. Larsson, T.; Koppang, E.O.; Espe, M.; Terjesen, B.F.; Krasnov, A.; Moreno, H.M.; Rørvik, K.A.; Thomassen, M.; Mørkøre, T. Fillet quality and health of Atlantic salmon (*Salmo salar* L.) fed a diet supplemented with glutamate. *Aquaculture* **2014**, *426*, 288–295.
98. Langley-Evans, S.C. Nutrition in early life and the programming of adult disease: A review. *J. Hum. Nutr. Diet.* **2014**, *28*, 1–14.
99. Fellous, A.; Favrel, P.; Guo, X.M.; Riviere, G. The Jumonji gene family in *Crassostrea gigas* suggests evolutionary conservation of Jmj-C histone demethylases orthologues in the oyster gametogenesis and development. *Gene* **2014**, *538*, 164–175.

CHAPTER 18

Genetic Identification of Anisakid Nematodes Isolated From Largehead Hairtail (*Trichiurus Japonicus*) in Korea

Jeong-Ho Kim[1]*, Woo-Hwa Nam[1] and Chan-Hyeok Jeon[2]

[1] *Department of Marine Bioscience, Gangneung-Wonju National University, Gangneung, Gangwon 210-702, South Korea*

ABSTRACT

Background

The nematode species belonging to genus *Anisakis* occur at their third larval stage in numerous marine teleost fish species worldwide and known to cause accidental human infection through the ingestion of raw or undercooked fish or squids. They may also draw the attention of consumers because of the visual impact of both alive and dead worms. Therefore, the information on their geographical distribution and clear species identification is important for epidemiological survey and further prevention of human infection.

Results

For identification of anisakid nematodes species isolated from largehead hairtail (*Trichiurus japonicus*), polymerase chain reaction-restriction fragment length polymorphism (PCR-RFLP) analysis of internal transcribed spacers of ribosomal DNA were conducted. Mitochondrial cytochrome c oxidase subunit 2 gene was also sequenced, and phylogenetic analysis was conducted. From the largehead hairtail ($n = 9$), 1259 nematodes were isolated in total. Most of the nematodes were found encapsulated throughout the viscera (56.2 %, 708/1259) or moving freely in the body cavity (41.5 %, 523/1259), and only 0.3 % (4/1259) was found in the muscles. By PCR-RFLP, three different nematode species were identified. *Anisakis pegreffii* was the most dominantly found (98.7 %, 1243/1259) from the largehead hairtail, occupying 98.7 % (699/708) of the nematodes in the mesenteries and 98.1 % (513/523) in the body cavity. Hybrid genotype (*Anisakis simplex* × *A. pegreffii*) occupied 0.5 %, and *Hysterothylacium* sp. occupied 0.2 % of the nematodes isolated in this study.

Conclusions

The largehead hairtail may not significantly contribute accidental human infection of anisakid nematode third stage larvae because most of the nematodes were found from the viscera or body cavity, which are not consumed raw. But, a high prevalence of anisakid nematode larvae in the largehead hairtail is still in concern because they may raise food safety problems to consumers. Immediate evisceration or freezing of fish after catch will be necessary before consumption.

Keywords: *Anisakis pegreffii, Hysterothylacium* sp., Largehead hairtail, *Trichiurus japonicas,* Cutlass fish

BACKGROUND

Anisakid nematodes (Nematoda: Anisakidae) infect animals belonging to almost all phyla and commonly found in aquatic vertebrates. Their life cycles involve various hosts at different levels in marine food webs. Crustaceans work as the first intermediate hosts; fish and cephalopods work as intermediate or paratenic hosts and marine mammals and fish-eating birds as definitive hosts (Mattiucci and Nascetti 2008). Since Van Thiel et al. (1960) reported that the third stage larvae (L3) of *Anisakis* from the Atlantic herring *(Clupea harengus)* can infect human, the number of clinical cases has been increasing worldwide, with the increase of knowledge about this parasite and improvement of diagnostics, as well as more globalized customs of eating raw marine fish (Cipriani et al. 2015).

The consumption of raw or undercooked fish harboring anisakid nematode L3 can lead to human infection (Audicana and Kennedy 2008). According to Fagerholm (1991), 10 genera exist in the family Anisakidae. However, the human infection is known to be caused frequently by the L3 of the genera *Anisakis* Dujardin, 1845 and less frequently by *Pseudoterranova* Krabbe, 1878 (Sakanari and McKerrow 1989). Of the genus *Anisakis*, *A. simplex* sensu stricto (s.s.) and *A. pegreffii* are recognized as zoonotic species causing anisakiasis in Asia and Europe, respectively (Mattiucci et al. 2013; Umehara et al. 2007). Thus, not all the nematode larvae found in fish are infective to human, and a clear diagnosis is important for proper epidemiological survey and further prevention of human infection.

Identification of anisakid nematodes has been traditionally conducted by morphological observation, but it is not always sufficient to make a clear identification at the species level because morphological characters of taxonomic significance in this group are very few, particularly for the larval stages lacking reliable diagnostic features at the species level (Mattiucci and Nascetti 2008). Moreover, morphological identification of nematodes is not always efficient when a huge number of the specimens have to be identified or the specimens are physically damaged and the morphological keys are missed. DNA-based diagnostic techniques such as polymerase chain reaction-restriction fragment length polymorphism (PCR-RFLP) of internal transcribed spacer (ITS) region, direct sequencing of ribosomal DNA (rDNA), and mitochondrial DNA cytochrome c oxidase subunit 2 (mtDNA cox2) gene analysis have solved these constraints and have been widely used (Mattiucci and Nascetti 2008 and the references therein). Current taxonomy of the genus *Anisakis*based on these techniques revealed that there are two clades including nine species in this genus, and *A. simplex*, the most frequently encountered *Anisakis* species in human infection belongs to clade I (Mattiucci and Nascetti 2008).

The largehead hairtail or cutlassfish (Trichiuridae: *Trichiurus* species) distributes in circumtropical and temperate waters of the world and has been considered as one of the economically important fish species in the western North Pacific including Korea (Nakamura and Parin 1993). The annual catch of the largehead hairtail in Korea has been fluctuating between 50,000 and 90,000 t, after a sharp decrease in mid-1990 (Kim et al. 2011). Thus, importation of frozen fish from several countries has been increasing to meet the domestic demand (Cha and Kim 2009). Juveniles feed mostly on euphausiids, small pelagic planktonic crustaceans and small fishes, and adults feed mainly on fishes and occasionally on squids and crustaceans

(Nakamura and Parin 1993). All of these food items are known to be the intermediate hosts of anisakid nematodes (Mattiucci and Nascetti 2008).

There have been many papers regarding anisakid nematodes isolated from *Trichiurus* species (Borges et al. 2012; Kong et al. 2015; Lee et al. 2009; Shih 2004; Umehara et al. 2010; Zhang et al. 2013). But, the identification of host species was not clearly conducted, and several scientific names (e.g., *Trichiurus lepturus*, *Trichiurus haumela*, *Trichiurus* sp.) have been indiscriminately used for the largehead hairtail in their studies. However, clear identification of host species, together with the information on their geographical distribution and feeding behavior are important to understand the occurrence of anisakid nematodes and predict the likelihood of infection in a given area. In particular, the largehead hairtails are often sold as fresh or frozen fillets, which are difficult to identify since most of the morphological features are no longer available. In this case, other techniques are necessary to identify these morphologically similar and closely related species in commercial products.

Although the largehead hairtail is commercially and economically important fish species in Korea, its anisakid nematode fauna has not been clearly investigated and only fragmentary information exists. Lee et al. (2009) reported *A. pegreffii* and *Anisakis typica* from several marine fish species including *T. lepturus*, but they did neither mention the origin of the fish nor the prevalence of infection in each host fish species. Moreover, there was no information in the microhabitat of anisakid nematodes in host fish. In this study, we investigated anisakid nematode fauna from the largehead hairtail caught from Korean waters. We collected nematodes in the largehead hairtail freshly caught from coastal waters around Jeju, Korea. The host species were identified by mitochondrial DNA (mtDNA) cytochrome oxidase I (COI) gene sequencing described by Hsu et al. (2009). The collected nematodes were identified by PCR-RFLP with subsequent sequencing and mtDNA cox2 gene sequencing, to provide information on the epidemiology of *Anisakis* infection in Korea.

METHODS

Nematodes collection

Fresh largehead hairtail ($n = 9$) caught by lines were purchased from local fisheries market located in Hallim, Jeju Island, off the southern coast of the Korean Peninsula, 2015. The freshly caught fish were placed on flake ice and immediately transported to the laboratory. All of the individual fish were measured and weighed. Then, they were dissected to collect nematodes. The body cavity was longitudinally opened, and the viscera were carefully examined for collecting nematodes. The muscles were sliced, placed, and pressed between transparent glass plates, then inspected with the naked eyes under the light. All of the collected live nematodes were washed repeatedly with 0.9 % NaCl solution and preserved individually in absolute alcohol for molecular analysis. Prevalence (the number of host infected with parasites divided by the number of hosts examined) and mean intensity (the average of parasite infection among the infected hosts) were used as quantitative descriptors of the parasite population as previously described by Bush et al. (1997).

DNA extraction and PCR

Before genomic DNA extraction, the nematodes were washed three times with PBS, then placed in 1.5 ml Eppendorf tube containing 400 μL TNES lysis buffer and crushed with a sterile pestle. Genomic DNA was extracted using the phenol-chloroform method with slight modification (Wasko et al. 2003). The extracted DNA was resuspended with 50 μL TE buffer (10 mM Tris-HCl, pH 7.5, 1 mM EDTA). DNA concentration and purity were measured using Nanodrop 1000 (Thermo Scientific, USA). The ITS region (ITS1-5.8S-ITS2) of rDNA was amplified using primers A (forward: 5′-GTCGAATTCGTAGGTGAACCTGCGGAAGGATCA-3′) and B (reverse: 5′-GCCGGATCCGAATCCTGGTTAGTTTCTTTTCCT-3′) and B (D'Amelio et al. 2000). Amplification was conducted using MyCyclerTM (BioRad, USA), with the following conditions. Denaturation at 94 °C for 5 min, then 35 cycles at 94 °C for 40 s, 54 °C for 40 s, 72 °C for 90 s, and post-amplification at 72 °C for 7 min.

Digestion of PCR products with restriction enzymes

Restriction enzymes *Hinf* I, *Hha* I and *Rsa* I were used in the RFLP analysis for the amplified PCR products (D'Amelio et al. 2000). Restriction endonuclease digestion was performed using 3 μL of PCR products, 0.5 μL of restriction enzymes, 1 μL of CutSmart® buffer (NEB, USA) and distilled water up to a final volume of 10 μL. The digestion with these restriction enzymes was performed at 37 °C for 90 min. The digested products were analyzed by electrophoresis in 1.5 % agarose gel containing ethidium bromide and photographed using Gel Logic 100 Imaging System (Biostep, Germany).

Mitochondrial DNA cox2 gene amplification and sequencing

For the amplification of mtDNA cox2 gene, the primers 210 (reverse: 5′-CACCAACTCTTAAAATTATC-3′) and 211 (forward: 5′-TTTCTAGTTATATA GATTGRTTYAT-3′) were used (Nadler and Hudspeth 2000). After initial denaturation at 94 °C for 5 min, 35 cycles were run at 94 °C for 40 s, 48 °C for 40 s, and 72 °C for 60 s followed by post-amplification step at 72 °C for 7 min. PCR products were purified by AccuPrep® Gel Purification Kit (Bioneer, Korea) according to the manufacturer's instructions. 10 ng/μL of purified PCR products were directly sequenced by ABI Prism 3730 XL DNA Analyzer (PE Applied Biosystems, USA).

Phylogenetic analysis

The sequence data of mtDNA cox2 gene was aligned with the published sequences in GenBank database (NCBI) using Clustal W method (Thompson et al. 1994). The phylogenetic tree was constructed by MEGA version 6 for comparing the genetic relationship among other published sequences by the neighbor-joining criteria (Tamura et al. 2013). The nucleotide sequences were registered in GenBank.

Host identification

Genomic DNA was extracted from the muscle tissue of largehead hairtail individuals according to the methods described by Wasko et al. (2003). PCR amplification of mtDNA COI gene was conducted with the primers and conditions described by Hsu et al. (2009). PCR products were purified and sequenced by the same protocols mentioned above. The obtained sequences were aligned with other published sequences including *T. lepturus*, *T. japonicus*, and *Trichiurus* sp. registered in NCBI using Clustal W method (Thompson et al. 1994). The phylogenetic tree was constructed by MEGA version 6 for comparing the genetic relationship among other published sequences of *Trichiurus* species by the neighbor-joining criteria (Tamura et al. 2013). The nucleotide sequences were registered in GenBank.

RESULTS

We examined nine individual largehead hairtail (mean whole body length = 90.7 ± 5.2 cm, mean body weight = 462.0 ± 47.3 g). All the examined individual largehead hairtail fish harbored nematodes (100 %, 9/9). In total, 1259 nematodes were found and mean intensity was 139.9 larvae/fish (1259/9). Most of the nematodes were found embedded in the surfaces of viscera including mesenteries (56.2 %, 708/1259) or existed freely in the body cavity (41.5 %, 523/1259). Two percent of nematodes were found in the intestines (0.6 %, 7/1259) or stomach (1.4 %, 17/1259). Only 0.3 % (4/1259) of nematodes was found in the muscles. All the information is summarized in Table 1.

Table 1 Infection levels of nematodes in the largehead hairtail in this study

P[a] (%)	MI[b]	% of nematodes (number of nematodes in each microhabitat/total number of nematodes)					
		Viscera			Body cavity	Muscles	Total
		Intestines	Stomach	Mesenteries			
100.0 (9/9)	139.9 (1259/9)	0.6 (7/1259)	1.4 (17/1259)	56.2 (708/1259)	41.5 (523/1259)	0.3 (4/1259)	100 (1259/1259)

[a]Prevalence of infection
[b]Mean intensity

The amplification of the rDNA region produced approximately 1 kb fragment (data not shown). PCR-RFLP analysis using *Hinf*I, *Hha*I, and *Rsa*I restriction enzymes revealed different banding patterns depending on the nematodes species. Digestion of the PCR product with *Hinf*I produced three different patterns, i.e., 350–300–250 bp, 700–350 bp, and 620–350–300–250 bp, respectively (Fig. 2a). Digestion using *Hha*I produced two different patterns, i.e., 550–430 bp and 350–250–200–150 bp (Fig. 2b), and digestion using *Rsa*I produced two different patterns, i.e., 550–300 bp and 650–220 bp, respectively (Fig. 2c). These banding patterns corresponded to the known patterns of *A. pegreffii* (lane 1, 2 in Fig. 1a–c), *Hysterothylacium* sp. (lane 3, 4 in Fig. 1a–c), and hybrid genotype (*A. simplex* × *A. pegreffii*) (lane 5, 6 in Fig. 1a–c), respectively.

Of 1259 nematodes, most of the nematodes (98.7 %, 1243/1259) were identified as *A. pegreffii*. The rest of them were identified as hybrid genotype, occupying 0.5 % (6/1259) and *Hysterothylacium* sp., occupying 0.2 % (2/1259) of the nematodes. PCR amplification was failed for eight individuals (0.6 %, 8/1259), probably due to the failure of DNA extraction. The results are summarized in Table 2. For *A. pegreffii*, 20 individuals were randomly selected and sequenced for mtDNA cox2 gene because a huge number of samples were collected. For other species, all the individuals were sequenced. All of the samples generated 629 bp size products (data not shown). When the phylogenetic tree was constructed with the mtDNA cox2 sequences of these samples and those previously registered in NCBI, taxonomic position of the *Anisakis* species identified in this study corresponded with the results obtained by PCR-RFLP; The mtDNA cox2 sequences of *A. pegreffii* (Genbank accession number: KU921231-921250) were clustered with those isolated from chub mackerel in Korea and Japan (Fig. 2), with 97.9–100.0 % homology. The mtDNA cox2 sequences of *Hysterothylacium* sp. showed the highest homology (92.0 %) with *Hysterothylacium* sp. isolated from chub mackerel (Genbank accession number: KC633443). The mtDNA cox2 sequences of all the hybrid genotype specimens were found to be *A. pegreffii* (data not shown).

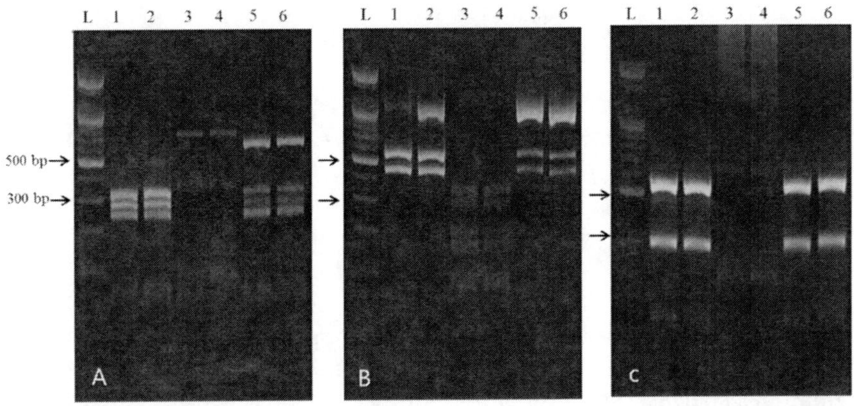

Figure 1 PCR-RFLP profiles of the nematode species obtained by digestion of PCR products with *Hinf*I (**a**), *Hha*I (**b**), and *Rsa*I (**c**) restriction enzymes, respectively (*L* ladder, *1–2*: *A. pegreffii*, *3–4*: *Hysterothylacium* sp., *5–6*: hybrid genotype)

Table 2 Anisakid nematodes assemblage of the largehead hairtail examined in this study

Microhabitat/ species name		*A. pegreffii*	Hybrid	*Hysterothylacium* sp.	Fail	Total
Viscera	Intestines	7	0	0	0	7
	Stomach	17	0	0	0	17
	Mesenteries	701	3	0	4	708
Body cavity		514	3	2	4	523
Muscle		4	0	0	0	4
	Total	1243	6	2	8	1259

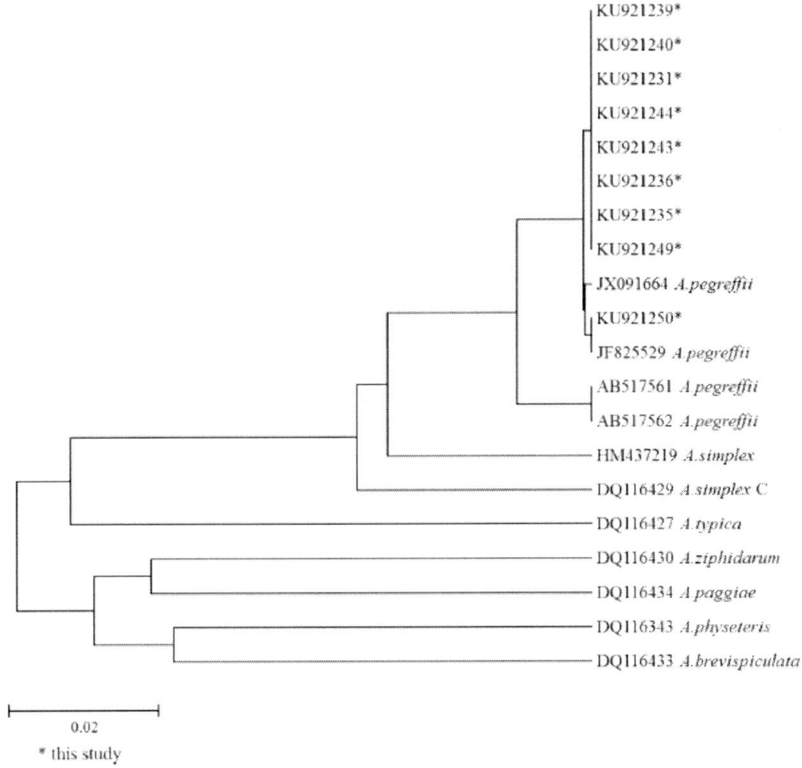

Figure 2 Molecular phylogenetic trees showing the genetic relationships among *Anisakis* spp. based on mtDNA cox2 gene sequences. Analysis was performed using the MEGA 6 program. The *scale bar* indicates evolutionary distance

PCR amplification and sequencing of mtDNA COI sequences from fish samples revealed that all of the nine samples produced one single band of 630 bp (data not shown). The obtained sequences (Genbank accession number: KU963588, 963589) were clustered with those of *T. japonicus* previously registered in GenBank, with 98.4–99.8 % homology (data not shown).

DISCUSSION

Human infection with *A. simplex* L3 accidentally occur when raw or undercooked fish contaminated with the L3 is ingested. Although live L3 can rarely develop to L4 or adults in the human gastrointestinal tracts, they can cause mild to severe epigastric pain, nausea, vomiting, and diarrhea (Audicana and Kennedy 2008). In addition, allergic reactions occur in some patients, which may occur even by the ingestion of dead *A. simplex*larvae (Audicana and Kennedy 2008). In fish, the larvae are mainly located freely in the visceral cavity or encapsulated as flat, tight spirals in and on the visceral organs; however, a minor proportion may migrate into the muscles, which may draw attention of consumers (Levsen and Berland 2012). Thus, it is necessary to have knowledge about the host range of each *Anisakis* species, together with the clear identification and their geographical distribution to avoid accidental human infection and warrant food safety issue.

In this study, most of the nematodes were found throughout the viscera (56.2 %) or from the body cavity of the largehead hairtail (41.5 %) (Table 1). Those from the viscera were embedded in or stick to the surfaces of the mesenteries and intestines, whereas those from the body cavity were freely moving in the body cavity. On the other hand, only 0.4 % (6/1,420) of the nematodes was found in the muscles.

Of 1259 nematodes, most of the nematodes (98.7 %, 1243/1259) were identified as *A. pegreffii* by PCR-RFLP and direct sequencing. Other species, *Hysterothylacium* sp. and hybrid genotype (*A. simplex* × *A. pegreffii*) were rarely found.

A. pegreffii Campana-Rouget and Biocca, 1955 is the dominant species of *Anisakis* in the Mediterranean Sea and also widely distributed in the austral region between 30° N and 55° S. Many pelagic and demersal fish species are known to be the intermediate/paratenic hosts caught in this area, and toothed whales (Delphinidae, Ziiphidae, Physeteridae) and baleen whales (Neobalaenidae) are known to be the definitive hosts (Mattiucci and Nascetti 2008). *A. pegreffii* has been also reported frequently at the larval stage in many fish species in East Asian waters (Bak et al. 2014 and the references therein). Recent reports in Italy suggest that *A. pegreffii* can be the etiological agent of human infection (Mattiucci et al. 2013).

Genus *Hysterothylacium* Ward and Magath, 1917 is one of the common genera in the family Raphidascarididae (Fageholm 1991), in which there are approximately 70 species (Shamsi et al. 2013). The adult stages are usually found in the alimentary canal of predatory teleost, while the L3 are found encapsulated in the mesentery and viscera of various species of marine teleosts and also in various marine invertebrates (Keskin et al. 2015). Thus, it is not host specific in the larval stages, exhibiting a global distribution (Shamsi et al. 2013). Many species of the genus *Hysterothylacium* have been poorly described; thus, their genetic characteristics have been frequently deposited in Genbank without clear identification at the species level. Although the RFLP patterns in this study was in accordance with those of *Hysterothylacium*sp. described in Bak et al. (2014), we could not identify these specimens at the species level because the highest homology of the mtDNA cox2 sequences showed 92 % with that of *Hysterothylacium* sp. registered in GenBank. It would be necessary to conduct morphological observation and further molecular analysis, for clear identification of these specimens.

The prevalence of infection and mean intensity for anisakid nematodes L3 in *T. japonicus* calculated in this study were extremely higher than those described by Lee et al. (2009). There can be several explanations for this difference. In our study, we collected freshly caught *T. japonicus* from Korean waters, whereas they collected *T. lepturus* in a fisheries market, without information on their geographical origin.

In marine food webs, the abundance of phytoplankton is strongly related to temperature and changes in phytoplankton biomass can affect all subsequent stages, first impacting on marine crustaceans, the predominant group in zooplanktons abundance and biomass, and thereafter migratory fish, cephalopods, and predatory marine mammals further up the food chain (Utaaker and Robertson 2015). The life cycles of *Anisakis* species involve various hosts at different levels in marine food webs. The high prevalence of *Anisakis*species in fish hosts in this study is therefore thought to reflect the richness of micro/macroinvertebrates and possibly subsequently have relations with the biomass of other predators such as pinnipeds and cetaceans in waters investigated. Several papers also reported relatively a high prevalence of anisakid nematodes larvae in the largehead hairtail (Borges et al. 2012; Kong et al. 2015; Umehara et al. 2010), as in this study. The low number of the nematodes larvae in the previous study (Lee et al. 2009) may just reflect the low abundance of micro/macroinvertebrates in the areas where the largehead hairtail were caught.

Several papers described anisakid nematodes L3 from the largehead hairtail (Borges et al. 2012; Kong et al. 2015; Lee et al. 2009; Shih 2004; Umehara et al. 2010; Zhang et al., 2013), but there was no report in which *T. japonicus* was mentioned as host species, except for our study. It is not clear if the host species mentioned was in fact *T. japonicus* or other *Trichiurus* species in their studies because the identification of host species was not clearly conducted. But except for the paper by Borges et al. (2012), they probably sampled *T. japonicus* because the known geographical distribution of *T. japonicus*, the northwestern Pacific (Hsu et al. 2009; Tzeng et al. 2007) includes their sampling areas.

After *Anisakis* L3 in the invertebrate or primary fish hosts are ingested by predatory fish, they penetrate in the intestinal wall and then may remain within the body cavity or migrate into the musculature or internal organs of the fish. The different relative abundance of these L3 among these microhabitats in fish can be affected by many factors such as the parasites species, the fish species, the fish age, and the environmental conditions (Lymbery and Cheah 2007). Although the mechanisms of larval penetration within fish hosts are still unknown, some studies revealed that L3 migrate from the visceral organs to the muscle after the death of the fish hosts and this postmortem migration is thought to be also affected by interactions among parasites, host, and external environments (Cipriani et al. 2016; Lymbery and Cheah 2007). Thus, the risk of human infection can be enhanced if the fish are ingested whole or if the fish have been kept without evisceration for a period of time.

Several methods such as postharvest handling, freezing, and adequate cooking are suggested for prevention or control measures of the human infection by ingestion of raw or undercooked fish harboring *Anisakis* L3 (Audicana and Kennedy 2008). Deep freezing and/or adequate cooking remain the most effective measures to prevent human infection (Audicana and Kennedy 2008; Lymbery and Cheah 2007). In addition, immediate evisceration or freezing of fish after catch will be necessary for the prevention of human infection (Lymbery and Cheah 2007). In case of our study, most of the *Anisakis* L3 on the largehead hairtail was found from the body cavity and viscera, which are generally not consumed raw and removed before cooking. In addition, *A. simplex* (s.s.), the most commonly involved species in human infection, was not found in this study. Thus, the largehead hairtail may not be the common source of human infection in Korea. But care should be still taken because *A. pegreffii* is also considered as a possible human pathogen (Mattiucci et al. 2013), and a high prevalence of anisakid nematode larvae in the largehead hairtail may raise food safety problems to the consumers. Immediate evisceration or freezing of fish after catch will be necessary before consumption.

CONCLUSIONS

Information on the anisakid nematodes fauna of commercially important fish species is important because the L3 of anisakid nematodes, mostly *A. simplex*, can cause accidental human infection in countries where the customs of eating raw or undercooked fish or cephalopods exist. Given the growing popularity of these customs worldwide, the risk of human infection is widespread, together with the increased number of new host records and expansion of the geographical ranges of *Anisakis* spp. The largehead hairtail, one of the commercially important fish species in Korea, was investigated for anisakid nematodes fauna. Most of the nematodes were found throughout the viscera (58.1 %) and the body cavity (41.5 %), which is not consumed raw. Only 0.3 % was found in the muscles. *A. pegreffii* was the most frequently encountered nematode species (98.7 %) from the largehead hairtail in this study. Detailed information on the distribution of different *Anisakis* species in the largehead hairtail in this study has significant implications for the biology, ecology, and epidemiology of anisakid nematodes.

ABBREVIATIONS

COI, cytochrome c oxidase subunit 1; ITS, internal transcribed spacer; L3, third stage larvae; mtDNA cox2, mitochondrial DNA cytochrome c oxidase subunit 2; RFLP, restriction fragment length polymorphism

ACKNOWLEDGEMENTS

We thank Jeong-Ho Jang, Min-Woo Kim, Chan-Goo Kim, and Hae-Young Yun for their help during the experiments.

FUNDING

This research was supported by the Basic Science Research Program through the National Research Foundation of Korea (NRF) funded by the Ministry of Science, ICT, and future planning (2015R1A2A2A01004406).

REFERENCES

1. Audicana MT, Kennedy MW. *Anisakis simplex*: from obscure infectious worms to inducer of immune hypersensitivity. Clin Microbiol Rev. 2008;21:360–79. doi:10.1128/CMR.00012-07.
2. Bak TJ, Jeon CH, Kim JH. Occurrence of anisakid nematode larvae in chub mackerel (*Scomber japonicus*) caught off Korea. Int J Food Microbiol. 2014;191:149–56. doi:10.1016/j.ijfoodmicro.2014.09.002.
3. Borges JN, Cunha LFG, Santos HLC, Monteiro-Nato C, Santos CP. Morphological and molecular diagnosis of anisakid nematode larvae from cutlassfish (*Trichiurus lepturus*) off the coast of Rio de Janeiro, Brazil. PLOS One. 2012;7:e40447. doi:10.1371/journal.pone.0040447.
4. Bush AO, Lafferty KD, Lotz JM, Shostak A. Parasitology meets ecology on its own terms: Margolis et al., revisited. J Parasitol. 1997;83:575–83. doi:10.2307/3284227.

5. Cha YG, Kim KS. A causality analysis of the prices between imported fisheries and domestic fisheries in distribution channel. J Fish Business Admin. 2009;40:105–26 (in Korean with English summary).
6. Cipriani P, Smaldone G, Acerra V, D'Angelo L, Anastasio A, Bellisario B, et al. Genetic identification and distribution of the parasitic larvae of *Anisakis pegreffii* and *Anisakis simplex* (s. s.) in European hake *Merluccius merluccius* from the Tyrrhenian Sea and Spanish Atlantic coast: implications for food safety. Int J Food Microbiol. 2015;198:1–8. doi:10.1016/j.ijfoodmicro.2014.11.019.
7. Cipriani P, Acerra V, Bellisario B, Sbaraglia GL, Cheleschi R, Nascetti G, et al. Larval migration of the zoonotic parasite *Anisakis pegreffii* (Nematoda: Anisakidae) in European anchovy, *Engraulis encrasicolus*: implications to seafood safety. Food Control. 2016;59:148–57. doi:10.1016/j.foodcont.2015.04.043.
8. D'Amelio S, Mathiopoulos KD, Santos CP, Pugachev ON, Webb SC, Picanço M, et al. Genetic markers in ribosomal DNA for the identification of members of the genus *Anisakis* (Nematoda: ascaridoidea) defined by polymerase-chain-reaction-based restriction fragment length polymorphism. Int J Parasitol. 2000;30:223–6. doi:10.1016/S0020-7519(99)00178-2.
9. Fageholm HP. Systematic implications of male caudal morphology in ascaridoid nematode parasites. Syst Parasitol. 1991;19:215–28. doi:10.1007/BF00011888.
10. Hsu KC, Shih NT, Ni IH, Shao KT. Speciation and population structure of three *Trichiurus* species based on mitochondrial DNA. Zool Studies. 2009;48:835–49.
11. Keskin E, Koyuncu CE, Genc E. Molecular identification of *Hysterothylacium aduncum* specimens isolated from commercially important fish species of Eastern Mediterranean Sea using mtDNA cox1 and ITS rDNA gene sequences. Parasitol Int. 2015;64:222–8. doi:10.1016/j.parint.2014.12.008.
12. Kim YH, Yoo JT, Lee EH, Oh TY, Lee DW. Age and growth of largehead hairtail *Trichiurus lepturus* in the East China Sea. Kor J Fish Aquat Sci. 2011;44:695–700. doi:10.5657/KFAS.2011.0695 (in Korean with English Summary).
13. Kong QM, Fan LF, Zhang JH, Akao N, Dong KW, Lou D, et al. Molecular identification of *Anisakis* and *Hysterothylacium*larvae in marine fishes from the East China Sea and the Pacific coast of Japan. Int J Food Microbiol. 2015;199:1–7. doi:10.1016/j.ijfoodmicro.2015.01.007.
14. Lee MW, Cheon DS, Choi CS. Molecular genotyping of *Anisakis* species from Korean sea fish by polymerase chain reaction restriction fragment length polymorphism (PCR-RFLP). Food Control. 2009;20:623–6. doi:10.1016/j.foodcont.2008.09.007.
15. Levsen A, Berland B. *Anisakis* species. In: Woo PTK, Buchmann K, editors. Fish parasites: pathobiology and protection. CAB International: USA; 2012. p. 298–309.
16. Lymbery AJ, Cheah FY. Anisakid nematodes and anisakiasis. In: Murrell KD, Fried B, editors. Food-borne parasitic zoonoses: fish and plant-borne parasites. New York: Springer Science; 2007. p. 185–207.
17. Mattiucci M, Nascetti G. Advances and trends in the molecular systematics of anisakid nematodes, with implications for their evolutionary ecology and host-parasite co-evolutionary process. Adv Parasitol. 2008;66:47–148. doi:10.1016/S0065-308X(08)00202-9.
18. Mattiucci M, Fazii P, De Rosa A, Paoletti M, Megna AS, Glielmo A, et al. Anisakiasis and gastroallergic reactions associated with *Anisakis pegreffii* infection, Italy. Emer Inf Dis. 2013;19:196–9. doi:10.3201/eid1903.121017.
19. Nadler SA, Hudspeth DSS. Phylogeny of the Ascaridoidea (Nematoda: Ascaridida) based on three genes and morphology: hypotheses of structural and sequence evolution. J Parasitol. 2000;86:380–93. doi:10.1645/0022-3395(2000)086[0380:POTANA]2.0.CO;2.
20. Nakamura I, Parin NV. FAO Species Catalogue. Vol. 15. Snake mackerels and cutlassfishes of the world (Families Gempylidae and Trichiuridae). An annotated and illustrated catalogue of the snake mackerels, snoeks, escolars, gemfishes,

sackfishes, domine, oilfish, cutlassfishes, scabbardfishes, hairtails, and frostfishes known to date. FAO Fish Synop. 1993;125:99–106.
21. Sakanari JA, McKerrow JH. Anisakiasis. Clin Microbiol Rev. 1989;2:278–84. doi:10.1128/CMR.2.3.278.
22. Shamsi S, Gasser R, Beveridge I. Description and genetic characterisation of *Hysterothylacium* (Nematoda: Raphidascarididae) larvae parasitic in Australian marine fishes. Parasitol Int. 2013;62:320–8. doi:10.1016/j.parint.2012.10.001.
23. Shih HH. Parasitic helminth fauna of the cutlass fish, *Trichiurus lepturus* L., and the differentiation of four anisakid nematode third-stage larvae by nuclear ribosomal DNA sequences. Parasitol Res. 2004;93:188–95. doi:10.1007/s00436-004-1095-7.
24. Tamura K, Stecher G, Peterson D, Filipski A, Kumar S. MEGA6: Molecular Evolutionary Genetics Analysis Version 6.0. Mol Biol Evol. 2013;30:2725–9. doi:10.1093/molbev/mst197.
25. Thompson JD, Higgins DG, Gibson TJ. CLUSTAL W: improving the sensitivity of progressive multiple sequence alignment through sequence weighting, position-specific gap penalties and weight matrix choice. Nucleic Acids Res. 1994;22:4673–80. doi:10.1093/nar/22.22.4673.
26. Tzeng CH, Chen CS, Chiu TS. Analysis of morphometry and mitochondrial DNA sequences from two *Trichiurus* species in waters of the western North Pacific: taxonomic assessment and population structure. J Fish Biol. 2007;70(B):165–76. doi:10.1111/j.1095-8649.2007.01368.x.
27. Umehara A, Kawakami Y, Araki J, Uchida A. Molecular identification of the etiological agent of the human anisakiasis in Japan. Parasitol Int. 2007;56:211–5. doi:10.1016/j.parint.2007.02.005.
28. Umehara A, Kawakami Y, Ooi HK, Uchida A, Ohmae H, Sugiyama H. Molecular identification of *Anisakis* type I larvae isolated from hairtail fish off the coasts of Taiwan and Japan. Int J Food Microbiol. 2010;143:161–5. doi:10.1016/j.ijfoodmicro.2010.08.011.
29. Utaaker KS, Robertson LJ. Climate changes and food borne transmission of parasites: a consideration of possible interactions and impacts for selected parasites. Food Res Int. 2015;68:16–23. doi:10.1016/j.foodres.2014.06.051.
30. Van Thiel PH, Kuipers FC, Roskam RT. A nematode parasitic to herring, causing atue abdominal syndromes in man. Trop Geogr Med. 1960;2:97–113.
31. Wasko AP, Martins C, Oliveira C, Foresti F. Non-destructive genetic sampling in fish. An improved method for DNA extraction from fish fins and scales. Hereditas. 2003;138:161–5. doi:10.1034/j.1601-5223.2003.01503.x.
32. Zhang LP, Du XJ, An RY, Li L, Gasser RB. Identification and genetic characterization of *Anisakis* larvae from marine fishes in the South China Sea using electrophoretic-guided approach. Electrophoresis. 2013;34:888–94. doi:10.1002/elps.201200493.

CHAPTER 19

p53 and TAp63 Promote Keratinocyte Proliferation and Differentiation in Breeding Tubercles of the Zebrafish

Boris Fischer[1], Manuel Metzger[1], Rebecca Richardson[1], Philipp Knyphausen[1], Thomas Ramezani[1], Rainer Franzen[2], Elmon Schmelzer[2], Wilhelm Bloch[3], Thomas J. Carney[4], Matthias Hammerschmidt[1,5,6*]

[1] *Institute of Developmental Biology, University of Cologne, Cologne, Germany*

[2] *Cell Biology, Max Planck Institute for Plant Breeding Research, Cologne, Germany*

[3] *Institute of Cardiology and Sports Medicine, German Sport University Cologne, Cologne, Germany*

4 Institute of Molecular and Cell Biology, Proteos, Singapore

[5] *Center for Molecular Medicine Cologne, University of Cologne, Cologne, Germany*

[6] *Cologne Excellence Cluster on Cellular Stress Responses in Aging-Associated Diseases, University of Cologne, Cologne, Germany*

ABSTRACT

p63 is a multi-isoform member of the p53 family of transcription factors. There is compelling genetic evidence that ΔNp63 isoforms are needed for keratinocyte proliferation and stemness in the developing vertebrate epidermis. However, the role of TAp63 isoforms is not fully understood, and TAp63 knockout mice display normal epidermal development. Here, we show that zebrafish mutants specifically lacking TAp63 isoforms, or p53, display compromised development of breeding tubercles, epidermal appendages which according to our analyses display more advanced stratification and keratinization than regular epidermis, including continuous desquamation and renewal of superficial cells by derivatives of basal keratinocytes. Defects are further enhanced in TAp63/p53 double mutants, pointing to partially redundant roles of the two related factors. Molecular analyses, treatments with chemical inhibitors and epistasis studies further reveal the existence of a linear TAp63/p53->Notch->caspase 3 pathway required both for enhanced proliferation of keratinocytes at the base of the tubercles and their subsequent differentiation in upper layers. Together, these studies identify the zebrafish breeding tubercles as specific epidermal structures sharing crucial features with the cornified mammalian epidermis. In addition, they unravel essential roles of TAp63 and p53 to promote both keratinocyte proliferation and their terminal differentiation by promoting Notch signalling and caspase 3 activity, ensuring formation and proper homeostasis of this self-renewing stratified epithelium.

Author Summary

The mammalian epidermis is a stratified self-renewing epithelium, in which cell loss at the surface is properly balanced by cell proliferation in basal layers to ensure tissue homeostasis. But how is this balance genetically controlled? Here, we address this question in zebrafish breeding tubercles, epidermal appendages in which keratinocytes undergo more advanced differentiation processes than in regular fish epidermis, sharing crucial features with the cornified mammalian skin. We identify a linear pathway consisting of the transcription factor p53 and its close relative TAp63, which activate Notch signalling and thereby caspase 3 to promote terminal differentiation and eventual shedding of keratinocytes in upper tubercle layers, while at the same time employing non-cell autonomous mechanisms to promote keratinocyte proliferation at the tubercle base, thereby ensuring proper development and homeostasis of this self-renewing tissue. Such a two-fold function of the pathway is consistent with the formerly reported dual role of a caspase during wing regeneration in the fruitfly. Our findings will help to better understand the seemingly contrary effects described for TAp63 in different mammalian systems, while demonstrating partial functional redundancy between p53 and TAp63 during epidermal development in fish.

INTRODUCTION

The mammalian epidermis is a self-renewing stratified epithelium on the outer surface of the skin. During embryogenesis, it develops from the surface ectoderm, which is initially a single-layered epithelium. Stratification is initiated (E10 in mouse) with the formation of the outer periderm, leading to a bi-layered epidermal organization in which peridermal cells are attached to each other via tight junctions to protect the embryo against the amniotic fluid [1], [2]. Further epidermal maturation (E12.5–E17.5 in mouse) takes place beneath with the consecutive formation of the spinous, granular and cornified layers, establishing the later epidermal barrier, while the periderm is sloughed off [3]–[6]. Crucial contributions to this later epidermal barrier come from the granular layer, in which cells are sealed to each other via tight junctions, and from the outer cornified layers, providing physical resistance and preventing dehydration. Formation of this water barrier is essential for the adaptation to terrestrial life both during the ontogeny [7] and the evolution [8], [9] of land-based vertebrates. Accordingly, fish lack epidermal cornification.

In zebrafish, the embryonic and larval epidermis is bi-layered, consisting of an outer enveloping layer (EVL) segregating from inner cells before surface ectoderm specification [10], [11], and a basal keratinocyte layer, resembling the bi-layered organization of the mammalian epidermis at midgestation stages. Further stratification of the zebrafish epidermis only commences with the onset of metamorphosis (after three weeks of development), during which the epidermis becomes multi-layered. However, keratinocytes remain metabolically active throughout, including the outer-most layer, and lack morphological signs of cornification [12], [13]. This is in contrast to the organization of breeding tubercles, contact and secondary sex organs on the head and pectoral fin rays. Based on morphological studies in several other fish species, they were described as "keratinized epidermal appendages covered by a layer of dead cells with altered content" [14], suggesting that in breeding tubercles, keratinocytes might undergo more advanced, cornification-like differentiation processes.

Mammalian keratinocyte cornification is a multi-step process initiated by a switch in the expression of particular keratin genes, followed by the expression of the keratin-bundling protein filaggrin and proteins such as involucrin and loricrin, which together with keratins become cross-linked by transglutaminases (Tgm1-4) to reinforce the formation of a cornified envelope [15]–[17]. Furthermore, lipids stored in lamellar bodies are extruded into the extracellular space to form a lipid envelope. During these later steps, cells enter an apoptotic-like phase, lose cytoplasmic organelles including the nucleus, and are eventually sloughed from the skin surface. This loss of corneocytes by desquamation is tightly balanced by keratinocyte proliferation in basal layers, allowing constant self-renewal during epidermal homeostasis.

A key regulator of mammalian epidermal stratification and keratinocyte proliferation and differentiation is p63, a homolog of the tumour suppressor and transcription factor p53 [18]–[24]. Use of alternative promoters of the *Tp63* gene gives rise to two isoform categories: those with an N-terminal transactivation domain (TAp63 isoforms), as also present in p53, and those lacking this domain (ΔNp63 isoforms) and acting as inhibitory competitors of their TA counterparts. In addition, alternative splicing gives rise to at least three different C-terminal isoforms (α,β,γ) in each category, and to different N-terminal isoforms of TAp63 [25]. Recent analyses of mice specifically lacking the ΔN isoforms indicate that ΔNp63 is required for maintaining the proliferative potential of basal keratinocytes in embryonic epidermis while preventing their premature entry into terminal differentiation [26], consistent with findings obtained in cultured keratinocytes [27]–[29]. In contrast, TAp63 and p53 might promote keratinocyte differentiation. Thus, in inducible cell lines, TAp63 activates genes involved in keratinocyte differentiation, including different *tgms* [30]. Also p53, via direct and indirect activation of Notch signalling, has been shown to be required for keratinocyte differentiation in cell culture systems and during squamous cell carcinoma suppression in mouse tumour models [31], [32]. siRNA-mediated knockdown studies in organotypic cultures of human keratinocytes further suggest that it is primarily p53 that antagonizes the proliferation-stimulating effect of ΔNp63, and that the contribution of TAp63 to keratinocyte differentiation is minor compared to that of ΔNp63 [29], [33]. Furthermore, TA-specific p63 mutant mice lack an abrogation of keratinocyte differentiation [34]–[36], leaving the *in vivo* role of TAp63 during keratinocyte development unclear.

Notch signalling promotes different steps of mammalian keratinocyte differentiation *in vivo* [37]–[41]. Upon binding of Delta or Jagged ligands, the Notch receptor is cleaved and releases its intracellular domain (NICD), which binds to the transcriptional repressor RBP-J, enabling it to activate target genes that are repressed in the absence of Notch signals. Findings concerning the interconnection between p53/p63 and Notch signalling are controversial. Thus, *Jagged1/2*and *Notch1* have been reported to be positively regulated by p53 and TAp63γ [31], [42]–[46], but also by ΔNp63 [26], [43], while in other systems, the Notch target *Hes1* was negatively regulated by both TAp63 and ΔNp63 [43]. How Notch signalling promotes keratinocyte differentiation is also incompletely understood [38], [47]. One described Notch/RBP-target gene in embryonic mouse keratinocytes encodes caspase 3 [39]. This cysteine protease not only executes apoptosis, but also promotes terminal differentiation processes in a range of cell types [48], leading to delayed basal keratinocyte differentiation in *caspase 3* mutant mouse embryos [39].

Here, we identify the zebrafish breeding tubercles as sites with higher keratinocyte proliferation in basal layers as well as more advanced, cornification-like keratinocyte differentiation in upper layers, including transglutaminase expression and stronger keratinization, exclusive presence of tight junctions in second-tier keratinocytes, and rudimentary lipid envelope formation and constant desquamation and renewal of surface keratinocytes. In addition, we identify zebrafish TAp63 isoforms and a TA-specific p63 loss-of-function mutant, revealing essential and partially redundant roles of TAp63 and p53 to promote both keratinocyte proliferation at the base and terminal keratinocyte differentiation in upper layers of breeding tubercles. Both effects are mediated via Notch signalling and activated caspase 3, although these mediators are restricted to upper breeding tubercle layers, pointing to a combination of cell autonomous and non-cell autonomous effects. These findings will help to better understand the seemingly controversial roles described for TAp63 and p53 in different systems.

RESULTS

Regular epidermis and breeding tubercles display different patterns of superficial cell renewal

Even after metamorphosis (approximately 30 days post fertilization; dpf), when the zebrafish epidermis has become multi-layered, superficial epidermal cells display crucial similarities to the embryonic enveloping layer (EVL), like the presence of tight junctions and distinct microridges at their outer surface [12]. To determine the developmental origin of superficial cells of the adult zebrafish epidermis, we carried out transgenic lineage tracing experiments, using promoter elements that drive embryonic expression confined to the outer EVL (*krt4*) [49] or the basal layer (*krt19*) [50], respectively (Figure 1A; 5 dpf). Employing these promoters in an inducible binary transgenic Cre/Lox-system, embryonic EVL cells or basal keratinocytes were stably labeled via tamoxifen application from 1–4 dpf (Figure 1B–F). At 60 dpf, superficial cells of the regular body epidermis consisted of a mixture of derivates of embryonic EVL cells (Figure 1B) and basal keratinocytes (Figure 1C,D). However, both expressed the same marker genes (Figure 1B–D; and data not shown) and displayed identical morphological characteristics (Figure 1E). This suggests that in regular epidermis, embryonic EVL cells can persist beyond metamorphosis and are only slowly replaced by derivatives of basal keratinocytes. However, renewed superficial cells seem to have the same properties and functions as persisting EVL cells.

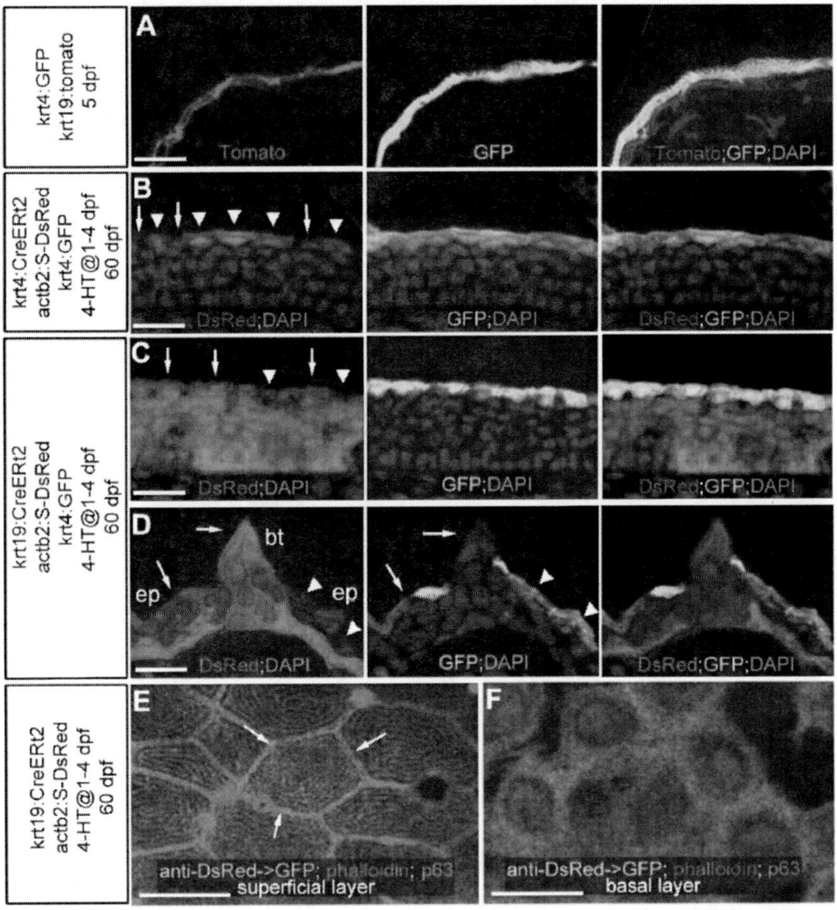

Figure 1. Transgenic lineage tracing of enveloping layer cells and basal keratinocytes in regular epidermis and breeding tubercles. (A–D) show transgene-encoded GFP and Tomato/dsRed fluorescence; transverse cryosections stained with DAPI; red channel on the left, green channel in the middle, merged channels on the right; (A) larva at 5 dpf; (B–D) juvenile fish at 60 dpf; (B,C) regular epidermis; (D) breeding tubercle on lower jaw. In (B–D), superficial cells deriving from the embryonic enveloping layer (EVL) are indicated with arrowheads, superficial cells deriving from the embryonic basal layer with arrows. (A) At 5 dpf, krt4-driven transgene expression is confined to EVL (in green), krt19-driven transgene expression to basal layer (in red). (B,C) At 60 dpf, derivatives of EVL cells (red; B) and basal keratinocytes (red, C) are present in the outer layer of regular

epidermis, intermixed with unlabelled cells, but expressing the same marker gene (tg(krt4:GFP) in green). (D) At 60 dpf, the breeding tubercle (bt), including its superficial layer, consists solely of derivatives of basal cells (in red). However, superficial cells lack expression of krt4:GFP (green). (E,F) anti-DsRed (in green) and anti-p63 (in pink) immunofluorescences, combined with phalloidin-staining of cytoskeletal actin (in red); confocal images of superficial (E) and basal layer (F) of regular epidermis; 60 dpf. A clone of labelled basal cells (F; green) has given rise to a single superficial cell (marked by arrows in E), which displays the same microridge pattern as non-labelled neighbours. Scale bars: 20 µm (A–D), 10 µm (E). Abbreviations: 4-HT, 4-hydroxytamoxifen; act2b:S-DsRed, actb2:loxP-STOP-loxP-dsREDEx; bt, breeding tubercle; ep, regular epidermis.

A different pattern was observed in spike-like epidermal structures that according to their location and fine structure (see below) were identified as breeding tubercles [14]. Here, superficial cells consisted solely of derivatives of basal keratinocytes and failed to express markers of the embryonic EVL and of the superficial cells of the regular epidermis (Figure 1D; n=13/13), suggesting that they have been renewed completely and have acquired a different fate.

Breeding tubercles of adult zebrafish display more advanced stratification and keratinocyte differentiation

In adult zebrafish, breeding tubercles were exclusively present in a disc- and row-like structure on each side of the lower jaw (Figure 2A), as well as in rows along the bony rays of the pectoral fins of males (Figure 2B), but not females (Figure 2C). Scanning electron microscopy (SEM) revealed that superficial cells of breeding tubercles, also called cap cells [14], lacked the microridges present at the outer surface of regular epidermis (Figure 2D,E). In addition, they lacked expression of *tg(krt4:GFP)* (Figure 2G) and other specific markers of superficial cells of regular epidermis (data not shown), and displayed a different cytoplasmic composition in AFOG trichrome stainings (Figure 2F). Furthermore, immunofluorescence revealed a higher keratin content of tubercle keratinocytes compared to regular epidermis (Figure 2H). Correspondingly, expression levels of the type II keratin gene *krt8* were significantly higher in breeding tubercles than in regular epidermis (Figure 2I). Furthermore, the type I keratins *krtt1c11a* (ZFIN: zgc:136902) and *krt17* (ZFIN: zgc:92061) were exclusively expressed in lower tubercle layers, but not in regular epidermis (Figure 2J and data not shown), while expression of type II keratin *krt5* and type I cytokeratin *cki* [51] was shared by regular epidermis and lower tubercle layers, but absent in upper layers of the tubercles (Figure 2K and data not shown). Strikingly, *tgm1*[52], encoding the cross-linking enzyme transglutaminase 1 involved in cornification of the mammalian epidermis [53], was exclusively expressed in upper layers of the breeding tubercles (Figure 2L).

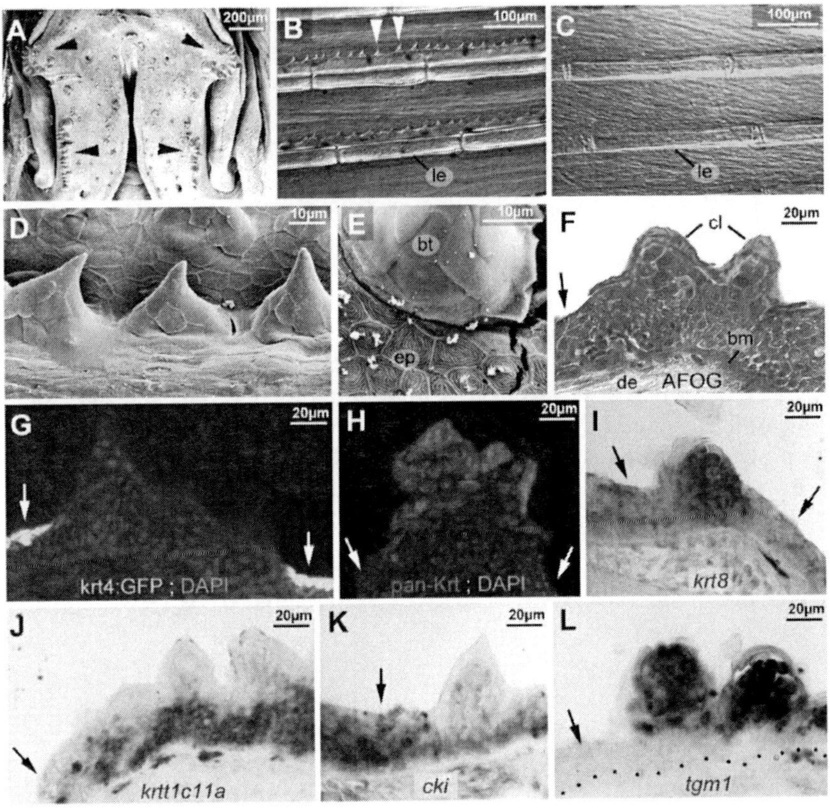

Figure 2. Breeding tubercle keratinocytes undergo more advanced keratinization. All images show breeding tubercles in wild-type zebrafish of one year of age; (A–E) views on body surface; (F–L) transverse sections through lower jaw. (A) Scanning electron micrograph (SEM) of the lower jaw; breeding tubercles are indicated by arrowheads. (B,C) Nomarski micrographs of pectoral fins of male (B) and female (C) fish. Bony rays (lepidotrichia) are marked (le). (D,E) SEM images of breeding tubercles (bt) on a male pectoral fin. (F) Trichrome (AFOG) staining; breeding tubercles consist of more cell layers than adjacent epidermis (arrow). The tubercle cap layer (cl) is stained in lighter red, resulting from its stronger keratinization. (G–L) GFP fluorescence of tg(krt4:GFP) line (G, green) and pan-type II Keratin immunofluorescence (H; red), in combination with DAPI staining. (I–L) in situ hybridization with indicated probes. Epidermis is indicated by arrows; dotted line in L marks basement membrane.

Abbreviations: bm, basement membrane; bt, breeding tubercle; cl, cap layer; de, dermis; ep, regular epidermis; le, lepidotrichia.

Striking differences were also observed at the structural level. In regular epidermis, basal and intermediary keratinocytes were of similar shapes and organized in a rather irregular pattern. In contrast, basal cells of tubercles were more regularly aligned to each other, while intermediary cells displayed a looser and spinous-like organization (Figures 2F and 3A,B), connected to each other via local desmosomes, but with wide extracellular spaces in between (Figure 3C). Tight junctions were only found in the second tier layer directly beneath the cap layer (Figure 3D), similar to their confinement to granular cells beneath the cornified layers in mammals. Desmosomes were also present between second tier and outer cap cells. However, in many samples, they were in the process of regression, leading to a partial dissociation of the two layers (Figure 3E). In addition, cap cells contained large vesicles reminiscent of lamellar bodies [54], extruding their content into the space between cap and second tier cells (Figure 3F), which was filled with material resembling the extracellular lipid lamellae in cornifying layers of the mammalian epidermis [54] (Figure 3G). Gradual loss of desmosomes was also observed between adjacent cap cells, but accompanied by cell membrane deterioration and cellular fusion (Figure 3H). Intracellular, cap and second tier cells contained large amounts of electron-dense granules (Figure 3G,H), while apart from nuclei (see below), cell organelles like mitochondria were largely absent (data not shown).

In sum, these data indicate that breeding tubercles display a more pronounced stratification than regular epidermis, consisting of different layers with distinct morphological properties and some cornification-like features in the superficial layer.

Breeding tubercles are formed during metamorphosis and undergo regular desquamation and renewal of superficial layers

Regular epidermis becomes multi-layered during metamorphosis [12], with final thicknesses between 3 and 10 layers, depending on the position on the body (data not shown). Breeding tubercles develop at the same time. At 21 dpf (6.0–6.5 mm body length) and 24 dpf (6.5–7.0 mm), the future breeding tubercle domain on the lower jaw was still covered by microridge-bearing peridermal cells (Figure 4A,B). At 28 dpf (7.5–9.0 mm), first elevations with microridge-free outer cells were present (Figure 4C), which had acquired the mature spiky shape at 31 dpf (10.0–11.0 mm) (Figure 4D).

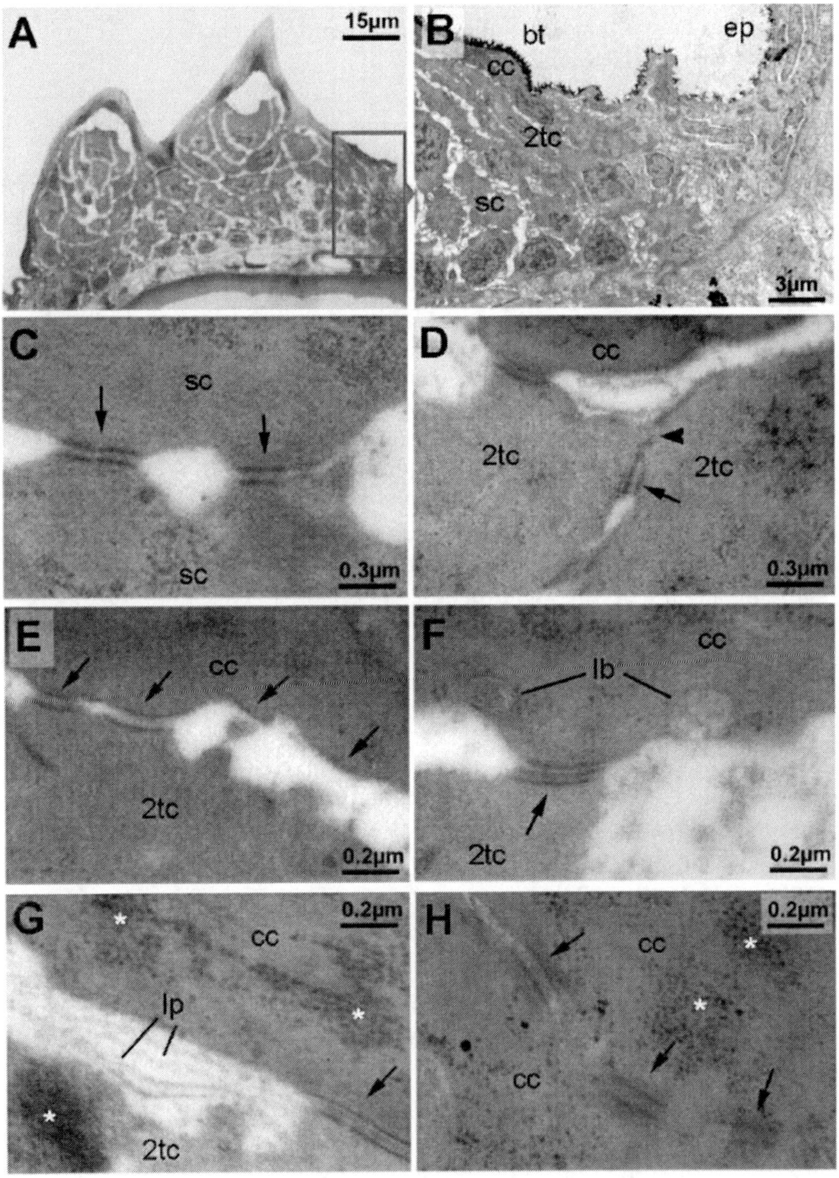

Figure 3. Ultrastructure of different tubercle cell layers. (A) Semi-thin section through breeding tubercles of lower jaw of a 1 year old wild-type, stained with methylene blue. For schematic overview, see also Figure 9A. Parts of the region shown in (B) are framed in red. (B–H) Transmission electron micrograph of ultrathin sections of same specimen as in (A). (B) Marginal zone of breeding tubercle. In the breeding tubercle (bt), basal cells

are more regularly aligned (red asterisks) than in the adjacent epidermis (ep; yellow asterisks). Spinous (sc), second-tier (2tc) and outer cap cells (cc) are indicated. (C) Interphase between two spinous cells. Arrows point to desmosomes. (D) Interphase between two second-tier cells. At their apical side, cells are sealed to each other via tight junctions (arrowhead), directly followed by a desmosome (arrow), displaying the same spatial organization as in peridermal cells of regular epidermis (not shown) [12]. (E–G) Interphase between second-tier and outer cap cell, displaying progressive desmosomal regression (arrows) (E), extrusion of lamellar-body-like vesicles (lb) into the extracellular space (F), which is filled with material reminiscent of lipid lamellae (lp; G). Arrows point to desmosomes, asterisks mark highly abundant protein aggregates in cap and second tier cells (also in H). (H) Interphase between two outer cap cells, displaying progressive desmosomal regression and cell membrane deterioration. Abbreviations: 2tc, second-tier cell; cc, cap cell; lb, lamellar body; lp, lipid lamella; sc, spinous cell.

Already at early elevation stages, tubercles were observed in which some of the smooth outer cells had been lost, exposing second tier cells with a rougher surface (Figure 4E), possibly reflecting former contact points to the lost outer cells. During later stages, entire sheets of outer cells were lifted up (Figure 4F) and shed. Corresponding spiky discs (Figure 4G–I) or rows were found in the water, in particular after spawning. Confocal analyses after DAPI-staining revealed that the spikes of these sloughed structures were hollow, indicating that they consisted solely of cap layers. To investigate whether shed cap layers are renewed, we took advantage of their unique property to be readily and permissively stained with externally applied dyes like calcein (Figure 4J) or methylene blue, reflecting their loss of cell membrane integrity, a hallmark of cell death [55]. In contrast, second tier and deeper cells remained unstained. In daily analyses of the same fish over several weeks, calcein permeability of individual discs or rows was randomly lost within one day and only regained after 7–14 days (Figure 4K,L; n=12), suggesting that shed cap cells are replaced by second tier cells and that it takes the latter 7–14 days to terminally differentiate.

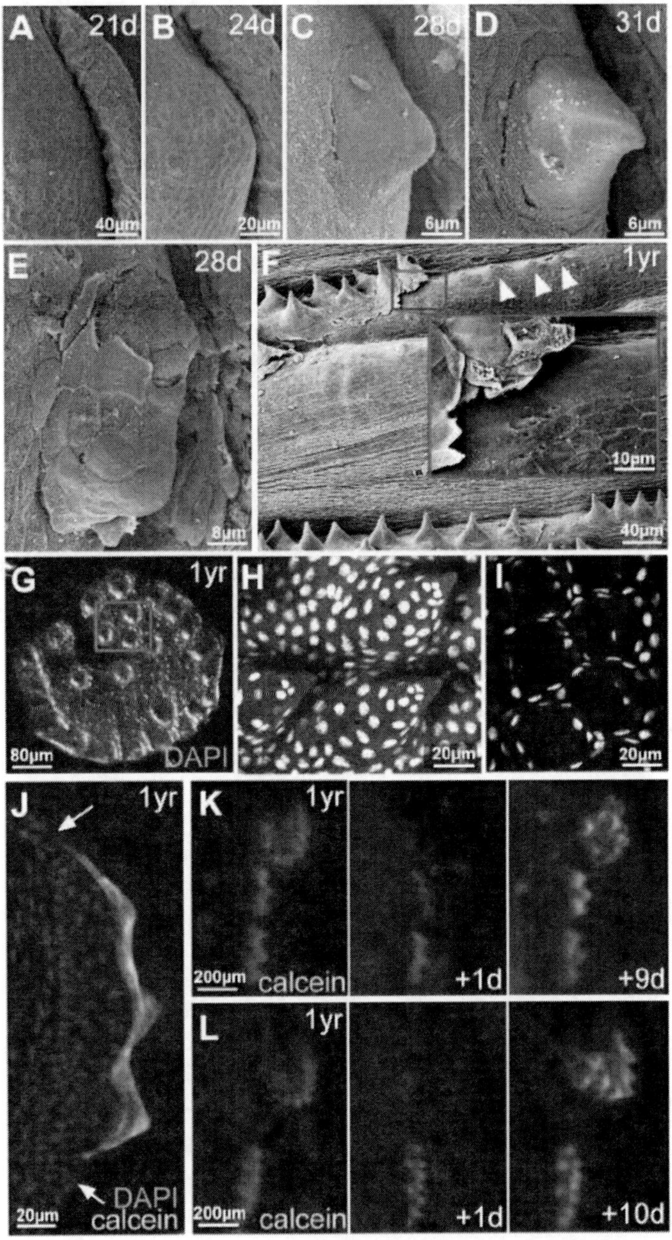

Figure 4. Outer cells of breeding tubercles undergo desquamation and renewal. (A–D) SEM images showing developmental time course of breeding tubercle formation on lower jaw between 21 and 31 days of development. (E) SEM image of a breeding tubercle on the lower jaw at 28

days of development. In the upper part of the image superficial cells have been lost, and former second tier cells are exposed. (F) SEM image of breeding tubercles on the pectoral fin of a male fish; 1 year of age. Inset shows magnification of boxed region. Entire rows of breeding tubercle cap layers are lifted up. Directly after desquamation of the cap layer, tubercles have a more shallow and non-spiky shape (arrowheads). (G–I) Shed cap layers of tubercles of disc-like structure on lower jaw, found in the water after spawning of 1 year old fish; stained with DAPI. Region boxed in (G) is shown in (H–I) as confocal micrographs at higher magnification; (H) maximal projection; (I) single plane. (J) Transverse cryosection through the breeding tubercles of the lower jaw of a 1 year old fish that had been incubated in calcein solution; calcein fluorescence in green; nuclei stained with DAPI in blue. Arrows point to outer layer of regular epidermis. (K,L) Calcein fluorescence of breeding tubercle region on the lower jaw of live fish, 1 year of age, repetitively stained with calcein and monitored daily. Days after first image acquisition are indicated (+1 d, +9 d, +10 d).

Spinous and cap layers of breeding tubercles display high levels of activated caspase 3, whereas basal layers exhibit high proliferation rates

In contrast to mammalian corneocytes, cap cells of breeding tubercles still contain their nuclei. However, compared to the nuclei of basal and spinous keratinocytes (Figure 5A,B), their chromatin was strongly condensed (Figure 5C), similar to the pyknosis apparent during cell death [56]. Nevertheless, cap cells were TUNEL (Terminal deoxynucleotidy transferease dUTP nick end labelling)-negative (Figure 5D,E), pointing to the absence of DNA fragmentation, whereas they showed high levels of activated caspase 3 (aCasp3; Figure 5F). This cysteine protease, which does not only execute apoptosis, but also promotes differentiation of embryonic keratinocytes [39], [48], was also present in spinous cells in intermediary tubercle layers (Figure 5F), which had normal nuclei (Figure 5B) and excluded externally applied dyes (Figure 4J), thus lacking all hallmarks of cell death. In contrast, aCasp3 was absent from basal tubercle layers and regular epidermis (Figure 5F). This aCasp3 distribution pattern was largely complementary to that of p63, which was present in all layers of regular epidermis, but confined to lower layers of breeding tubercles (Figure 5G). Cell proliferation, assayed via BrdU incorporation, displayed the same pattern like p63, complementary to aCasp3 (Figure 5H). However, proliferation rates in lower tubercle layers were significantly higher than in regular epidermis, both in fully grown (Figure 5H,J) as well as in developing tubercles (Figure 5I,J).

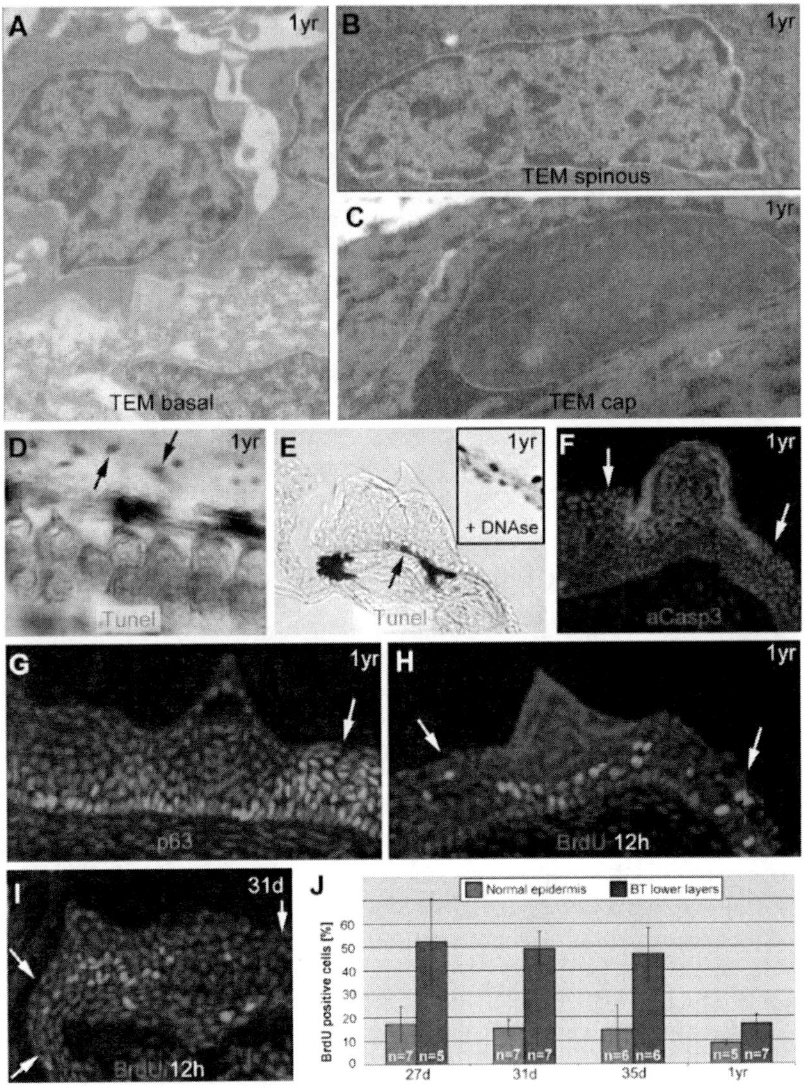

Figure 5. Basal tubercle cells display increased proliferation rates, upper cells increased levels of activated caspase 3 in the absence of other hallmarks of cell death. (A–C) TEM images of same specimen as shown in Figure 3, revealing nuclei with normal chromatin in basal and spinous cells of breading tubercles (A,B), whereas the chromatin of outer cap cells is more condensed (C). (D,E) TUNEL staining of pectoral fins of male fish; 1 year of age; whole mount (D) and transverse sections (E). Single TUNEL-positive cells are present in tubercle-free, inter-ray region (D, arrows) and in basal layer of breeding tubercles (E; arrow), whereas upper

layers of breeding tubercles are TUNEL-negative (E). Inset in (E) shows a section through the inter-ray region after DNase treatment, with TUNEL signals in all nuclei (positive control). (F–I) anti-activated caspase 3 immunofluorescence (IF) (F), anti-p63 IF (G), anti-BrdU iIF after BrdU incorporation for 12 hours (H,I), co-stained with DAPI; transverse section through breeding tubercle of lower jaw; 1 year old fish (F–H); 31 day old fish (I). Arrows point to regular epidermis. (J) Percentages of BrdU-positive nuclei of specimens as shown in (H,I) in regular epidermis (red) and lower (BrdU-positive) layers of breeding tubercles (blue). Standard deviations and numbers of evaluated samples (n) are indicated. For all stages, differences between regular epidermis and breeding tubercles are statistically significant: 27 d, $p=0.000431$; 31 d, $p=0.00000006$; 35 d, $p=0.000178$; 1 yr, $p=0.000295$. Abbreviations: d, days; yr, year.

In sum, in contrast to regular epidermis, where keratinocytes of all layers display comparable p63 and proliferation levels while lacking activated caspase 3, at least two domains can be distinguished in breeding tubercles: aCasp3-positive cells in the upper layers that are postmitotic and undergo more advanced differentiation, and aCasp3-negative keratinocytes at the base, with proliferation rates even higher than in regular epidermis. Furthermore, aCasp3 in upper layers is not correlated with apoptosis.

TAp63 and p53 mutants display compromised breeding tubercle development

In light of the described observations, we next studied breeding tubercle development in zebrafish loss-of-function mutants in p53 and TAp63, potential regulators of keratinocyte differentiation and proliferation in mammalian cell culture systems [29], [31]–[33]. The *p53*^{zdf1}allele bears an (M214K) exchange of a conserved amino acid residue in the DNA-binding domain that compromises p53 activity [57]. For zebrafish *p63*, only ΔN isoforms had been described thus far [58], [59]. However, via exon prediction of genomic sequences upstream of the ΔNp63-specific exon (Ensembl accession number ENSDARG00000044356) and validation via RT-PCR analyses and cDNA sequencing, two N-terminal TAp63 isoforms, TA1 and TA4, were identified (Figures 6A and S1) that are similar to the corresponding isoforms in mammals [25] (39% aa identity; 55% aa similarity; Figure S2A,B; GenBank accession numbers KF682365, KF682366). TA1 corresponds to mammalian full-length TAp63, TA4 to Δ40TAp63, which was the initially described human isoform [18]. Furthermore, the exon-intron organization is conserved between mammals and fish (Figure S1), and sequences from shark to human segregate in the expected phylogenetic pattern (Figure S2C).

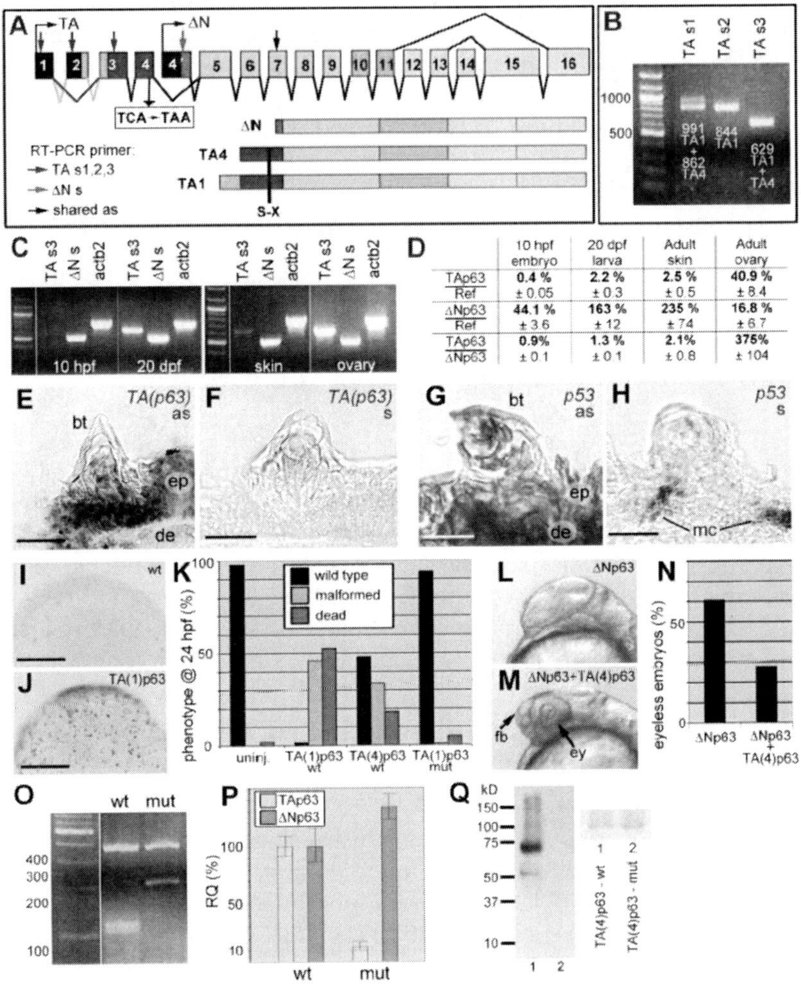

Figure 6. Structure, expression and biological activity of zebrafish TAp63, and molecular features of the mutant hu2525 allele. (A) Schematics of structure of zebrafish *p63* gene and of the encoded TAp63 (TA1 and TA4) and ΔNp63 isoforms. The different C-terminal isoforms (a,b,g) are not considered. The TA-specific transactivation domain is in dark (TA4) or light and dark (TA1) blue, the ΔN-specific N-terminal domain in red. Positions of primers used for RT-PCR analyses in (B–D) are indicated. Skipping of TA-specific exon 2 (indicated by dark blue lines) leads to the formation of the shorter TA4 isoform. The position of the TCA->TAA nonsense mutation (hu2525), and the resulting truncations of all TAp63 isoforms are indicated. (B) RT-PCR analyses of adult zebrafish skin, using indicated alternative TA-specific sense primers (for exact sequences and

positions, see Figure S1), together with an antisense primer in exon 7 shared between TA and ΔNp63. TAs1 generates two bands, reflecting the presence or absence of exon2-encoded sequences and representing either TA1 (0.991 kb band; exon 2 present) or TA4 (0.862 kb band; exon 2 absent). TAs2 positioned in exon2 generates one band, representing TA1. TAs3 positioned in exon 3 shared by TA1 and TA4 generates one band. (C) RT-PCR analyses of TA1+TA4 and ΔNp63 expression at different stages of zebrafish development (10 hpf=end of grastrulation; 20 dpf=onset of metamorphosis) and in skin and ovaries of adult zebrafish (1 year). Identical results were obtained in three independent experiments. (D) Quantitative RT-PCR analyses of *ΔNp63* and *TAp63* transcript levels relative to standard *rps23*transcript, and ratios of *ΔNp63* and *TAp63* isoforms in whole embryos/larvae at 10 hpf and 20 dpf and in adult skin and ovaries. (E–H) in situ hybridization with TA-specific TAp63 (E,F) and p53 (G,H) antisense (E,G) and sense control (F,H) RNA probes; transverse sections through breeding tubercles of lower jaw of wild-type fish, 1 year old. Breeding tubercles (bt), regular epidermis (ep), dermis (de) and melanocytes (mc) are indicated. (I,J) Tunel staining of apoptotic cells in uninjected embryo (I) and embryo injected with *TA(1)p63γ* mRNA (J); mid-gastrula stage (8 hpf). View on animal pole. (K) Quantification of TAp63-induced phenotypes (severe malformations and death) at 24 hpf, resulting from apoptosis during mid-gastrula stages, as shown in panels I,J (ratio between affected and total (n) embryos; in %; 3 independent injections). TA(1)p63γ(wt), n=240; TA(4)p63γ(wt), n=314; TA(1)p63γ(hu2525 mut), n=234. (L,M) Representative examples of embryos at 32 hpf, injected with ΔNp63α1 mRNA (L) or co-injected with ΔNp63α1 and TA(4)p63γ mRNA (M); lateral views on head region. Over-expression of ΔNp63 leads to the loss of forebrain (fb) and eyes (ey), resulting from ventral shifts during early patterning of the embryonic ectoderm (L) [58], while these structures are rescued in the embryo over-expressing both ΔNp63 and TAp63 (M). (N) Quantification of headless phenotypes (in % of wild-type plus headless embryos (n); 3 independent injections; ΔNp63α1, n=277; ΔNp63α1+TA(4)p63γ(wt), n=354). For simplicity, dead and malformed embryos in the co-injection are not considered. However, frequencies were significantly lower (25%) compared to embryos only injected with TA(4)p63γ(wt) (52%; see panel K), indicating that ΔNp63 also alleviates the pro-apoptotic effect of TAp63. (O) *Mbo*I RFLP analysis of TAp63 cDNA fragment amplified from skin of adult wild-type or *TAp63*$^{hu2525/hu2525}$ mutant fish. The mutant only contains transcripts harbouring the hu2525 mutation that destroys an *Mbo*I site (GATC; see Figure S3), resulting in a 208 bp instead of the two wild-type cleavage products of 101 bp and 107 bp. (P) Quantitative RT-PCR analyses of TAp63 and ΔNp63 mRNA levels in adult skin of wild-type and *TAp63*$^{hu2525/hu2525}$ mutant fish (2 independent experiments), revealing that in the mutant, TAp63 levels are reduced to

13.3%, and ΔNp63 are elevated to 134% of the wild-type levels. (Q) Anti-Myc-tag Western blot of lysates from embryos injected with mRNA encoding wild-type (lane 1) or hu2525 mutant N-terminally Myc-tagged zebrafish TA(4)p63γ. Ponceau red staining of membrane is shown as loading control. Calculated protein masses of 6xMyc-TA(4)p63γ proteins are: wt, 65.3 kDa; mut, 14.4 kDa. Identical results were obtained in 3 independent experiments. Scale bars are: 20 μm (E–H), 200 μm (I,J).

Both predicted TAp63 isoforms (TA1, TA4) were expressed in adult zebrafish skin (Figure 6B). Comparative regular (Figure 6C) and quantitative real-time RT-PCR (Figure 6D) analyses further revealed almost exclusive presence of ΔNp63 transcripts in early zebrafish embryos, similar to the situation in mouse [60]. At the onset of metamorphosis (20 dpf), TAp63 levels were approximately 5fold increased, but still constituted only 1.3% of the total p63 levels (Figure 6D). Also in adult zebrafish skin, TAp63 was expressed at much lower levels than ΔNp63 (approx. 2%), whereas higher TAp63 transcript levels were found in the ovary, the site of essential TAp63 function in mouse [34] (Figure 6D). In situ hybridizations with an isoform-specific RNA probe further revealed that TAp63 was expressed in regular epidermis as well as in all layers of the breeding tubercles (Figure 6E,F), similar to the expression pattern of p53 (Figure 6G,H).

To investigate whether the zebrafish TAp63 transcripts are translated into biologically functional proteins, we carried out over-expression studies in zebrafish embryos. Injected synthetic mRNA encoding the TA1 or TA4 isoform of zebrafish TAp63γ caused widespread apoptosis during gastrulation stages (Figure 6I,J), leading to embryonic death or severe malformations during further development (Figure 6K), similar to the previously reported effects of injected zebrafish p53 mRNA [61]. In addition, zebrafish TAp63γ mRNA significantly rescued the (headless) phenotype caused by ΔNp63α over-expression (Figure 6L–N), as previously reported for mouse TAp63 and p53 [58]. This indicates that zebrafish TAp63 transcripts give rise to a protein with p53-like pro-apoptotic and ΔNp63-antagonizing activities.

Using target-selected mutagenesis [62], a TA-specific zebrafish p63 mutant (*Tp63*^{hu2525}) mutant was isolated that bears a TCA->TAA nonsense mutation (S48X) in the last TA-specific exon truncating the N-terminal isoforms of TAp63, while leaving ΔNp63 isoforms unaffected (Figures 6A and S3). RT-PCR followed by restriction fragment length polymorphism (RFLP) analysis of skin samples confirmed the presence of the premature stop codon in *TAp63*transcripts of *TAp63*$^{hu2525/hu2525}$ animals (Figure 6O). Furthermore, qRT-PCR revealed that in *TAp63*$^{hu2525/hu2525}$ mutant skin, TAp63 mRNA levels were more than 7.5 fold reduced compared to wild-type siblings, whereas ΔNp63 mRNA levels were slightly up-regulated (Figure 6P), suggesting that the mutant TAp63 transcripts undergo nonsense-mediated mRNA decay and that TAp63 normally has a subtle negative effect on ΔNp63 transcription. Unfortunately, we could not investigate TAp63 proteins in *TAp63*$^{hu2525/hu2525}$ mutants, as all tested p63 antibodies failed to detect p63 proteins after Western blotting of zebrafish extracts (data not shown). However, upon injection of synthetic mRNAs encoding N- or C-terminally Myc-tagged zebrafish TAp63γ, wild-type mRNA gave rise to full-length fusion protein, whereas hu2525 mutant mRNA yielded no products (Figure 6Q and data not shown), suggesting that the stop codon cannot be by-passed, that no internal start site is used, and that the truncated protein is unstable. Furthermore, in contrast to wild-type TAp63γ, hu2525 mutant mRNA lacked pro-apoptotic activity upon over-expression in early zebrafish embryos (Figure 6K; compare columns 2 and 4). Together, this suggests that the hu2525 mutant is a TAp63 null.

Like *p53*$^{zdf1/zdf1}$ mutants, *TAp63*$^{hu2525/hu2525}$ mutants were viable and fertile. Anti-p63 immunofluorescence analysis revealed normal staining in regular epidermis of TAp63 mutants (Figure 7A,B; Figure S4E–G), suggesting that ΔNp63 is the major p63 isoform of the adult zebrafish skin, consistent with our RT-PCR data. However, both TAp63 and p53 mutants displayed specific breeding tubercle deficiencies of variable strength (C1,C2). SEM (Figure 7C–E), calcein stainings (Figure 7F) and histological sections (Figure 7G–I; Figure S4A–D) revealed complete absence (C2) in some individuals and reduced numbers and/or sizes of tubercles (C1) in others, while frequencies of strongest phenotypes were significantly increased in TAp63/p53 double mutants (Figure 7F). At the molecular level, even tubercle remnants of more weakly affected (C1) TAp63 and/or p53 mutants displayed significant alterations compared to wild-type tubercles, such as reduced *tgm1* (Figure 7J–L) and ectopic (ΔN)p63 expression (Figure 7A,B) in upper layers, as well as reduced proliferation rates at the base, which again was most prominent in TAp63/p53 double mutants (Figure 7M–P). In sum, this makes the molecular signature of breeding tubercle remnants of weaker TAp63/p53 mutants more similar to that of the regular epidermis.

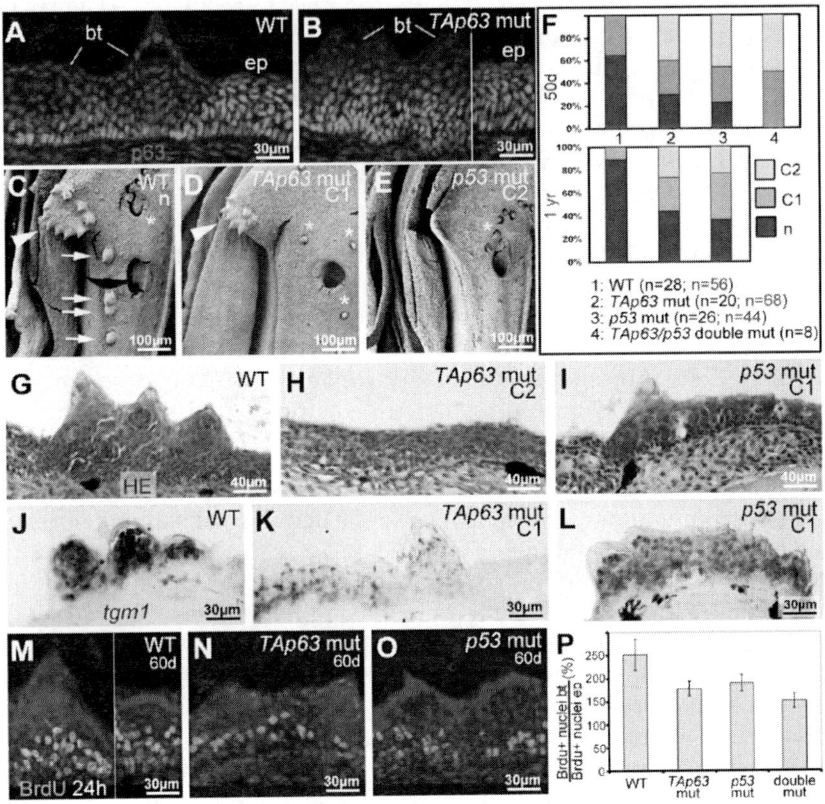

Figure 7. TAp63 and p53 mutants display reduced proliferation in basal layers and reduced differentiation in upper layers of tubercle remnants. (A,B) anti-p63 immunofluorescence; transverse sections through lower jaw of wild-type (A) and TAp63 mutant (B), 1 year of age. Breeding tubercles (bt) and adjacent regular epidermis (ep) are indicated. For larger field images and p53 mutant, see Figure S4E–G. (C,D,E) SEM images of lower jaw, ventral view. The TAp63 mutant (D; intermediate phenotype; C1) has a smaller disc (arrowheads in C,D) with smaller and fewer tubercles, while the posterior row of tubercles (arrows in C) is missing. The p53 mutant (E; strong phenotype; C2) lacks all breeding tubercles. Asterisks mark smaller, more randomly distributed appendages, which most likely are unculi [85] that are unaffected in the mutants. (F) Distribution of phenotypic strengths (n, normal; C1, intermediate; C2, strong) in lower jaw tubercles of wild-type, TAp63 mutant, p53 mutant and TAp63/p53 double mutant fish at an age of 50 days (upper panel) and 1 year (lower panel), determined via calcein staining (for example, see Figure 8V,X). Numbers of evaluated specimens are indicated. Similar results were obtained for the pectoral fins of males (not shown). (G–I) Hematoxylin &

eosin staining; transverse sections through breeding tubercle disc region of lower jaw of wild-type (G), TAp63 mutant (H; strong phenotype; C2), and p53 mutant (I; intermediate phenotype; C1) fish at 1 year. For larger field images, see Figure S4A–D. (J–L) *tgm1 in situ* hybridization; transverse sections through breeding tubercle disc region of lower jaw at an age 1 year. In TAp63 mutant (K), *tgm1*levels in upper layers of breeding tubercles are strongly reduced compared to wild type (J), whereas in the p53 mutant (L), the effect is weaker. Identical results were obtained in three independent experiments. (M–O) anti-BrdU immunofluorescence, combined with nuclear DAPI staining; transverse sections through breeding tubercles on the lower jaw of wild-type (M), TAp63 mutant (N), and p53 mutant (O) fish at an age of 60 days and directly after incubation with BrdU for 24 hours. (P) Ratios between keratinocyte proliferation rates in lower layers of breeding tubercles and regular epidermis (epidermal rates were unaltered in all cases; BrdU+ $nuclei_{bt}$/total $nuclei_{bt}$//BrdU+ $nuclei_{ep}$/total $nuclei_{ep}$; in %). Standard deviations are indicated. Numbers of evaluated samples were: WT, 36 sections from 4 fish; TAp63 mut, 28 sections from 4 fish; p53 mutant, 31 sections from 4 fish; TAp63/p53 double mutant, 21 sections from 3 fish. Observed decreases in keratinocyte proliferation in tubercles of mutants are statistically significant (Student's t-test; WT - TAp63 mut, $p=0.0041$; WT - p53 mut, $p=0.008$; WT – TAp63/p53 double mut, $p=0.0033$).

Together, our data suggest that TAp63 and p53 are required for proper keratinocyte proliferation at the base of breeding tubercles, as well as for cornification-like differentiation processes in upper layers.

TAp63 and p53 act via Notch signalling and caspase 3

In light of their known roles in keratinocyte differentiation in mouse [39], we also explored the involvement of Notch signalling and caspase 3 during zebrafish tubercle development, as well as their epistatic relationships to TAp63/p53. Differentiating tubercle keratinocytes displayed strong Notch signalling, as revealed by *Tg(TP1bglob:eGFP)*^{um13}, a transgenic in vivo reporter with 12 RBP-J_k binding sites [63], which was strongly expressed in upper layers of breeding tubercles, but not in lower layers and regular epidermis, complementary to the distribution of (ΔN)p63 (Figure 8A), but overlapping with aCasp3 (Figure 8B). Double labelling with calcein blue as a marker of differentiated cap cells revealed that during development, Notch signalling in the tubercle domain is initiated several days before the first cap cells have differentiated (Figure 8C,D). Furthermore, TAp63 and p53 mutants displayed strongly reduced Notch signalling and aCasp3 levels in their (significantly smaller) tubercles at 50 dpf (Figure 8F–H,L–N), although wild-type tubercles of similar sizes at 30 dpf were strongly positive for both (Figure 8E,K).

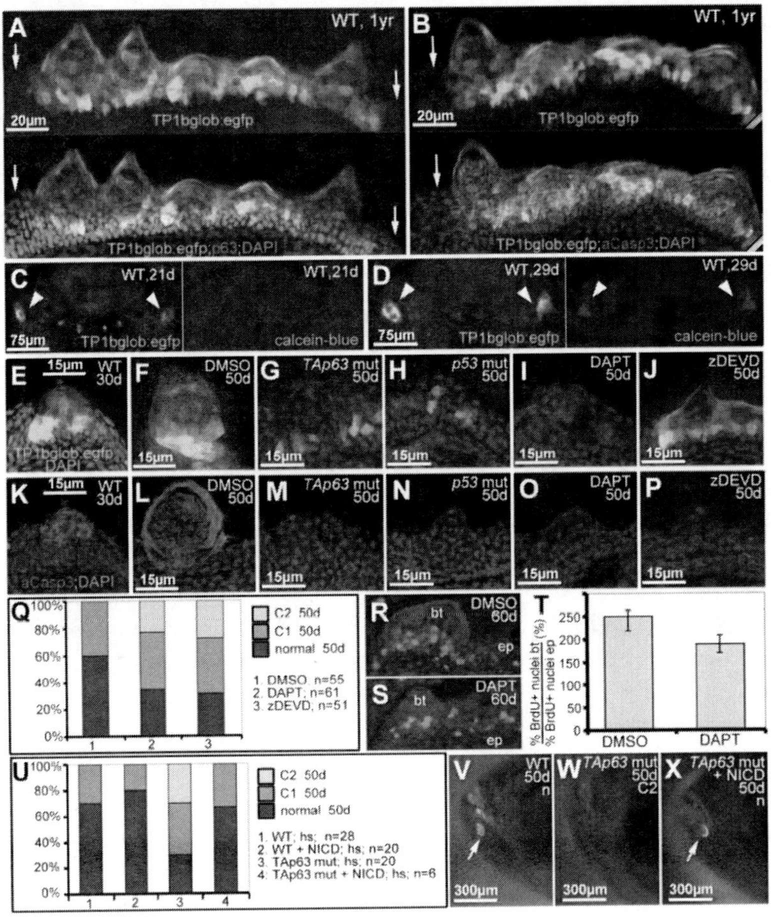

Figure 8. TAp63 and p53 act via Notch signalling and activated caspase 3. (A,B) Double anti-p63 (in red) and anti-GFP (in green) (A) or anti-aCasp3 (in red) and anti-GFP (in green) (B) immunofluorescences of Notch signalling reporter fish at 1 year of age; transverse sections through breeding tubercles in the disc region of the lower jaw; green channel in upper panels; merged images including DAPI staining in lower panels. Regular epidermis is indicated by arrows. (C,D) GFP and calcein-blue *in vivo*imaging of lower jaw of Notch signalling reporter fish at 21 dpf (C) and 28 dpf (D); ventral views; green channel to the left, blue channel to the right; the bilateral breeding tubercle disc regions are indicated by arrowheads. (E–P) GFP in vivo imaging of Notch signalling reporter (in green; E–J) and anti-aCasp3 immunofluorescence (in red; K–P) in untreated

controls at 30 days (E,K), DMSO-treated controls (F,L), TAp63 mutants (G,M) p53 mutants (H,N), DAPT-treated wild types (I,O) or zDEVD-treated wild types (J,P), all at 50 days of age; transverse sections through breeding tubercles in disc region of lower jaw. For each condition, identical results were obtained in all of at least 10 investigated individuals (see Figure S5). (Q) Distribution of phenotypic strengths (normal; intermediate, C1; strong, C2) in lower jaw tubercles of DMSO-, DAPT- or zDEVD-treated wild-type fish; all fish (numbers indicated; n) were evaluated at 50 days of age after calcein staining (not shown). (R,S) anti-BrdU immunofluorescence, combined with nuclear DAPI staining; transverse sections through breeding tubercles on the lower jaw of DAPT- (M), and DMSO-treated control fish (N) at an age of 60 days and directly after incubation with BrdU for 24 hours. Scale is as in panels (E–P). (T) Ratios between keratinocyte proliferation rates in lower layers of breeding tubercles and regular epidermis (epidermal rates were identical; BrdU+ $nuclei_{bt}$/total $nuclei_{bt}$//BrdU+ $nuclei_{ep}$/total $nuclei_{ep}$; in %). Standard deviations are indicated. Numbers of evaluated samples were: DMSO, 30 sections from 5 fish; DAPT, 19 sections from 4 fish. Difference is statistically significant (Student's t-test; p=0.0008). (U) Distribution of phenotypic strengths (normal; intermediate, C1; strong, C2) in lower jaw tubercles of TAp63 mutant and wild-type siblings versus *Tg(5xUAS-E1b:6xMYC-notch1a); Tg(-1.5hsp70l:Gal4)*double transgenic TAp63 mutants or wild-type siblings after heatshock-treatments from 20–50 days of development; all fish (numbers indicated; n) were evaluated at 50 days of age after calcein staining (see V–X for examples). (V–X) 50 day old non-transgenic control (WT; V), non-transgenic TAp63 mutants (W) and double transgenic TAp63 mutant after transgenic NICD expression from 20–50 days of development (X); calcein in vivo staining, lateral views on head. Arrows in (V,X) point to stained breeding tubercles in disc region of lower jaw. Abbreviations: ep, regular epidermis; bt, breeding tubercle.

To study whether Notch signalling and caspase 3 are required for breeding tubercle formation, we treated wild-type fish from 20 to 50 dpf with the specific Notch/g-secretase inhibitor DAPT or the caspase 3 peptide inhibitor zDEVD-fmk. While both treatments did not affect the general conditions and growth of the fish (Figure S5), DAPT-treatment caused significant reductions in the numbers and sizes of tubercles, in the activity of the Notch reporter and in aCasp3 levels (Figure 8I,O,Q), as also seen in p53 and TAp63 mutants (Figures 7F and 8G,H,M,N). zDEVD-fmk treatment had similar effects on tubercle numbers and sizes and on aCasp3 levels (Figure 8P,Q), while leaving Notch signalling unaffected (Figure 8J). Furthermore, despite the restriction of Notch signalling to upper breeding tubercle layers (see above), DAPT treatment led to a significant reduction of keratinocyte proliferation at the base of the tubercles (Figure 8R–T), comparable to the effects caused by loss of TAp63 and p53 function (Figure

7P). Finally, we re-introduced Notch signalling into TAp63 mutants, using a binary transgenic approach for temporally controlled expression of the constitutively active intracellular domain of Notch1 (NICD) [64]. While heatshock-induced NICD expression from 20–50 dpf had little effect in wild-type fish, it significantly elevated tubercle numbers in TAp63 mutants back to wild-type conditions (Figure 8U–X).

Together, these data suggest that TAp63/p53, Notch signalling and aCasp3 constitute a linear pathway required for proper breeding tubercle formation.

DISCUSSION

Both the *in vivo* role of TAp63 isoforms during vertebrate keratinocyte development and the stratification status of the epidermis of adult zebrafish were not fully understood. Here, we present data shedding new light onto both questions, identifying the zebrafish breeding tubercles as epidermal appendages with more advanced stratification and keratinocyte differentiation, and with regular desquamation and self-renewal, while presenting genetic evidence for an *in vivo* involvement of a TAp63/p53->Notch->caspase 3 pathway in these processes.

Breeding tubercles as sites of more advanced keratinocyte differentiation and regular self-renewal

Thus far, it had been unclear whether the epidermis of adult zebrafish undergoes self-renewal as in mammals. Our transgenic lineage analyses indicate that both in regular epidermis and in breeding tubercles, derivatives of basal keratinocytes can be found in the outer cell layer (Figure 1). Furthermore, long-term observations of individual fish show that outer layers of breeding tubercles are shed off, and that it takes cells of the next layer several days to fully develop outer layer properties before they undergo desquamation themselves (Figure 4).

Electron microscopy and marker analyses further revealed striking differences in the differentiation of keratinocytes in breeding tubercles versus regular epidermis. In regular epidermis, basal and intermediary cells display largely identical ultrastructural features [12]. In addition, they are (ΔN)p63-positive and mitotically active throughout all layers, and express the same keratin genes (Figures 2 and 5). Only when ending up in the superficial layer do they become strikingly different, express specific keratins and other structural proteins, including components of tight junctions, which are only present in this layer. Strikingly, basal keratinocyte-derived superficial cells of the regular epidermis are indistinguishable from persisting cells of the early embryonic enveloping layer, with which they form a uniform and continuous periderm-like sheet (Figure 1E).

In contrast, breeding tubercles exhibit a much more pronounced stratification (see Figure 9A for schematic drawing). They consist of more epidermal cell layers and display a keratin gene expression pattern strikingly different from that of regular epidermis (Figure 2). Cell proliferation is confined to basal layers, while the expression of transglutaminase 1, a crucial cross-linking enzyme during mammalian cornification [53], which is absent in regular zebrafish epidermis, is confined to upper layers (Figures 2 and 5). Structurally, at least 4 different layer types can be distinguished. The basal layer, in which cells are organized in a much more regular and columnar fashion than in regular epidermis, several spinous layers, a second-tier layer, in which cells are sealed to each other via tight junctions, and a heavily keratinized outer cap layer (Figures 3 and 9A). This organization is similar to that of the epidermis of adult mammals. In addition, the presence of extruding lamellar body-like vesicles and of lipid deposits at the second tier – cap layer interphase is reminiscent of lipid envelope formation during mammalian cornification [54], while the progressive degradation of desmosomes between second tier and cap layer cells resembles the fate of corneodesomosomes in mammalian corneocytes [65]. However, there are also differences between zebrafish breeding tubercles and the cornifying mammalian epidermis. Thus, despite their high keratin content, tubercle keratinocytes lack obvious intermediary filament bundles and a cornified envelope (CE), in line with the absence of genes encoding the keratin-bundling protein filaggrin and the CE components loricrin and involucrin in teleost genomes [66]. Furthermore, in contrast to mammalian corneocytes, sloughed cap cells of tubercles still contain their nuclei and display some (loss of cell membrane integrity, nuclear pyknosis), but not all hallmarks of apoptosis [55], [56] (Figures 4and 5).

Overall, this makes differentiated zebrafish tubercle keratinocytes more similar to the "immature horny cells with nuclei" observed in fetal human epidermis, which are supposed to represent a "transition phase of keratinisation" [2]. In line with such a more basal state, keratinocyte differentiation in zebrafish tubercles requires Notch signalling and caspase 3, regulators that are also needed for early steps of epidermal cornification in mammals [37], [39].

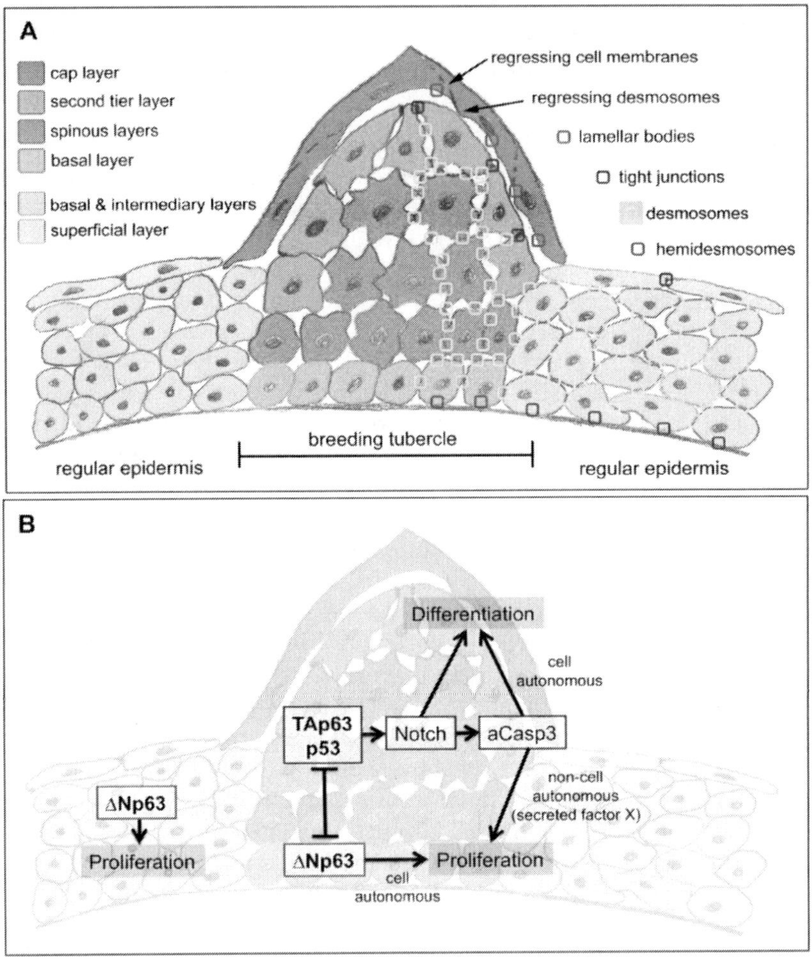

Figure 9. Schematics of breeding tubercle organization and regulation. (A) Schematic of cellular composition of breeding tubercles and regular epidermis. The different layers and cell types are colour-coded. On the right side, the sites of cellular structures and junctions shown in Figure 3 (TEM) are indicated. (B) Model for the dual role of TAp63/p53 on keratinocyte proliferation and differentiation, and its connection to ΔNp63. For details, see text.

A linear TAp63->Notch->caspase 3 pathway drives advanced keratinocyte development in breeding tubercles

There is compelling evidence in mouse and zebrafish that ΔNp63 is required during early steps of epidermal development, promoting the proliferation and stemness of basal keratinocytes, while possibly blocking differentiation processes [26], [58], [59]. However, the *in vivo* role of its counterpart TAp63 during keratinocyte development and differentiation remained elusive. Also, to our knowledge, no defects during regular keratinocyte development have been reported for p53 mutants as yet [31], [32].

Here, we revealed specific defects during keratinocyte development in breeding tubercles of zebrafish p53 and TA-specific TAp63 mutants. The used p53 allele, zdf1 (also called e7), bears a missense mutation in the DNA binding domain. Although the strongest of all available p53 mutants [57], [67], it might not be a complete functional null. However, inferences with other members of the p53/63/73 family seem very unlikely, as zdf1 homozygous embryos lack the phenotypes caused by loss of p63 or p73 [58], [59], [61], [67]. Antimorphic effects are also unlikely for the used TAp63 allele, hu2525, as the mRNA and the resulting truncated protein seem to be unstable (Figure 6P,Q). Furthermore, the protein would only contain part of the transactivating domain, while lacking DNA binding and oligomerization domains (Figure 6A). Therefore, we conclude that both TAp63 and p53 are per se essential for normal breeding tubercle development.

Similar tubercle defects as in TAp63 and p53 mutants were obtained upon chemical inhibition of Notch signalling or caspase 3 activity (Figure 8). aCasp3 levels are strongly reduced in TAp63 and p53 mutants, as well as after loss of Notch signalling, whereas Notch signalling is only lost in TAp63 and p53 mutants, but not after inhibition of caspase 3. In addition, re-introduction of Notch signalling into TAp63 mutants rescues their tubercle deficiencies. Together with their shared expression in differentiating keratinocytes of wild-type tubercles (Figures 6E,G and 8B), this provides *in vivo* evidence for the presence and requirement of a linear TAp63/p53->Notch->caspase 3 pathway.

Future studies have to elucidate the genetic control of breeding tubercle formation upstream of TAp63 and p53. As previously suggested [21], we believe that TAp63/p53 activity in zebrafish tubercles is indirectly promoted via negative interferences with its antagonist ΔNp63, which in skin is present in vast excess over TAp63 (Figure 6D), but, in contrast to TAp63 (Figure 6E), most likely restricted to the base of breeding tubercles (Figure 7A). This initial inhibition of ΔNp63 production in upper layers of the tubercle anlage might be reinforced by various negative feedbacks between TAp63/Notch/caspase and ΔNp63. Thus, in mammalian epidermal keratinocytes, Notch1 has been shown to repress ΔNp63 expression [43], while TAp63 induces caspase-dependent ΔNp63 degradation [68], consistent with the observed increased ΔNp63 transcript levels in the skin (Figure 6P) and the presence of ectopic ΔNp63 protein in upper tubercle layers of TAp63 mutant zebrafish (Figure 7A,B).

The dual effect of TAp63->Notch->Caspase 3 on keratinocyte proliferation and differentiation

It might appear paradoxical that in addition to reduced keratinocyte differentiation in upper layers of breeding tubercles (Figure 7J–L), loss of TAp63 or p53 function also leads to reduced keratinocyte proliferation at the base of the tubercles (Figure 7M–P), pointing to both differentiation- and proliferation-promoting effects of these regulators. Reduced tubercle growth and basal keratinocyte proliferation was also obtained upon inhibition of the TAp63/p53 mediators Notch or caspase 3 (Figure 8E–T), although both are only active in post-mitotic keratinocytes in upper tubercle layers (Figure 8A,B), pointing to non-cell autonomous mitogenic effects. Interestingly, in regenerating wings of the fruitfly *Drosophila melanogaster*, the enzymatic activity of the caspase Dronc generated in apoptotic cells does not only execute cell death in a cell-autonomous manner, but also promotes proliferation of adjacent cells even when Dronc-positive cells are prevented from dying, pointing to the presence of a Dronc-dependent mitogenic signal that acts in an apoptosis-independent and non-cell autonomous manner [48], [69]. It is tempting to speculate that a similar mechanism might be at play in breeding tubercles, in which a TAp63/p53->Notch->caspase 3 pathway active in upper epidermal layers, while cell-autonomously promoting terminal keratinocyte differentiation, enhances proliferation of cells in lower layers in a paracrine fashion (via an unknown secreted factor X; Figure 9B), accounting for proper tubercle growth during development and for proper balancing between cell loss via desquamation and cell renewal during tissue homeostasis [70]. Such a proliferation-stimulating role seems in contrast to the known functions of Notch and p53 as tumour suppressors [71], [72]. However, it is in line with the initial identification of p53 as an oncogene [73] and its more recently described function in the context of metabolic control [74].

Why TAp63 mouse mutants might lack overt defects during epidermal development

The breeding tubercle phenotype of zebrafish TAp63 and p53 mutants is not fully penetrant and variable in strength. Interestingly, however, phenotypic penetrance and average strength are significantly higher in TAp63/p53 double mutants, suggesting that the two structurally related transcription factors play partially redundant roles. Although other reasons cannot be ruled out, this suggests that a similar, possibly even more pronounced functional redundancy might also account for the apparent absence of epidermal defects in TA-specific p63 mutant mice [34]–[36]. Of note, we even identified a few TAp63/p53 double mutant zebrafish in which breeding tubercles were not completely lost. This could be due to some remaining p53 activity (see above). In addition, it might point to the existence of further partially redundant factors. TAp73, the third member of the family, is a candidate, which in mouse is expressed in all examined tissues and required side by side with TAp63 in oocytes to prevent genomic instability and female infertility [24], [34], [75]. Particular ΔNp63 isoforms might also be involved. Although in most cases they have dominant negative effects on TAp63 isoforms and p53, acting as transcriptional repressors, cases have been reported where they transactivate target genes [27]and positively cooperate with p53 [29].

Together, this demonstrates the complexity of the p53/p63/p73 system of transcriptional regulation, while revealing that it can be helpful to perform genetic analyses in different *in vivo*model systems, taking advantage of variations that have occurred during vertebrate evolution.

MATERIALS AND METHODS

Zebrafish lines and genotyping

Unless stated otherwise, wild-type fish from a mixture of TL and EK were used. The mutant *Tp63*^{hu2525} (S48X) line was generated upon our request in the Hubrecht Institute, NL, using target-selected mutagenesis (TILLING) [62]. The stable transgenic lines *Tg(krt4:creERt2)*^{fr33}, *Tg(krt19: dTomato)*^{fr34} and *Tg(krt19:creERt2)*^{fr35} were generated using the Tol2 kit [76], [77] with the described *krt4* [49] or *krttc19e* [50] (here abbreviated as *krt19*) promoter fragments for construct generation, followed by standard injection and screening procedures.

The mutant line *Tp53*^{zdf1} (M214K) [57] and the transgenic lines *Tg(krt4: GFP)*^{gz7} [49], *Tg(actb2:loxP-STOP-loxP-*

dsREDEx)^{sd5} [78], *Tg(TP1bglob:eGFP)^{um13}* [63], *Tg(5xUAS-E1b:6xMYC-notch1a)^{kca3}* [64] *and* *Tg(-1.5hsp70l:Gal4)^{kca4}* [63] have been previously described. For NICD expression, *Tg(5xUAS-E1b:6xMYC-notch1a)*, *Tg(-1.5hsp70l:Gal4)^{kca4}*double transgenic fish were heat-shocked from 20–50 dpf once a day for 1 hour at 40°C.

The *TAp63^{hu2525}* allele was genotyped using the dCAPS (derived Cleaved Amplified Polymorphism Sequence) method [79] with PCR primers CTGACCCCGAGGTTGTCTAA(sense) and TGCTAATCTGTATAGTA TTGGA AGCT (antisense) and subsequent *Hind*III digest. The *Tp53^{zdf1}* allele was identified via an RFLP (Restriction fragment length polymorphism) genotyping assay with PCR primers CCAGAGTATGTTCTGTCCA (sense) and TGATTGTGAGGATGGGCCTGCGGAATC (antisense) and subsequent *Bsty*I restriction digest.

Fish carrying the *Tg(-1.5hsp70l:Gal4)^{kca4}* or *Tg(5xUAS-E1b:6xMYC-notch 1a)^{kca3}* transgene were identified by PCR transgene amplification with the primers CGGGCATTTTACTTTTATGTTGC (gal4, sense), CATCATTAGCG TCGGTGAG (gal4 antisense), CATCGCGTCTCAGCCTCAC (NICD sense), CGGA AT CG TTTATTGGTGTCG primer (NICD antisense), yielding a 1.2 or 0.3 kb amplification product, respectively.

All zebrafish experiments were approved by the national animal care committees (LANUV Nordrhein-Westfalen; 8.87-50.10.31.08.129; 84-02.04.2012.A251; City of Cologne; 576.1.36.6.3.01.10 Be) and the University of Cologne

Lineage tracing

Tg(krt19:creERT2)^{fr35}, *Tg(actb2:loxP-STOP-loxP-dsREDEx)^{sd5}* double transgenic, *Tg(krt19:creERT2)^{fr35}*, *Tg(actb2:loxP-STOP-loxP-dsREDEx)^{sd5}*, *Tg(krt4:GFP)^{gz7}* triple transgenic, or *Tg(krt4: creERT2)^{fr33}*, *Tg(actb2:loxP-STOP-loxP-dsREDEx)^{sd5}*; *Tg(krt4:GFP)^{gz7}*triple transgenic embryos were treated with 5 μM 4-Hydroxytamoxifen (Sigma Aldrich; H7904) in the dark at 28°C from 24 hpf to 96 hpf before being returned to normal system conditions for growing up. For Figure 1A–D, larvae or adult fish were fixed with 4% paraformaledehyde (PFA)/PBS overnight at 4°C, followed by cryosectioning, mounting of sections in Mowiol (Carl Roth) containing DAPI and fluorophore analysis with a Zeiss Apotome. For Figure 1E fish were stained with rabbit anti-RFP (1:100; MBL, PM005) (secondary=Alexa Fluor-488 anti-rabbit (1:100; Invitrogen, A11008)), mouse anti-p63 (1:100, Santa Cruz, sc-8431) (secondary=Alexa Fluor-647 anti-mouse (1:100; Invitrogen, A21240)) and Rhodamine-Phalloidin (1:100, Invitrogen, R415), and analyzed via confocal microcopy (Zeiss LSM710 Meta).

Tissue-labelling procedures

Epidermal cell proliferation was assessed by BrdU incorporation after incubating adult fish in 100 μg/ml BrdU (Sigma) in fish system water for 12 or 24 hours, followed by anti-BrdU immunolabelling. TUNEL assay was performed using the *in situ* Cell Death Detection Kit, POD (Roche) according to the manufacturers recommendations.

For *in vivo* calcein staining, fish were incubated for two hours in calcein green or calcein blue solution (100 mg/l; Sigma Aldrich). After extensive washings, fish were anaesthetized with Tricaine (ethyl-3-aminobenzoate methanesulfonate, Fluka) for fluorescence analysis of live whole mounts or after PFA fixation and cryosectioning.

For histological, immunofluorescence and in situ hybridization analyses, adult zebrafish were sacrificed by Tricaine overdose and fixed in 4% PFA overnight at 4°C. Samples for paraffin embedding were decalcified in 0.5 M EDTA (pH 7.4) at room temperature for 5 days, dehydrated in a graded series of alcohols, cleared in Roti-Histol (Carl Roth) and embedded in paraffin wax. 10 μm sections were cut using a Leica RM2255 microtome. Samples for cryosections were orientated in 15% sucrose with 1% agarose in PBS and mounted in tissue freezing medium (Leica). 10 or 12 μm sections were obtained using a Leica CM1850 cryostat.

Paraffin sections were stained with hematoxylin & eosin or acidic fuchsin orange G (AFOG) trichrome (Gennova) according to standard protocols. For immunofluorescence analysis of paraffin- or cryosections, antigen retrieval was performed with 10 mM sodium citrate (pH 6.0) at 70°C for two hours, followed by washes and primary and secondary antibody incubations in PBS supplemented with 10% fetal calf serum (FCS), and mounting of sections in Mowiol containing DAPI. Primary antibodies other from the ones described above were: rabbit anti-activated caspase3 (1:1000, abcam ab-13847), mouse anti-BrdU (1:200, Roche 1170376), mouse anti-pan Keratin Type II (1:200, Progen 61006). Secondary antibodies used were: anti-mouse Cy3 (1:1000, Invitrogen), anti-rabbit Cy3 (1:1000, Invitrogen), Alexa Fluor-488 anti-rabbit (1:1000, Invitrogen)

In situ hybridization on paraffin sections was performed according to [80]. Antisense RNA probes were generated via in vitro transcription with Dig RNA labelling mix (Roche) and the following templates and conditions: *krt8* (GenBank BI875660): 1.8 kb cDNA fragment cloned from EST into pBluescript SK, linearization with *Hin*dIII, transcription with T3 RNA polymerase; *cki* (GenBank AF197880): 0.6 kb fragment in pSPORT, *Eco*RI, SP6 RNA pol; *krt5* (GenBank AF197909): 0.4 kb fragment in pBluescript SK, *Kpn*I, T3 RNA pol; *krt17* (ZFIN-ID zgc:92061; GenBank BI850052): 1.5 kb fragment in pSPORT, *Eco*RI, SP6 RNA pol.

For *krtt1c11a* and *tgm1*, 1.0 kb fragments were amplified via RT-PCR and cloned into pGMTeasy (Promega) (*krtt1c11a: Spe*I, T7 RNA pol) or pCRII (*tgm1*: *Xho*I, SP6 RNA pol). For a TA-specific TAp63 probe, a 408 bp TA1 cDNA fragment was amplified via RT-PCR with the primers 5′-CAGGGGCTAGCTTCTAGTGG-3′ (sense) and 5′-TGTAAGGGGCTCCTCAGGCTC-3′ (antisense) and cloned into pGEMTeasy. The plasmid was digested with *Spe*I and transcribed with SP6 RNA pol for antisense, and with *Nco*I and T7 RNA pol for sense probe. For p53, EST clone MPMGp609B127Q8 with the full-length p53 cDNA in pSPORTI was linearized with *Eco*RI and transcribed with SP6 RNA pol for antisense, and with *Bam*HI and T7 RNA pol for sense probe.

Images were captured on a Zeiss Axiophot, Zeiss Apotome, Zeiss Confocal (LSM710 META) or Leica M165 FC stereo microscope.

Electron microscopy

Transmission electron microscopy (TEM) of adult zebrafish was carried out as described [81]. For Scanning electron microscopy (SEM), adult fish were sacrificed and fixed overnight in 4% PFA at 4°C, dehydrated and either cryo-fixed, sputter-coated (gold/palladium) and transferred onto the SEM cryo-stage while still frozen, or critical point dried (CPD), sputter-coated and evaluated at room temperature.

Chemical treatments

Fish were raised from 20–50 dpf in E3-medium containing 100 μg/ml of the γ-secretase inhibitor DAPT (N-N-(3,5-difluorophenacetyl)-L-alanyl)-S-phenylglycien t-butylester; Sigma-Aldrich 208255) [82], 5 μg/ml of the caspase 3 peptide inhibitor z-DEVD-fmk (Calbiochem 264155-80) [83] or 0.2% DMSO as control. Standard length (SL) of the fish was used to control equal development of each group.

RT-PCR and RFLP analysis

RNA of whole zebrafish embryos at different developmental time points or from isolated tissues or organs of adults was isolated using the trizol reagent (Invitrogen). cDNA was generated using random hexamer primers. Regular PCR was carried out with the TA-specific sense primers TAS1-3 or the ΔN-specific sense primer ΔNs (see Figure S1), combined with a shared reverse primer 5-GTGACTGGGTGGGGCTATTT-3. Zebrafish *actb2* (GenBank: BC0675676) specific primers were used as control (sense, 5′-AGTTTGAGTCGGCGTGAAGT-3′; antisense, 5′-AGGCTGTGCTGTCCCTGTAT-

3′). PCR reactions were performed with an annealing temperature of 55°C for 35 cycles.

For cDNA RFLP analysis, the 629 bp fragment shown in Figure 6B was amplified with primers TA3 and the reverse primer 5′-GTGACTGGGTGGGGCTATTT-3′, followed by overnight digest with *Mbo*I (NEB) and electrophoresis in 4% agarose gel, revealing 379, 107, 101 and 42 bp cleavage products in wild-type, but only 279, 208 and 42 bp products in hu2525 mutant cDNA.

Quantitative RT-PCR was performed in triplicates (2 experiments each) with TaqMan primers (see below) and an Applied Biosystems 7500 Fast Real-Time PCR System under default PCR conditions, resulting in specific 65 bp (TAp63; shared by TA1 and TA4) and 72 bp (ΔNp63) products. Used primers were: TA-forward, 5′-GCCTGAGGAGCCCCTTACA-3′; ΔN-forward, 5′-CCAATGCTCCCTCATCCTACA-3′, TA-reverse and ΔN-reverse, 5′-CATTTTGA TCCATGCTGTTGAGA-3′; TA-TaqMan probe, 5′-CTCAGT ATACAAGCCTGGG-3′; ΔN-TaqMan probe, 5′-AGCCTCAGTATACAAGCC-3′; standard, rps23 (ribosomal protein S23; standard; Applied Biosystems; Dr.0343030371m1). Amplification efficiencies were determined with a dilution series of cDNA from adult skin, and were above 95% for all three amplificants (TAp63, 99,4%; ΔNp63, 95.9%; rps23, 95.2%). Data were analyzed using Biosystems Prism SDS and Excel software, applying ΔCT and ΔΔCT calculations.

Generation of expression constructs, mRNA synthesis and microinjection

To generate the TAp63 expression constructs pCS2-TA(1)p63γ and pCS2-TA(4)p63γ, a replacement strategy was used, amplifying the N-terminal fragments of the TA1 and TA4 isoforms of TAp63 via RT-PCR from adult skin of wild-type and hu2525 mutants with forward primers 5′-TTGGAT CCACCATGACCTCTCCTTATGCAGC-3′ (TA1) or 5′-TTGGATCCAC CATG TCACAGG GCCAGGGCTC-3′ (TA4), and reverse primer 5′-GTGACTG GGTG GGGCTATTT-3′, followed by *Bam*H1/*Bsp*M1 digest and cloning into *Bam*H1/*Bsp*M1-digested pCS2-ΔNp63γ [58]. To generate expression constructs for TAp63 with six N-terminal Myc tags, TAp63γ coding sequences were amplified from wild-type and hu2525 mutant pCS2-TA(4)p63γ plasmids with primers 5′-CGAATTCAACCATGT CACAGGGC CAGGGCTC-3′ (sense) and 5′- TTTCTAGATCACACTGATTGAGAACTCTTTTT G-3′ (antisense), digested with *Eco*RI and *Xba*I, and cloned into *Eco*RI/*Xba*I digested pCS2-MT (www.addgene.org/vector-database/2296/). For expression constructs with six C-terminal Myc tags, amplification was performed with primers 5′-TTGGATCCACCATGTCACAGGGCCAGGGCTC-3′ (sense) and 5′-CGATCGATTCACTGATTGAGAACTCTTTTTGTC-3′, followed by digestion with *Bam*HI and *Cla*I, and cloning into *Bam*HI/*Cla*I, digested pCS2-MT.

Capped RNA was prepared after restriction digest of these expression constructs or pCS2-ΔNp63α1 [58] with *Kpn*I, using the Message Machine kit (Ambion, Austin, TX). RNA was dissolved in water, and 1 nl per embryo injected. TAp63γ mRNAs were injected at a concentration of 10 ng/μl, 6xMyc-TAp63γ mRNAs at a concentration of 5 ng/μl, and ΔNp63α1 mRNA at a concentration of 25 ng/μl. Apoptosis and resulting embryonic death or embryonic malformations were scored at 8 hpf and 24 hpf, respectively, ΔNp63α1-induced loss of eyes at 32 hpf, as described [58].

Immunoblotting

Zebrafish embryos were dechorionated and deyolked, and cells were collected as described [84]. Cell pellets or adult tissues were either directly dissolved in SDS loading buffer as described [84], or first lysed in chilled CSH buffer (50 mM Tris-HCl (pH 7.5), 250 mM NaCl, 1 mM EDTA, 1% Triton-X100, supplemented with cOmplete Protease Inhibitor Cocktail, Roche), followed by protein concentration determination. 10–12% SDS-PAGE, blotting on nitrocellulose membrane, Ponceau staining and immunodetection were carried out as described [84]. Used primary antibodies were: anti-Myc, 9B11 (mouse, Cell Signaling Technology; 1:2000); anti-p63, 4A4 (mouse, Santa Cruz Technologies, against aa 1–205 of human ΔNp63), D-9 (mouse, Santa Cruz Biotechnology, against aa 15–151 of human ΔNp63), H-137 (rabbit, Santa Cruz Biotechnology, against aa 15–151 of human ΔNp63), H-129 (rabbit, Santa Cruz Biotechnology, against aa 513–641 at C-terminus of human TAp63α).

ACKNOWLEDGMENTS

We are very grateful to Julia Hatzold, Mojgan Ghilav and Iris Riedl-Quinkertz for experimental help, to Edwin Cuppen and colleagues at the Hubrecht Institute, NL, for isolating and shipping the TA-specific p63 mutant, and to Zhiyuan Gong, Wiebke Herzog, Arndt Siekmann, Thomas Look, Gilbert Weidinger and Alexander Reugels for mutant and transgenic lines.

REFERENCES

1. Nakamura H, Yasuda M (1979) An electron microscopic study of periderm cell development in mouse limb buds. Anat Embryol (Berl) 157: 121–132.
2. Holbrook KA, Odland GF (1975) The fine structure of developing human epidermis: light, scanning, and transmission electron microscopy of the periderm. J Invest Dermatol 65: 16–38.
3. Byrne C, Tainsky M, Fuchs E (1994) Programming gene expression in developing epidermis. Development 120: 2369–2383.
4. Hardman MJ, Moore L, Ferguson MW, Byrne C (1999) Barrier formation in the human fetus is patterned. J Invest Dermatol 113: 1106–1113.
5. Koster MI, Roop DR (2004) The role of p63 in development and differentiation of the epidermis. J Dermatol Sci 34: 3–9.
6. Koster MI, Roop DR (2007) Mechanisms regulating epithelial stratification. Annu Rev Cell Dev Biol 23: 93–113.
7. Williams ML, Hanley K, Elias PM, Feingold KR (1998) Ontogeny of the epidermal permeability barrier. J Investig Dermatol Symp Proc 3: 75–79.
8. Alibardi L (2003) Adaptation to the land: The skin of reptiles in comparison to that of amphibians and endotherm amniotes. J Exp Zool B Mol Dev Evol 298: 12–41.
9. Wu P, Hou L, Plikus M, Hughes M, Scehnet J, et al. (2004) Evo-Devo of amniote integuments and appendages. Int J Dev Biol 48: 249–270.
10. Kimmel CB, Warga RM, Schilling TF (1990) Origin and organization of the zebrafish fate map. Development 108: 581–594.
11. Fukazawa C, Santiago C, Park KM, Deery WJ, Gomez de la Torre Canny S, et al. (2010) poky/chuk/ikk1 is required for differentiation of the zebrafish embryonic epidermis. Dev Biol 346: 272–283.
12. Le Guellec D, Morvan-Dubois G, Sire JY (2004) Skin development in bony fish with particular emphasis on collagen deposition in the dermis of the zebrafish (Danio rerio). Int J Dev Biol 48: 217–231.
13. Richardson R, Slanchev K, Kraus C, Knyphausen P, Eming S, et al. (2013) Adult zebrafish as a model system for cutaneous wound healing research. J Invest Dermatol 133: 1655–1665.
14. Whitear M (1986) The skin of fishes including Cyclostomes: epidermis. In: Bereiter-Hahn J, Matoltsy AG, Richards KS, editors. Biology of the Integument 2: Vertebrates. Berlin, Heidelberg: Springer-Verlag,: 8–38.
15. Candi E, Schmidt R, Melino G (2005) The cornified envelope: a model of cell death in the skin. Nat Rev Mol Cell Biol 6: 328–340.
16. Mack JA, Anand S, Maytin EV (2005) Proliferation and cornification during development of the mammalian epidermis. Birth Defects Res C Embryo Today 75: 314–329.
17. Proksch E, Brandner JM, Jensen JM (2008) The skin: an indispensable barrier. Exp Dermatol 17: 1063–1072.
18. Yang A, Kaghad M, Wang Y, Gillett E, Fleming MD, et al. (1998) p63, a p53 homolog at 3q27-29, encodes multiple products with transactivating, death-inducing, and dominant-negative activities. Mol Cell 2: 305–316.
19. Mills AA, Zheng B, Wang XJ, Vogel H, Roop DR, et al. (1999) p63 is a p53 homologue required for limb and epidermal morphogenesis. Nature 398: 708–713.
20. Yang A, Schweitzer R, Sun D, Kaghad M, Walker N, et al. (1999) p63 is essential for regenerative proliferation in limb, craniofacial and epithelial development. Nature 398: 714–718.
21. Yang A, McKeon F (2000) P63 and P73: P53 mimics, menaces and more. Nat Rev Mol Cell Biol 1: 199–207.
22. King KE, Weinberg WC (2007) p63: defining roles in morphogenesis, homeostasis, and neoplasia of the epidermis. Mol Carcinog 46: 716–724.

23. Crum CP, McKeon FD (2010) p63 in epithelial survival, germ cell surveillance, and neoplasia. Annu Rev Pathol 5: 349–371.
24. Dötsch V, Bernassola F, Coutandin D, Candi E, Melino G (2010) p63 and p73, the ancestors of p53. Cold Spring Harb Perspect Biol 2: a004887.
25. Bamberger C, Schmale H (2001) Identification and tissue distribution of novel KET/p63 splice variants. FEBS Lett 501: 121–126.
26. Romano RA, Smalley K, Magraw C, Serna VA, Kurita T, et al. (2012) DeltaNp63 knockout mice reveal its indispensable role as a master regulator of epithelial development and differentiation. Development 139: 772–782.
27. King KE, Ponnamperuma RM, Yamashita T, Tokino T, Lee LA, et al. (2003) deltaNp63alpha functions as both a positive and a negative transcriptional regulator and blocks in vitro differentiation of murine keratinocytes. Oncogene 22: 3635–3644.
28. King KE, Ponnamperuma RM, Gerdes MJ, Tokino T, Yamashita T, et al. (2006) Unique domain functions of p63 isotypes that differentially regulate distinct aspects of epidermal homeostasis. Carcinogenesis 27: 53–63.
29. Truong AB, Kretz M, Ridky TW, Kimmel R, Khavari PA (2006) p63 regulates proliferation and differentiation of developmentally mature keratinocytes. Genes Dev 20: 3185–3197.
30. Candi E, Rufini A, Terrinoni A, Dinsdale D, Ranalli M, et al. (2006) Differential roles of p63 isoforms in epidermal development: selective genetic complementation in p63 null mice. Cell Death Differ 13: 1037–1047.
31. Yugawa T, Handa K, Narisawa-Saito M, Ohno S, Fujita M, et al. (2007) Regulation of Notch1 gene expression by p53 in epithelial cells. Mol Cell Biol 27: 3732–3742.
32. Guinea-Viniegra J, Zenz R, Scheuch H, Jimenez M, Bakiri L, et al. (2012) Differentiation-induced skin cancer suppression by FOS, p53, and TACE/ADAM17. J Clin Invest 122: 2898–2910.
33. Truong AB, Khavari PA (2007) Control of keratinocyte proliferation and differentiation by p63. Cell Cycle 6: 295–299.
34. Suh EK, Yang A, Kettenbach A, Bamberger C, Michaelis AH, et al. (2006) p63 protects the female germ line during meiotic arrest. Nature 444: 624–628.
35. Guo X, Keyes WM, Papazoglu C, Zuber J, Li W, et al. (2009) TAp63 induces senescence and suppresses tumorigenesis in vivo. Nat Cell Biol 11: 1451–1457.
36. Su X, Paris M, Gi YJ, Tsai KY, Cho MS, et al. (2009) TAp63 prevents premature aging by promoting adult stem cell maintenance. Cell Stem Cell 5: 64–75.
37. Blanpain C, Lowry WE, Pasolli HA, Fuchs E (2006) Canonical notch signaling functions as a commitment switch in the epidermal lineage. Genes Dev 20: 3022–3035.
38. Watt FM, Estrach S, Ambler CA (2008) Epidermal Notch signalling: differentiation, cancer and adhesion. Curr Opin Cell Biol 20: 171–179.
39. Okuyama R, Nguyen BC, Talora C, Ogawa E, Tommasi di Vignano A, et al. (2004) High commitment of embryonic keratinocytes to terminal differentiation through a Notch1-caspase 3 regulatory mechanism. Dev Cell 6: 551–562.
40. Rangarajan A, Talora C, Okuyama R, Nicolas M, Mammucari C, et al. (2001) Notch signaling is a direct determinant of keratinocyte growth arrest and entry into differentiation. Embo J 20: 3427–3436.
41. Estrach S, Cordes R, Hozumi K, Gossler A, Watt FM (2008) Role of the Notch ligand Delta1 in embryonic and adult mouse epidermis. J Invest Dermatol 128: 825–832.
42. Sasaki Y, Ishida S, Morimoto I, Yamashita T, Kojima T, et al. (2002) The p53 family member genes are involved in the Notch signal pathway. J Biol Chem 277: 719–724.
43. Nguyen BC, Lefort K, Mandinova A, Antonini D, Devgan V, et al. (2006) Cross-regulation between Notch and p63 in keratinocyte commitment to differentiation. Genes Dev 20: 1028–1042.
44. Lefort K, Mandinova A, Ostano P, Kolev V, Calpini V, et al. (2007) Notch1 is a p53 target gene involved in human keratinocyte tumor suppression through negative regulation of ROCK1/2 and MRCKalpha kinases. Genes Dev 21: 562–577.

45. Candi E, Rufini A, Terrinoni A, Giamboi-Miraglia A, Lena AM, et al. (2007) DeltaNp63 regulates thymic development through enhanced expression of FgfR2 and Jag2. Proc Natl Acad Sci U S A 104: 11999–12004.
46. Yugawa T, Narisawa-Saito M, Yoshimatsu Y, Haga K, Ohno S, et al. (2010) DeltaNp63alpha repression of the Notch1 gene supports the proliferative capacity of normal human keratinocytes and cervical cancer cells. Cancer Res 70: 4034–4044.
47. Okuyama R, Tagami H, Aiba S (2008) Notch signaling: its role in epidermal homeostasis and in the pathogenesis of skin diseases. J Dermatol Sci 49: 187–194.
48. Lamkanfi M, Festjens N, Declercq W, Vanden Berghe T, Vandenabeele P (2007) Caspases in cell survival, proliferation and differentiation. Cell Death Differ 14: 44–55.
49. Gong Z, Ju B, Wang X, He J, Wan H, et al. (2002) Green fluorescent protein expression in germ-line transmitted transgenic zebrafish under a stratified epithelial promoter from keratin8. Dev Dyn 223: 204–215.
50. Lee RTH, Asharani PV, Carney TJ (2014) Basal keratinocytes contribute to all strata of the adult zebrafish epidermis. PLoS One e84858
51. Padhi BK, Akimenko MA, Ekker M (2006) Independent expansion of the keratin gene family in teleostean fish and mammals: an insight from phylogenetic analysis and radiation hybrid mapping of keratin genes in zebrafish. Gene 368: 37–45.
52. Deasey S, Grichenko O, Du S, Nurminskaya M (2011) Characterization of the transglutaminase gene family in zebrafish and in vivo analysis of transglutaminase-dependent bone mineralization. Amino Acids 42: 1065–1075.
53. Matsuki M, Yamashita F, Ishida-Yamamoto A, Yamada K, Kinoshita C, et al. (1998) Defective stratum corneum and early neonatal death in mice lacking the gene for transglutaminase 1 (keratinocyte transglutaminase). Proc Natl Acad Sci U S A 95: 1044–1049.
54. Madison KC (2003) Barrier function of the skin: "la raison d'etre" of the epidermis. J Invest Dermatol 121: 231–241.
55. Pyati UJ, Look AT, Hammerschmidt M (2007) Zebrafish as a powerful vertebrate model system for in vivo studies of cell death. Semin Cancer Biol 17: 154–165.
56. Galluzzi L, Vitale I, Abrams JM, Alnemri ES, Baehrecke EH, et al. (2012) Molecular definitions of cell death subroutines: recommendations of the Nomenclature Committee on Cell Death 2012. Cell Death Differ 19: 107–120.
57. Berghmans S, Murphey RD, Wienholds E, Neuberg D, Kutok JL, et al. (2005) tp53 mutant zebrafish develop malignant peripheral nerve sheath tumors. Proc Natl Acad Sci U S A 102: 407–412.
58. Bakkers J, Hild M, Kramer C, Furutani-Seiki M, Hammerschmidt M (2002) Zebrafish DeltaNp63 is a direct target of Bmp signaling and encodes a transcriptional repressor blocking neural specification in the ventral ectoderm. Dev Cell 2: 617–627.
59. Lee H, Kimelman D (2002) A dominant-negative form of p63 is required for epidermal proliferation in zebrafish. Dev Cell 2: 607–616.
60. Laurikkala J, Mikkola ML, James M, Tummers M, Mills AA, et al. (2006) p63 regulates multiple signalling pathways required for ectodermal organogenesis and differentiation. Development 133: 1553–1563.
61. Rentzsch F, Kramer C, Hammerschmidt M (2003) Specific and conserved roles of TAp73 during zebrafish development. Gene 323: 19–30.
62. Wienholds E, van Eeden F, Kosters M, Mudde J, Plasterk RH, et al. (2003) Efficient target-selected mutagenesis in zebrafish. Genome Res 13: 2700–2707.
63. Parsons MJ, Pisharath H, Yusuff S, Moore JC, Siekmann AF, et al. (2009) Notch-responsive cells initiate the secondary transition in larval zebrafish pancreas. Mech Dev 126: 898–912.
64. Scheer N, Campos-Ortega JA (1999) Use of the Gal4-UAS technique for targeted gene expression in the zebrafish. Mech Dev 80: 153–158.

65. Caubet C, Jonca N, Brattsand M, Guerrin M, Bernard D, et al. (2004) Degradation of corneodesmosome proteins by two serine proteases of the kallikrein family, SCTE/KLK5/hK5 and SCCE/KLK7/hK7. J Invest Dermatol 122: 1235–1244.
66. Vanhoutteghem A, Djian P, Green H (2008) Ancient origin of the gene encoding involucrin, a precursor of the cross-linked envelope of epidermis and related epithelia. Proc Natl Acad Sci U S A 105: 15481–15486.
67. Sidi S, Sanda T, Kennedy RD, Hagen AT, Jette CA, et al. (2008) Chk1 suppresses a caspase-2 apoptotic response to DNA damage that bypasses p53, Bcl-2, and caspase-3. Cell 133: 864–877.
68. Wu NL, Lee TA, Tsai TL, Lin WW (2011) TRAIL-induced keratinocyte differentiation requires caspase activation and p63 expression. J Invest Dermatol 131: 874–883.
69. Huh JR, Guo M, Hay BA (2004) Compensatory proliferation induced by cell death in the Drosophila wing disc requires activity of the apical cell death caspase Dronc in a nonapoptotic role. Curr Biol 14: 1262–1266.
70. Blanpain C, Fuchs E (2009) Epidermal homeostasis: a balancing act of stem cells in the skin. Nat Rev Mol Cell Biol 10: 207–217.
71. Ratushny V, Gober MD, Hick R, Ridky TW, Seykora JT (2012) From keratinocyte to cancer: the pathogenesis and modeling of cutaneous squamous cell carcinoma. J Clin Invest 122: 464–472.
72. Nicolas M, Wolfer A, Raj K, Kummer JA, Mill P, et al. (2003) Notch1 functions as a tumor suppressor in mouse skin. Nat Genet 33: 416–421.
73. Reich NC, Levine AJ (1984) Growth regulation of a cellular tumour antigen, p53, in nontransformed cells. Nature 308: 199–201.
74. Maddocks OD, Berkers CR, Mason SM, Zheng L, Blyth K, et al. (2013) Serine starvation induces stress and p53-dependent metabolic remodelling in cancer cells. Nature 493: 542–546.
75. Tomasini R, Tsuchihara K, Wilhelm M, Fujitani M, Rufini A, et al. (2008) TAp73 knockout shows genomic instability with infertility and tumor suppressor functions. Genes Dev 22: 2677–2691.
76. Kwan KM, Fujimoto E, Grabher C, Mangum BD, Hardy ME, et al. (2007) The Tol2kit: a multisite gateway-based construction kit for Tol2 transposon transgenesis constructs. Dev Dyn 236: 3088–3099.
77. Petersen LK, Stowers RS (2011) A Gateway MultiSite recombination cloning toolkit. PLoS One 6: e24531.
78. Bertrand JY, Chi NC, Santoso B, Teng S, Stainier DY, et al. (2010) Haematopoietic stem cells derive directly from aortic endothelium during development. Nature 464: 108–111.
79. Neff MM, Turk E, Kalishman M (2002) Web-based primer design for single nucleotide polymorphism analysis. Trends Genet 18: 613–615.
80. Moorman AF, Houweling AC, de Boer PA, Christoffels VM (2001) Sensitive nonradioactive detection of mRNA in tissue sections: novel application of the whole-mount in situ hybridization protocol. J Histochem Cytochem 49: 1–8.
81. Laue K, Pogoda HM, Daniel PB, van Haeringen A, Alanay Y, et al. (2011) Craniosynostosis and multiple skeletal anomalies in humans and zebrafish result from a defect in the localized degradation of retinoic acid. Am J Hum Genet 89: 595–606.
82. Geling A, Steiner H, Willem M, Bally-Cuif L, Haass C (2002) A gamma-secretase inhibitor blocks Notch signaling in vivo and causes a severe neurogenic phenotype in zebrafish. EMBO Rep 3: 688–694.
83. Parng C, Anderson N, Ton C, McGrath P (2004) Zebrafish apoptosis assays for drug discovery. Methods Cell Biol 76: 75–85.
84. Link V, Shevchenko A, Heisenberg CP (2006) Proteomics of early zebrafish embryos. BMC Dev Biol 6: 1.
85. Roberts TR (1984) Unculi (Horny projections arising from single cells), an adaptive feature of the epidermis of ostariophysan fishes. Zool Scripta 11: 55–76.

Index